10th Anniversary of Water

10th Anniversary of Water

Special Issue Editor

Arjen Y. Hoekstra†

MDPI • Basel • Beijing • Wuhan • Barcelona • Belgrade

Special Issue Editor
Arjen Y. Hoekstra†
University of Twente
The Netherlands

Editorial Office
MDPI
St. Alban-Anlage 66
4052 Basel, Switzerland

This is a reprint of articles from the Special Issue published online in the open access journal *Water* (ISSN 2073-4441) from 2018 to 2020 (available at: https://www.mdpi.com/journal/water/special_issues/10th_anniversary).

For citation purposes, cite each article independently as indicated on the article page online and as indicated below:

LastName, A.A.; LastName, B.B.; LastName, C.C. Article Title. *Journal Name* **Year**, *Article Number*, Page Range.

ISBN 978-3-03936-340-7 (Pbk)
ISBN 978-3-03936-341-4 (PDF)

Contents

About the Special Issue Editor

Arjen Y. Hoekstra† (1967–2019) was Professor in Water Management at the University of Twente, and Visiting Professor at the Lee Kuan Yew School of Public Policy at the National University of Singapore. He held an M.S. degree in Civil Engineering and a Ph.D. degree in Policy Analysis, both from Delft University of Technology. Hoekstra has led a variety of interdisciplinary research projects and advised governments, civil society organizations, companies, and multilateral institutions like UNESCO and the World Bank. Hoekstra has received various awards, including an ERC Advanced Grant, Europe's most prestigious research grant.

Hoekstra pioneered in quantifying the water volumes virtually embedded in trade, thus showing the relevance of a global perspective on water use and scarcity. As creator of the water footprint concept, Hoekstra introduced supply chain thinking in water management. With the development of Water Footprint Assessment, he laid the foundation of a new interdisciplinary research field, addressing the relations between water management, consumption and trade. Hoekstra was founder of the Water Footprint Network (2008), co-initiator of the Water Footprint Research Alliance (2015), and founding member of the Planetary Accounting Network (2018).

Hoekstra's scientific publications cover a wide range of topics related to water, food, energy, and trade, and include a large number of highly cited articles and book chapters. His books have been translated into several languages and include The Water Footprint of Modern Consumer Society (Routledge, 2013, 2019), The Water Footprint Assessment Manual (Earthscan, 2011), and Globalization of Water (Wiley-Blackwell, 2008). Hoekstra was Editor-in-Chief of *Water*, an interdisciplinary open access journal covering all aspects of water, including water science, technology, management, and governance.

Hoekstra has taught in a variety of subjects, including, sustainable development, water management, river basin management, hydrology, water quality, water footprint assessment, natural resources valuation, environmental systems analysis, and policy analysis. He developed various educational tools, including the River Basin Game and the Globalization of Water Role Play.

His colleagues will miss him as a passionate human, always working and socially engaged. He had an open mind, was intrigued by other cultures, and visited many countries over the years. To-the-point, no-nonsense, and starting from a base of trust characterized Arjen. He enjoyed giving autonomy to the people he with whom he worked. He was socially active in our team and was working from his sustainable office, a green oasis filled with plants. He was an inspiration to us all.

water

MDPI

Editorial

10th Anniversary of *Water*

Jeroen C. J. H. Aerts

Institute for Environmental Studies (IVM), VU University Amsterdam, De Boelelaan 1087, 1081HV Amsterdam, The Netherlands; jeroen.aerts@vu.nl

Received: 28 April 2020; Accepted: 10 May 2020; Published: 12 May 2020

Abstract: This Special Issue was set up to mark the 10th anniversary of *Water*. The contributions to this Special Issue of *Water* were carefully selected by the late Guest Editor Prof. Dr. Arjen Hoekstra. Arjen was devoted to conducting excellent science and was motivated to create this Special Issue to be something 'special'. It was therefore dedicated to the publication of 11 comprehensive papers and reviews encompassing the most significant developments in the realm of water sciences in the last decade.

Keywords: Governance; flood adaptation costs; hydro-informatics; water management; water use

1. Introduction

Water is essential to all life on earth, but its management is facing increasing challenges due to socio-economic pressures such as population growth and the unsustainable use of water resources. Climate change will further exacerbate the water risk for society and the environment, and water-related extremes such as floods and droughts will increase in the future. The long-term aspects of these future trends and the inherent uncertainty within future projections present water managers with considerable challenges. The technocratic approach of working with fixed design standards for engineering seems insufficient; the environment is constantly changing, and, as a consequence of this, so too are the boundary conditions on which basis engineered water management solutions are developed. Therefore, water management is increasingly developing into an adaptive form of decision making, where flexibility, robustness and resilience are key [1]. Moreover, societal processes and the physical water system are increasingly interwoven, and there are little natural water systems not influenced by human activity. These developments require novel approaches in decision sciences, data processing, modelling techniques, catalyzed by the integration of social and natural sciences [2]. The international journal *Water* addresses these challenges as an outlet for cutting-edge inter-disciplinary approaches in water science. In this Special Issue, topics cover broad aspects of water systems, including water science, water quality, management, and governance.

2. Overview of the Special Issue

Grafton, R.Q., et al. [3] demonstrate the use of a new water governance reform framework (WGRF), which may help authorities to reform their governance frameworks to achieve convergence between water supply and demand and ensure freshwater ecosystem services are sustained. The importance of water governance is further illustrated in a paper by Purkey, D.R., et al. [4] that tries to answer the question of whether the best option even exists. Although such existential questions are not common in the water management community, they are not new to political theory. This paper explores early classical writing related to issues of knowledge and governance, as captured in the work of Plato and Aristotle, and then attempts to place a novel, analysis-supported, and stakeholder-driven water resources planning and decision-making practice within this philosophical discourse, referencing current decision theory.

Koech, R., and Langat, P.'s, [5] paper on irrigation reviews the advancements toward improving irrigation water use efficiency (WUE), with a focus on irrigation in Australia, but with some examples from other countries. The review shows that improvements in irrigation infrastructure through modernization and automation have led to water savings, and that the future is likely to see increased use of remote sensing techniques, wireless communication systems, and more versatile sensors to improve WUE.

A review paper by Aerts, J.C.J.H., [6] on the cost of flood adaptation measures provides the most recent empirical data regarding the cost of flood management by compiling peer-reviewed literature and research reports. The focus is on construction costs and expenses for operation and maintenance, including: (1) the flood-proofing of buildings, (2) flood protection, (3) beach nourishment and dunes, (4) nature-based solutions for coastal ecosystems, (5) channel management and nature-based solutions for riverine systems, and (6) urban drainage.

Related to urban drainage in the previous review, another contribution by Antunes, L.N., et al. [7] presents an overview of permeable pavements and studies of life cycle assessment that compare the environmental performance of permeable pavements with traditional drainage systems. Life cycle assessment studies are essential to guide planning and decision making, leading to systems that consider increasing water resources and reducing natural disasters and environmental impacts.

Further within the context of river management options, a paper on river relocation by Flatley, A., et al. [8] discusses shortcomings in current practice for river relocation and highlights areas for future research for the successful rehabilitation of relocated rivers. Relocations are common through history, carried out for a wide range of purposes, but most commonly to construct infrastructure and for mining. However, many assessment studies do not include the effects of climate change.

Another paper by Makropoulos, C., and Savić, D.A., [9] provides a comprehensive overview of the history of hydroinformatics. Hydroinformatics has arguably been able to mobilize wide ranging research and development and align the water sector more with the digital revolution of the past 30 years. In this context, this paper attempts to trace the evolution of the discipline from its computational hydraulics origins to its present focus on the complete socio-technical system and by providing a framework to highlight the links between different strands of the state-of-art hydroinformatic research and innovation.

A paper by Iaccarino, M. [10] on population growth and water use touches upon a very timely issue: the influence of water contamination on the human population. The paper considers historic data and shows that after a huge population increase between the years 25,000 and 5000 Before Common Era (BCE), the number of people did not change appreciably after 200 Common Era (CE), and increased only slowly in the period 1000 to 1500 CE. The authors show that the main cause of this observed slow-down in population growth was the increase in population density, which caused the appearance and spreading of infectious diseases, often due to the use of contaminated water.

Another paper by Destouni, G. and Prieto, C. [11] develops a data-driven approach to robustly assess freshwater changes due to climate change and/or human irrigation developments using the overarching constraints of catchment water balance. The results show that the resulting uncertainties in the water-balance constrained estimates of runoff and evapotranspiration (*ET*) changes are smaller than the input data uncertainties.

A paper by Hynds, P. et al. [12] on socio-hydrogeology, an extension of socio-hydrology, emphasises a bottom-up methodology involving non-expert end-users in hydrogeological investigations. The authors consider that multiple actors should be identified and incorporated using stakeholder network analysis, which may include policymakers, media and communications experts, mobile technology developers, and social scientists, to appropriately convey demographically focused bi-directional information, with the hydrogeological community representing the communication keystone.

The final paper by Clifford, C.C. and Heffernan, J.B., [13] analyzes the importance of artificial aquatic systems in modern landscapes. The provisioning of ecosystem services by these systems is underexplored and likely underestimated. Instead of accepting that artificial ecosystems have intrinsically low values, environmental scientists should determine what combination of factors, including setting, planning, and construction, affects these values. Scientists, social scientists, and policymakers should more thoroughly evaluate whether current study and management of artificial aquatic systems are based on the actual ecological condition of these systems, or judged differently, due to artificiality, and consider the possible resultant changes in goals for these systems.

3. Conclusions

The papers in this Special Issue are, of course, not reflecting all the recent developments in water sciences. However, it provides a snap-shot of important scientific developments required to support water managers in their endeavor to deal with increasing complexity and uncertainty. As such, Arjen would be proud of the result of this Special Issue, as all articles contribute to the much needed debate around the fair and sustainable allocation of fresh water resources [14].

Acknowledgments: All acknowledgements for making this special issue a success are for the late Arjen Hoekstra, guest editor of this Special Issue, and a devoted and inspiring researcher.

Conflicts of Interest: The author declares no conflict of interest.

References

1. Becker, G.; Huitema, D.; Aerts, J.C. Prescriptions for adaptive co-management: the case of flood management in the German Rhine basin. *Ecology and Society* **2015**, *20*. [CrossRef]
2. Aerts, J.C.J.H.; Botzen, W.J.; Clarke, K.; Cutter, S.L.; Michel-Kerjan, E.; Surminski, S.; Mysiak, J.; Merz, B.; Hall, J.; Kunreuther, H. Including Human Behavior in Flood risk assessment. *Nat. Clim. Chang.* **2018**, *8*, 193–199.
3. Grafton, R.Q.; Garrick, D.; Manero, A.; Do, T.N. The Water Governance Reform Framework: Overview and Applications to Australia, Mexico, Tanzania, U.S.A and Vietnam. *Water* **2019**, *11*, 137. [CrossRef]
4. Purkey, D.R.; Escobar Arias, M.I.; Mehta, V.K.; Forni, L.; Depsky, N.J.; Yates, D.N.; Stevenson, W.N. A Philosophical Justification for a Novel Analysis-Supported, Stakeholder-Driven Participatory Process for Water Resources Planning and Decision Making. *Water* **2018**, *10*, 1009. [CrossRef]
5. Koech, R.; Langat, P. Improving Irrigation Water Use Efficiency: A Review of Advances, Challenges and Opportunities in the Australian Context. *Water* **2018**, *10*, 1771. [CrossRef]
6. Aerts, J.C.J.H. A Review of Cost Estimates for Flood Adaptation. *Water* **2018**, *10*, 1646. [CrossRef]
7. Antunes, L.N.; Ghisi, E.; Thives, L.P. Permeable Pavements Life Cycle Assessment: A Literature Review. *Water* **2018**, *10*, 1575. [CrossRef]
8. Flatley, A.; Rutherfurd, I.D.; Hardie, R. River Channel Relocation: Problems and Prospects. *Water* **2018**, *10*, 1360. [CrossRef]
9. Makropoulos, C.; Savić, D.A. Urban Hydroinformatics: Past, Present and Future. *Water* **2019**, *11*, 1959. [CrossRef]
10. Iaccarino, M. Water, Population Growth and Contagious Diseases. *Water* **2019**, *11*, 386. [CrossRef]
11. Destouni, G.; Prieto, C. Robust Assessment of Uncertain Freshwater Changes: The Case of Greece with Large Irrigation—and Climate-Driven Runoff Decrease. *Water* **2018**, *10*, 1645. [CrossRef]
12. Hynds, P.; Regan, S.; Andrade, L.; Mooney, S.; O'Malley, K.; DiPelino, S.; O'Dwyer, J. Muddy Waters: Refining the Way forward for the "Sustainability Science" of Socio-Hydrogeology. *Water* **2018**, *10*, 1111. [CrossRef]
13. Clifford, C.C.; Heffernan, J.B. Artificial Aquatic Ecosystems. *Water* **2018**, *10*, 1096. [CrossRef]
14. Vanham, D.; Mekonnen, M.M.; Chapagain, A.K. Arjen Y. Hoekstra 1967–2019. *Nat. Sustain.* **2020**, *3*, 80. [CrossRef]

water

MDPI

Article

The Water Governance Reform Framework: Overview and Applications to Australia, Mexico, Tanzania, U.S.A and Vietnam

R. Quentin Grafton [1,*], Dustin Garrick [2], Ana Manero [3] and Thang Nam Do [1]

1 Crawford School of Public Policy, The Australian National University, Lennox Crossing, Canberra, ACT 2601, Australia; thang.do@anu.edu.au

2 School of Geography and the Environment, Oxford University Centre for the Environment, University of Oxford, South Parks Road, Oxford OX1 3QY, UK; dustin.garrick@ouce.ox.ac.uk

3 School of Agriculture and Environment, University of Western Australia, Perth, WA 6009, Australia; ana.maneroruiz@uwa.edu.au

* Correspondence: quentin.grafton@anu.edu.au; Tel.: +61410680584

Received: 21 December 2018; Accepted: 8 January 2019; Published: 14 January 2019

Abstract: The world faces critical water risks in relation to water availability, yet water demand is increasing in most countries. To respond to these risks, some governments and water authorities are reforming their governance frameworks to achieve convergence between water supply and demand and ensure freshwater ecosystem services are sustained. To assist in this reform process, the Water Governance Reform Framework (WGRF) is proposed, which includes seven key strategic considerations: (1) well-defined and publicly available reform objectives; (2) transparency in decision-making and public access to available data; (3) water valuation of uses and non-uses to assess trade-offs and winners and losers; (4) compensation for the marginalized or mitigation for persons who are disadvantaged by reform; (5) reform oversight and "champions"; (6) capacity to deliver; and (7) resilient decision-making. Using these reform criteria, we assess current and possible water reforms in five countries: Murray–Darling Basin (Australia); Rufiji Basin (Tanzania); Colorado Basin (USA and Mexico); and Vietnam. We contend that the WGRF provides a valuable approach to both evaluate and to improve water governance reform and, if employed within a broader water policy cycle, will help deliver both improved water outcomes and more effective water reforms.

Keywords: Murray–Darling Basin; Colorado; water scarcity; IWRM; equity

1. Introduction

World freshwater extractions (surface and groundwater) increased by about 2.5 times from 1960 to 2010. As a result, some four billion people live in conditions of severe water scarcity where levels of water consumption are more than twice that of the readily available water, at least, one month per year [1]. Further, up to 80% of the global population is exposed to high levels of water threats in relation to watershed disturbances, pollution, water resource development and biotic factors [2].

The on-going challenge is that water extractions are projected to increase by a further 50% by 2100 [3] and, with business as usual, about half of the world's population by 2050 is projected to reside in water basins where more than 40% of the available water is extracted [4]. Using a Blue Water Sustainability Index (BlWSI), which measures the proportion of blue water use from non-sustainable water resources (including groundwater and uses that reduce environmental flows), Wada and Bierkens [3] estimate that currently some 30% of human water consumption (including groundwater) is non-sustainable in that it will result in either the degradation of surface water or depletion of groundwater resources. At a regional level, Nechifor and Winning [5] project that India will need to lower its projected 2050 total water demand by almost 40% (292 Billion Cubic Meters, BCM), the rest of South Asia by 43% (100 BCM), the Middle East by 42% (168 BCM) and North Africa by 17% (30 BCM), if water extractions are not to exceed the current total renewable water resources of these regions.

Aeschbach-Hertig and Gleeson [6] argue that current food production in key farming regions of India, China and the USA cannot be maintained unless groundwater levels are stabilized. Another concern is that Wada et al. [7] estimate that, in 2000, non-renewable groundwater extraction contributed to 20% of global irrigation water extraction. This has important implications for food production because, globally, irrigation accounts for about 70% of global freshwater extractions and provides 40% of total human food calories. This tension between water for food production and other purposes is likely to be exacerbated into the future. For instance, by 2050: (1) the current water supply is projected to be less than the projected water applied for irrigation in major food-producing countries with production methods, and (2) a plateau is projected in terms of crop food production from water extractions if there are no further increases in the global irrigated agriculture area [8]. The key point is that in the absence of reform that includes: (1) better water governance and (2) how water is currently extracted and consumed, there are large, and with climate change, increasing risks to future food security [9].

Government responses to the global water crises have, typically, been to adopt a "hard" infrastructure or engineering solutions to increase water supply that has sometimes been part of "water nationalism" [10] and even as a "hydraulic mission" as part of a foreign policy tool [11]. Arguably, since at least the 1990s, and certainly since the Report of the World Commission on Dams was published in 2000 [12], there has been an increasing focus paid to "soft" infrastructure and governance [13]. A reprioritization towards governance and sustainability of water use is welcome as traditionally implemented water supply "solutions" have often been delivered with little regard to ecological impacts, to their negative consequences on freshwater ecosystems [14] or to the poor or marginalized who lack a voice in water planning [15]. While "hard" infrastructure is frequently needed to mitigate water insecurity, the pressures (such as population and per capita income growth, and urbanization) and states (such as per capita water availability, water variability, and climate change) of water challenges [16] require multiples responses (such as water demand management and also water justice which includes fairness, equity, participation and the democratization of water governance) [15].

Here, we provide guidance as to what should be strategic considerations in response to a growing water demand with a limited water resource and especially in terms of equity with regard to how water is allocated and used [17]. Our focus is on water governance reform noting that a change process is ongoing, must be context specific, and designed to respond to multiple challenges [18]. While there are already existing governance principles and frameworks [19–21], we contend that there is still a real need for practical guidance about how to apply key strategic considerations in relation to water reforms, and to do so in an integrative way. In Section 2, we briefly describe existing water governance frameworks and outline our own seven strategic considerations for water governance reform. In Section 3, we apply these strategic considerations in four different locations to show the added value of our approach and also discuss how our framework can be used to generate improved water outcomes. In Section 4, we offer our conclusions.

2. Water Governance Principles and Frameworks

Multiple frameworks and approaches exist in relation to governance, in general, and with respect to water, in particular. The Institutional Analysis and Development (IAD) framework provides a useful way of describing the broad governance space and includes: (1) the exogenous variables (biophysical constraints, community attributes and rules) and (2) the action arenas (situations and participants) that determine outcomes and feedbacks [22]. In the IAD framework, water governance reform operates within action arenas intended to influence or promote particular and desired outcomes. Within social, economic and political settings, the IAD framework also provides a means of describing and linking resource systems, governance systems, resource units, users, interactions, outcomes and related ecosystems [23].

2.1. Existing Water Governance Principles and Frameworks

The most long-standing water governance framework is the Integrated Water Resource Management (IWRM) that promotes coordinated management actions in relation to environmental sustainability, economic efficiency, and social equity [24]. Beyond the principles of integration across actions and their consequences, consideration of the "triple bottom line", an exhortation to ensure participatory approaches and the full inclusion of women in management, IWRM is non-prescriptive. This has allowed IWRM to be readily adapted to multiple contexts and applied in many different ways such that 80% of countries have already adopted its principles in their water laws [25]. Nevertheless, its usefulness has been questioned [26], especially in relation to what it fails to say in regard to water resource allocation, while its dominance as a water governance paradigm is challenged by the notions of water security [27,28], the nexus [29] and integrative approaches to water policy dilemmas [30].

At an intergovernmental level, the Organization for Economic Co-operation and Development (OECD) has developed a water governance framework in collaboration with its member governments and a range of stakeholders [20]. This framework is the emerging dominant water governance paradigm given its endorsement by OECD governments. It is based on 12 principles embedded around: (1) effectiveness, in relation to defining and achieving clear and sustainable water policy goals (Principles 1. Clear roles and responsibilities, 2. Appropriate scales within basin systems, 3. Policy coherence and 4. Capacity); (2) efficiency, to maximize the benefits of sustainable water management (Principles 5. Data and information, 6. Financing, 7. Regulatory frameworks and 8. Innovation); and (3) trust and engagement, build public confidence and inclusiveness with stakeholders (Principles 9. Integrity and transparency, 10. Stakeholder engagement, 11. Trade-offs across users and 12. Monitoring and evaluation) [31]. By contrast to IWRM, the OECD framework is highly prescriptive and features: (1) a "traffic light" of the current state of water governance; (2) detailed checklists of what should be done, with a series of "What, Who and How" questions for each Principle; and a (3) ten-point assessment that involves a diagnosis and the development of an action plan to resolve the "What, When, Who and How" of implementation.

Pegram et al. [21] developed a River Basin Planning framework intended to be strategic and multidisciplinary and to deliver improved economic, ecological and management solutions at a basin scale. This framework is described using multiple examples and actual cases to show how river basin planning can be practically delivered. In common with the OECD framework, River Basin Planning is highly prescriptive and has as its core ten golden rules. These rules include: 1. Develop a comprehensive understanding of the entire system; 2. Plan and act, even without full knowledge; 3. Prioritize issues and adopt a phased and iterative approach; 4. Enable adaptation; 5. Accept basin planning is an inherently iterative and chaotic process; 6. Develop relevant and consistent thematic plans; 7. Address issues at the appropriate scale; 8. Engage stakeholders; 9. Focus on implementation; and 10. Select the planning approach and methods to suit basin needs [21]. The key steps in River Basin Planning include: (1) situation assessment, including future trends and scenarios; (2) vision formulation, including goals and outcomes; (3) Basin strategies, including conservation, water use and development, disaster risk management and institutional management; and (4) detailed implementation, including activities and milestones, responsibilities and monitoring and review [21].

An alternative to the mainstream discourse on water governance is the Framework on Hydro-Hegemony (FHH) introduced by Zeitoun and Warner [32] and developed to analyze trans-boundary water conflicts. It has been widely employed at a river basin level [33] and provides an understanding of how three forms of power ("hard" or structural power; covert or bargaining power to shape agendas; and "ideational" power to shape perceptions and discourses) are used [34] and the nature of power asymmetries. While the FHH is not a governance framework per se, it does provide an effective means to better understand socio-political-economic relations in relation to water, and how these relations determine water outcomes.

2.2. The Water Governance Reform Framework (WGRF)

Given the existing water governance frameworks (IWRM, OECD, and River Basin Planning), why is there a need for a water governance reform framework? First, there is a "sweet spot" between the highly flexible, even vague, approach of IWRM and the highly prescriptive, even restrictive, rules and traffic lights frameworks of the OECD and River Basin Planning. Second, we contend the water governance reform framework (WGRF) offers both a more concise and easier-to-apply approach that complements the detailed water planning in the OECD and River Basin Planning approaches, among others, as well as more general policy frameworks. Third, as far as we are aware, the WGRF is the only water governance framework specifically developed and applied for strategic water reform. Fourth, it comprises three key strategies for integrative water security research that include: (1) linkage between the state of knowledge to decision-making; (2) an expanded water research agenda, such as comprehensive water accounting; and (3) a recognition of inequities in terms of water allocation and also the need for water justice [30].

The WGRF has as its core seven strategic considerations in relation to water reform and its implementation. Importantly, it is not a "checklist", but rather a set of strategic considerations that include: (1) well-defined and publicly available reform objectives; (2) transparency in decision-making and public access to available data; (3) water valuation of uses and non-uses to assess trade-offs and winners and losers; (4) compensation for the marginalized or mitigation for persons who are disadvantaged by reform; (5) reform oversight and "champions"; (6) capacity to deliver; and (7) resilient decision-making that is both beneficial and durable from a broad socio-economic perspective [35]. Of these strategic considerations; (3) in relation to water valuation, (5) reform oversight and "champions" and also (7) resilient decision-making are additional to the OECD and River Basin Planning frameworks. Importantly, the WGRF is also explicit about water equity in relation to (3) evaluation of winners and losers and (4) compensation for the marginalized and also those disadvantaged by reform. Thus, while the WGRF includes elements of existing frameworks, it is integrative, flexible and fit-for-purpose and, thus, a novel framework in its own right.

3. Applications of the WGRF to Australia, Tanzania, Mexico and USA, and Vietnam

The value of any policy framework is not in its principles or steps per se, but rather how they are applied and, importantly, whether the framework generates positive net public benefits to the alternatives. The tactical aspects of water reform must also be context-specific and, thus, the WGRF should not be applied as a step-by-step "How-To-Manual" because what is prioritized and the sequencing of water reform must differ according to values, capacity, hydrological constraints, institutions and other factors.

To show both how to apply the framework and to demonstrate its potential for decision makers, we provide four applications of the WGRF: Murray–Darling Basin (Australia); Rufiji Basin (Tanzania); Colorado Basin (Mexico and USA); and Vietnam. The choice of these applications is, in part, based on our respective knowledge and experiences. The applications were selected to ensure a large variation in terms of institutional context, history, financial resources and capacity, and biophysical differences so as to test the flexibility and applicability of the water governance reform framework. For each application, we present an overview of the biophysical and socio-economic environment, which is then followed by an evaluation of each of the seven strategic considerations.

3.1. Murray–Darling Basin, Australia

The Murray–Darling Basin (MDB) is located in Southeast Australia and covers an area of over one million km^2. The MDB suffers from highly variable rainfall and, sometimes, severe droughts. While there have been various supply-based strategies to respond to the risks of droughts, such as the construction of large up-stream water storages, there has also been a well-recognized need to undertake water reform in terms of how water is used within the basin [36].

The most recent water reform in the MDB began with the National Water Initiative (NWI), agreed to by Basin states and the Australian government in 2004 [37]. Article 5 of the NWI highlighted the need to " ... ensure the health of river and groundwater systems by establishing clear pathways to return all systems to environmentally sustainable levels of extraction." In addition, the NWI prioritized the establishment of consistent rules in relation to water rights and the need for comprehensive water accounting.

A lack of progress in the implementation of the NWI led to the *Water Act 2007* that reassigned the jurisdictional powers for governance of water in the MDB from the Basin states (Australian Capital Territory, New South Wales, Queensland, South Australia and Victoria) to the federal government. This act is being implemented through a ten-year Basin Plan that passed the Federal Parliament in November 2012 [38]. The 2012 Basin Plan specifies catchment and basin-level sustainable diversion limits (SDLs).

To encourage states to agree to the change water governance powers, initially opposed by the state of Victoria, and to give effect to key objects of the *Water Act 2007*, the federal government allocated A$10 billion (subsequently increased to A$13 billion in 2008) over ten years to compensate irrigators and ensure the success of water reform [39]. This financial reform package [36] included some A$8.9 billion to respond to over-allocation of water by buying back water entitlements from willing irrigators (A$3.1 billion) and also by modernizing irrigation infrastructure (A$5.9 billion) to increase the irrigation efficiency.

(1) Well-Defined Reform Objectives

The *Water Act 2007* has well-defined, but high-level objectives. The 2012 Basin Plan gives effect to this Act and key environmental targets to be achieved beyond 2019, as detailed in the Murray-Darling Basin Authority's (MDBA) basin-wide environmental watering strategy. Key environmental targets include: (1) maintain base flow levels at 60% of natural flows; (2) enforce environmentally sustainable limits on the quantities of surface water and groundwater that may be taken from the basin water resources; (3) increase overall flow by 10% more into the Barwon–Darling, 30% more into the River Murray and 30–40% more to the Murray mouth which opens to the sea 90% of the time to an average annual depth of one meter; (4) export 2 million tons of salt per year to the Southern Ocean; and (5) improve bird breeding with up to 50% more breeding events for colonial nesting species and a 30–40% increase in nests and broods for other waterbirds [40]. While there have been some environmental improvements in specific locations [41], none of these five key objectives at a basin scale, as of the start of 2019, have been realized [36,42–46].

(2) Transparency

The Australian government's own Productivity Commission, in a five-year review of the 2012 Basin Plan, raised serious concerns about the lack of transparency and accountability in terms of institutional and water governance arrangements [47]. It observed: 'This lack of transparency has resulted in stakeholders seeking information through other means, including Freedom of Information requests and orders for the production of documents in the Australian Parliament. The absence of transparency has engendered an environment of low confidence and trust in Governments' [47]. Further, despite multiples of billions of expenditures on water infrastructure by the Australian government, there has been no publicly available cost-benefit analysis in relation to these expenditures [46].

Notwithstanding these deficiencies, water reform funds allocated for water data collection have resulted in a substantial improvement in the quality and range of data available. These data include information on water flows within the basin, such as the Water Data Online and the National Water Account, both of which are available through the Bureau of Meteorology [48].

(3) Water Valuation

A number of water valuation studies have been undertaken in the MDB, including several non-market studies [49–54]. However, a major omission in terms of quantitative valuation studies is in terms of Indigenous water values [55], notwithstanding several important qualitative studies, such as Weir [56].

Hatton MacDonald et al. [54] provided a total economic value for environmental assets located at the Murray River Mouth of A$13 billion. Akter et al. [49] estimated marginal non-use values of water in a key Basin wetland that are comparable to the annual market price for water used for irrigation. While these non-market studies and hydrological-economic studies [57,58] have been important in making the possible trade-offs between use and non-use values of water transparent, there is no evidence they made any difference in the determination of SDLs in the 2012 Basin Plan. Indeed, sworn testimony to the Murray–Darling Basin Royal Commission (MDBRC), shows that the final determination about what should be the SDLs under the 2012 Basin Plan was dominated by political considerations with little or no regard to scientific or valuation studies [59].

(4) Compensation and Mitigation Mechanisms

Almost all the compensation paid for the reallocation of water as part of the 2012 Basin Plan has been provided to irrigators. The total compensation is A$8.9 billion, and expenditures to date include: (1) A$2.5 billion associated with the direct purchase of water entitlements from willing irrigators and (2) A$3.5 billion in subsidies and grants for water irrigation infrastructure [36]. The compensation already paid represents, on average, about A$750,000 per irrigator [60]. Given that there is no obligation for irrigators to either sell their water entitlements or to accept water infrastructure subsidies, the costs of reallocating water to the environment with the current water reform represents full compensation to irrigators for agreed to changes in their water diversions.

The approach taken to compensation and mitigation in the MDB has been to direct it to irrigators in the belief that benefits will "trickle down" to rural communities. An alternative would have been to invest up to several billions of dollars in "thriving communities" and still have had sufficient funds leftover to acquire the volumes of environmental water obtained under the actual water reform process [61]. Importantly, the Traditional Owners of the land and water of the basin, Australia's First Peoples, have received virtually no compensation for their land and water rights [62] and " … no material increase in water allocation for Indigenous—social, economic or cultural purposes" [63]. This is despite Article 25 of the 2004 NWI highlighting Indigenous needs in relation to water access and management, and NWI Articles 52–54 requiring water plans account for Indigenous access and native title rights to water.

(5) Reform Oversight and Champions

Reform oversight and a "champion" of change are critically important to deliver successful water reform, as shown in relation to water market reforms over the past 25 years [64]. A water reform "champion" did exist, the National Water Commission (NWC) that was created as part of the 2004 NWI, but it was abolished by an Act of Parliament in 2015. The Parliamentary Secretary responsible for announcing the NWC demise justified this decision on the basis that " … there is no longer a need for a stand-alone entity to undertake monitoring of Australia's progress on water reform" [65].

Despite the Parliamentary Secretary's claim that a body like the NWC was no longer needed, multiple failures in terms of implementation of the Basin Plan and delivery of key objects of the *Water Act 2007* have been identified. In particular, the Senior Counsel assisting the MDBRC, Richard Beasley, stated "The implementation of the Basin Plan has been marred by maladministration … The responsibility for that maladministration and mismanagement falls on both past and current executives of the MDBA and its board" [66]. This conclusion that the MDBA has not provided the reform oversight required, and that an alternative and truly independent water agency is required to audit progress on the Basin Plan and its delivery of the *Water Act 2007*, is supported by Grafton [46], Grafton and Wheeler [36] and the Australian Government's own Productivity Commission [47], among others.

(6) Capacity to Deliver

Australia is a world leader in various water disciplines that include hydrology, hydrogeology, water law, water and climate modeling, environmental flows, and water economics, among others. This expertise resides primarily within universities, but also within government research agencies that, at a federal level, include the Commonwealth Scientific and Industrial Research Organization (CSIRO), the Bureau of Meteorology and Geoscience Australia. Scientific and technical capacity also previously existed within the NWC, and some expertise resides in state agencies and in the private sector.

In a country well-endowed with scientific and social science capacity, much of this expertise has, unfortunately, not been used in relation to key decision making with respect to the basin, and even disregarded. Key implementing agencies, such as the MDBA, have relied heavily on the reports of paid consultants for key studies in relation to water reform yet, as noted by Wheeler et al. [67], many of the socio-economic consultant reports commissioned by the MDBA suffer from serious technical deficiencies. Of equal concern is sworn testimony that a key CSIRO report used in modeling the multiple benefits of the SDLs in the 2012 Basin Plan " ... was altered by the CSIRO management under pressure from people at the Basin Authority" and that " ... the report was altered in a way that made it misleading" [66]. This appears to be a systemic problem, not just limited to the MDBA, as Ken Matthews, previously the head of two Australian government departments and the CEO of the NWC, has observed that " ... current water decision-making processes have been designed with an assumption that good science and careful analysis will make its way up through the system, and that responsible ministerial decision makers will be at the helm to receive it. But it turns out that too often they are not" [68].

(7) Risk and Resilient Decision-Making

There are multiple risks that need to be managed in the context of water and the MDB. First, and foremost, is the very large variation in precipitation across seasons and years that can result in both floods and extended droughts. This risk has been managed by building very large water storages and allocating water to irrigators via water entitlements that are a share of a consumptive pool [69]. Thus, in periods of low water storages and inflows, holders of water entitlements, and in particular, those with low-reliability water entitlements, receive less than they would in "normal" years. This allocation of water also places a priority on providing water volumes to water rights with subsidiary priority on discretionary environmental flows that are not managed as water entitlements [70].

Despite a median climate change scenario developed by CSIRO for 2030 and 2050 that has, respectively, an 11% and 17% reduction in the runoff for the southern MDB [71], there was no consideration of climate change in the setting of SDLs as part of the Basin Plan [72,73]. Equally relevant, no consideration was made for the effects of climate change on the economic returns associated with the A$3.5 billion already spent, and billions to be spent, on upgraded irrigation infrastructure [73,74]. This is problematic because modeling suggests that a more resilient strategy, in relation to droughts, is to reduce water diversions rather than invest in water infrastructure [75].

3.2. *Rufiji Basin, Tanzania*

The Rufiji River Basin is the largest river basin in East Africa, covering almost 180,000 km^2—or one-fifth of Tanzania's area [76]. The basin varies widely in elevation (from sea-level to 2960 m) and climate, ranging from hot and humid along the coast, to cool and moderately dry in the highlands [77]. From an administrative viewpoint, the Rufiji River Basin comprises 26 districts; while, hydrologically, the basin is divided into four river catchments: Great Ruaha, Kilombero, Luwegu and Lower Rufiji. The Great Ruaha is the largest catchment, accounting for almost half of the basin's area and over 80%

of its consumptive water uses—although it only contributes one-fifth to the 31 BCM/year of the basin's flow at the delta.

Over the last decades, management of natural resources in Tanzania has undergone a significant transformation to a decentralized system. During pre-colonial times, natural resources were managed under customary rules [78], and it was not until the early 1900s that formal water law was introduced by German and British settlers. By the 1950s, and until two decades after independence in 1961, water governance remained vested under the national government [79]. During the 1980s, widespread and poor infrastructure performance, and also negative environmental outcomes, drove a shift towards decentralization and the introduction of river basins as water governance entities. The 1990s and 2000s were dominated by a comprehensive reform, culminating in the passage of the *2002 National Water Policy* and the *2009 Water Resources Management Act* [80].

The water-governing authorities are structured on five levels of management: (i) national; (ii) basin; (iii) catchment; (iv) district; and (v) community [81]. At the national level, the Ministry of Water and Irrigation (MoWI) is Tanzania's uppermost water authority and is responsible for formulating and updating the country-wide policies. Under the MoWI, there are nine Basin Water Boards that are responsible for the allocation and protection of water resources. Beneath the Boards are Catchment Water Committees (CWC) intended to coordinate Integrated Water Resource Management (IWRM) plans and to resolve regional water conflicts. Only a few CWC have been established [80] so to overcome this water governance gap, District Facilitation Teams have emerged from administrative District Councils and are responsible for conflict resolution, water infrastructure planning, and the formation of Water User Associations (WUAs) [82,83].

(1) Well-defined Reform Objectives

Tanzania's subsidiarity principles were adopted in line with the United Nations Agenda 21, calling for water resources management to be delegated at the lowest appropriate level. Nevertheless, not all objectives defined at the national level (e.g., cost-recovery, equity and efficiency) can effectively be devolved to WUAs. In an illustrative example in the upper Great Ruaha River catchment, van Koppen et al. [84] observed how a newly introduced fee system eroded customary water-sharing principles. Thus, informal self-supply in Tanzania remains a key feature of rural water provision [85] and must be considered, alongside formal rules, to deliver an effective IWRM framework [86].

(2) Transparency

Disputes over water are common at the local level, i.e., between and within WUAs. Local conflicts can impose severe consequences for the community and beyond, as they often result in eroded cooperation among users and a lack of adherence to water rules and mismanagement of the water resources. Bureaucratic and logistic constraints also pose major obstacles to access legal water institutions by remote, ill-resourced water users [82]. Thus, transferring authority to judge water-related disputes from regional courts to local WUAs could make the legal system more accessible. Further, transparent and accountable local institutions would fill in the current gap left by the limited reach of national institutions [87].

(3) Water Valuation

Decreasing stream flows and growing water demands for environmental flows, hydropower and agriculture have fuelled tensions over water resource allocation within the Rufiji River Basin. Tanzania relies on this basin for over four-quarters of its installed hydropower capacity and it is also home to the Ruaha National Park—the country's largest—which depends on the Rufiji River as its sole water source during the dry season.

Given its importance, numerous studies on the Great Ruaha catchment have been undertaken [87–91]. Most recently, Yang and Wi [92] observed that effective water policies should be articulated by a combination of strategies that enhance environmental outcomes and are socio-economically acceptable.

(4) Compensation and Mitigation Mechanisms

Irrigation is the largest user within the Rufiji River Basin, accounting for almost 80% of its consumptive water uses—mainly for traditional, smallholder schemes. The existing irrigation area (87,000 ha) is projected to almost quadruple by 2030 [77], in line with the national Agriculture Sector Development Program targeting rural growth and poverty reduction. By contrast, independent studies conclude that viable expansions are limited [93] and, instead, a priority should be to increase the productivity of high-value crops irrigated during the dry months [94]. Proposed measures to reduce consumptive water use in the Great Ruaha River could also be accompanied by targeted social interventions to minimize or avoid negative impacts on the local communities including non-irrigation economic activities, such as livestock or off-farm work [95,96].

(5) Reform Oversight and Champions

Water reform is embedded across three contexts: (1) a dichotomy between formal and customary laws; (2) multiple government bodies nested within each other—in a succession from national to local levels; and (3) water governance across two parallel lines of authority; (i) the Ministry of Water and Irrigation that follows natural water boundaries and (ii) Local Government Authorities, defined by administrative boundaries.

Overlapping mandates across multiple water-governing institutions have contributed to rivalries over horizontally (across lines of governance) and vertically (across hierarchical levels) [77]. Thus, a designated, independent body to provide oversight and to harmonize water management across all institutions is required [82].

(6) Capacity to Deliver

Tanzania's water policies of the early 2000s represented a step-change towards IWRM goals, including water-use efficiency, irrigated crop productivity and equity of water supply. Unfortunately, as van Koppen et al. [97] argue, these reforms have failed to achieve several of their objectives, in particular, cost-recovery and alleviation of basin-level water scarcity. This is attributed to a lack of scientific analysis and poor stakeholder consultation which means that IWRM principles have, as yet, to be translated into reality.

The national versus local-level policy dichotomy is highlighted by the principle of "water equity" that is mandated by the National Water Policy 2002, Water Sector Development Strategy 2006, Water Resources Management Act 2009, National Irrigation Policy 2010 and the National Irrigation Act 2013. Despite this equity goal, many WUAs lack the knowledge, physical and technical capacity to monitor and deliver "water equity" [82].

(7) Risk and Resilient Decision-Making

Rising aquifer levels due to over-irrigation are contributing to soil-salinization across the Rufiji River Basin [98] and other agricultural areas [99,100]. Moreover, groundwater is increasingly polluted as a result of human activities, notably by nitrates from poor sanitation and fertilizer use [101–103]. Data on hydrogeology is extremely limited [104], yet it is understood that abstractions for domestic uses are growing, despite the severe health risks resulting from polluted groundwater.

One of the most critical challenges faced by the Great Ruaha catchment is the need to adapt to a variable climate, population growth and institutional changes. This means that water reforms should be framed as part of broader-reaching policies such as the education of local communities about watershed protection [87] and the cultivation of drought and salt-resistant crop varieties.

3.3. *Colorado Basin, Mexico and USA*

The Colorado River (637,137 km^2) is an international river shared by two federal countries: Mexico and the USA. It straddles seven states in the USA and two states in Mexico, supporting 2.23 million hectares of irrigated agriculture and 40 million people, including several Tribal Nations [105]. The basin's multi-purpose reservoir system has the capacity to store approximately four years of annual average runoff (approximately 18.5 BCM), provides 4.2 GW of hydropower capacity, and supports a range of recreational uses, including rafting and boating. The upstream development of water resources has led to the decline of a once-vast delta ecosystem, which is now the focus of bi-national restoration efforts by the US and Mexico to secure water for base flows and pulse flows [18,106].

(1) Well-Defined Reform Objectives

Water allocation in the Colorado River Basin is governed by a complex mix of more than 100 laws, court decisions, operational guidelines, and technical rules known as the "Law of the River". The 1922 Colorado River Compact and the *1928 Boulder Canyon Project Act* established a fixed water allocation for downstream states within the US. This legal framework for interstate apportionment was confirmed in the Supreme Court decision on Arizona v. California in 1963; it requires "upper division" states (Wyoming, Colorado, Utah, and New Mexico) to deliver 92.5 BCM to the "lower division" states (Arizona, California, and Nevada) over a rolling 10-year period. It formally allocated an equivalent volume to the upper division states. Downstream delivery requirements from the upper division to lower division states are assessed on a rolling 10-year accounting period. In practice, the fixed allocation leaves the upper division states with residual flows and, hence, disproportionate exposure to hydro-climatic risks. Both divisions are responsible for Mexico's 1.85 BCM annual allocation secured under a 1944 international treaty.

The overarching reform responds to unsustainable water extractions, which are described as a "structural deficit" in which water use exceeds long-term renewable supplies. In 1999, long-term supply and demand intersected for the first time, coinciding with the beginning of an unprecedented 20-year sequence of dry years and increasing evidence from tree-rings and climate models of the potential for severe sustained drought and drying [107]. Despite over-allocation and a history of disputes, these pressures have prompted a reform period marked by institutional innovations. The states and other stakeholders in the basin have undertaken inter-related investments in institutions, infrastructure, and information to respond to the consequences of climate variability and change, starting with the development of interim guidelines for sharing surplus water among the states in 2001, and six years later, for sharing shortage.

(2) Transparency

Long-term planning and operational decision-making are guided by the Colorado River Simulation System (CRSS), a river system model that requires explicit and transparent assumptions regarding water availability, water deliveries and allocation rules [108,109]. The 2010–2012 Colorado River Basin Study used CRSS to engage stakeholders and establish a common language to navigate trade-offs associated with intensifying scarcity and shortage risks.

The reliance on CRSS for stakeholder engagement and water planning involves both strengths and weaknesses. On the one hand, the federal agency managing the reservoirs in the basin note that, as a result of the modeling system, 'transparency facilitated stakeholders being on relatively equal ground, rather than [certain parties] having an advantage' [108]. At the same time, the barriers to entry are substantial, which involves a steep learning curve that has required the capacity for building for historically marginalized groups, including environmental and indigenous stakeholders.

(3) Water Valuation

The development of water markets in parts of the Colorado River Basin has revealed the value of water in its competing uses, particularly between agriculture and urban uses. The Colorado-Big Thompson is home to the most active water market in the basin where shares of agricultural water are being leased or purchased by cities in the state of Colorado. In 2010, shares of water in the Colorado-Big Thompson project sold for approximately US$6,900/mL (approximately US$6.89/m^3, 2010 prices, [110]). More recent prices have been reported to be well over three times this amount, approaching US$24,300/mL [111] (2018 prices). Outside of the Colorado-Big Thompson, water markets are thin and constrained by high transaction costs and concerns about third-party effects [112]. As a consequence, the price of water established by markets or administrative decisions remains a poor reflection of the value of water in its competing uses, particularly for non-market benefits.

The non-market valuation of water in the Colorado River Basin has aimed to capture the economic value of water used for instream flows, recreational purposes and other types of ecosystem services. A 2014 report applied an ecosystem services framework to estimate the total economic value of the Colorado River, indicating economic benefits from US$56.6 to US$466.5 billion per year with an underlying asset value between US$1.5 and US$11.5 trillion [113]. The cultural values have proven difficult to integrate into decision-making, as illustrated by efforts to address lingering controversies over indigenous water justice and historical exclusion of indigenous groups from water planning and allocation [114].

(4) Compensation and Mitigation Mechanisms

Institutional mitigation and adaptations include interstate and bi-national agreements for coordinated operations of reservoir storage, together with new rules for managing surpluses and shortages, incentives for system efficiency improvements, and commitments to ecosystem restoration. Investments in infrastructure include the operation of desalination plants, conservation measures, and reservoir intakes. In the context of prolonged drought conditions, environmental flow requirements in the Delta have received additional attention, culminating in 2012 in "Minute 319" (updated in 2017 in "Minute 323"), an agreement made under the 1944 Treaty, coordinating US and Mexico's water storage and delivery options to enhance water supply reliability for Mexican water users and the Delta ecosystem [106].

Looking forward, the annual average cost of reducing shortage risks is projected to approach up to US$6 billion in 2060, reflecting a lingering bias toward supply-side solutions including the importation of water from other basins [105]. Much of these costs involves alternative water supplies (desalination) or importation, rather than improved water allocation, which is lower cost but concentrated on powerful agricultural user groups. Even restricting mechanisms to demand-side solutions will require substantial investment to sustain water security and safeguard the economic activities, urban centers, and ecosystems that depend on the river in a changing climate.

(5) Reform Oversight and Champions

The federal and bi-national structure of the basin involves multiple layers of oversight. It also creates ambiguity regarding authority and accountability [115]. The US Secretary of Interior serves as the "rivermaster" under the Colorado River Compact, and the federal government can, therefore, take unilateral action when the state governments lag. A credible threat has been issued in 2004 and 2018 as a spur to action by the state governments to negotiate rules for sharing the risks of shortage. Shortage rules issued in 2007 signaled this potential, noting that the Secretary 'shall evaluate and take additional necessary actions, as appropriate, at critical elevations in order to avoid Lower Basin shortage determinations as reservoir conditions approach critical thresholds ... ' ([115], p. 40). On 13 December 2018, the sitting commissioner of the Bureau of Reclamation, Brenda Burman, threatened federal action on drought contingency planning if the states fail to achieve an agreement by 31 January 2019.

Paralleling the experience in the Murray–Darling and other cases, the existence of champions has proven pivotal in overcoming resistance to reform, including key organizations as well as influential policy leaders to sustain progress. The nonprofit sector, led by environmental NGOs, such as Environmental Defense and Audubon, have played a key role, facilitated by established members of policy and planning networks, such as high ranking officials in the Bureau of Reclamation and within state agencies [106].

(6) Capacity to Deliver

The USA and Mexico have significant institutional capacity, technical knowledge and financial resources to deliver on reform. In terms of institutional capacity, the basin has a nested set of governance arrangements that link local users with state, federal and bi-national arrangements. The chief constraint stems from horizontal coordination challenges between states with flashpoints of conflict between Arizona and California and between the Upper Basin states, particularly Colorado, and those downstream, particularly Arizona. Increasingly, rural-urban conflicts are posing challenges both locally and within interstate negotiations. Vested interests in the agricultural sector represent a formidable barrier to sustained progress with powerful irrigation districts in Arizona and California mobilizing to thwart changes.

Technical and human capacity are substantial, as highlighted by the Colorado River Simulation System and the development of a multi-state research consortium, the Colorado River Governance Initiative. Thus, the scientific and technical understanding of the basin has been sufficient to support the reform process. The chief impediments have stemmed from the legal uncertainties and institutional coordination issues noted above with surprisingly limited dispute about the underlying science.

(7) Risk and Resilient Decision-Making

Information on climate change, paleo-hydrology and related climate risks have played an increasing role in planning decision-making, supported by a Bureau of Reclamation funded program [109]. The severity of climate change impacts are increasingly evident, but action has lagged [107] due to the lingering distributional conflicts regarding the risks of shortages. Despite these lags, efforts to address environmental flows have expanded during sustained drought and structural imbalances highlighted by two agreements which update the international treaty between the US and Mexico to include provisions for a flood pulse to restore the Colorado River Delta. These experiences show an increasing commitment to adaptive and flexible water allocation with growing recognition of the interdependence of water users and the need to progress toward systemic resilience.

3.4. *Vietnam*

Effective water resource management plays a crucial role in Vietnam's economic development, with 80% of its GDP generated in its key river basins [116]. Despite having, on average, about 2000 mm rainfall per year and about 3500 rivers in 16 major river basins, water availability is highly seasonal and unevenly distributed across the country [117]. In turn, this contributes to severe water scarcity in some regions at particular times of the year [118].

Vietnam has over 90 million people and is rapidly industrializing and urbanizing, which is contributing to greater extractions of both surface and groundwater [116,117]. Consequently, stark trade-offs are emerging in water allocation for agriculture, industry, and households. Further, about two-thirds of Vietnam's water resources are sourced beyond its borders, such that its internal water resource availability is 4200 m^3 per person compared to an average of 4900 m^3 for South East Asia [119].

The supply of water is affected by water pollution with only 12% of domestic wastewater [120] and 25% of industrial wastewater treated before being discharged into streams and rivers [121]. As a result, untreated wastewater has polluted rivers and lakes in and around big cities and industrial zones, undermining the health and livelihoods of millions of people [116,117,119].

Water scarcity and pollution are being exacerbated by climate change. It is estimated that a sea level rise of one meter by the end of this century will displace about 11% of the population, mainly in the Red River Delta and the Mekong River Delta and along 3000 km of the coast [122]. Recent severe typhoons and storm surges, as well as other extreme weather events, such as drought and flash floods, have cost around US$1.75 billion [118] and are projected to be a major challenge in the future in the Mekong Delta that produces half of Vietnam's rice [116,118].

(1) Well-Defined Reform Objectives

Recognizing the importance of water, the Vietnam Government is reforming its water governance. Key goals are defined in the National Strategy on Water Resources in five areas: (1) water resource protection; (2) water resource exploitation and usage; (3) water resource development; (4) mitigation of water-related damages; and (5) improvement of water resource management capacity [123]. The water reform objectives are legalized by the Law on Water Resources (LWR) issued in 1998, and revised in 2012, and the Law on Environmental Protection (LEP) issued in 1993, revised in 2005 and again in 2014. The object of the LWR is to provide a legal basis for the management, protection, exploitation and use of water resources, as well as for the prevention, control and remedy of harmful effects caused by water [124]. The LEP 2014 governs environmental protection activities, including water environment protection [125].

To achieve its water reform objectives, government agencies for managing water resources have been established at both the central and provincial level. At the central level, the Ministry of Natural Resources and Environment (MONRE) was established in 2002 to separate policy development and regulation of water resources from ministries overseeing the exploitation and use of water resources for economic development, such as the Ministry of Agriculture and Rural Development (irrigation and flood control), Ministry of Industry and Trade (hydropower) and Ministry of Construction (municipal water supply and drainage) [126,127]. All of these ministries also have their branches at the provincial level.

(2) Transparency

Water resource management in Vietnam faces several transparency challenges. In particular, limited public access to data and data sharing have led to inefficient policy implementation [119]. While water information and knowledge sharing do occur, they are often blocked due to limited coordination between different levels of administration [126]. In addition, individuals and organizations with data are reluctant to publish and share data. Further, the participation of civil society, research institutions and local communities in river basin decision-making is limited [128]. While public participation in environmental impact assessment and other water decision-making processes is legally permitted [124,125], insufficient guidance and a lack of enabling policy have hindered implementation. Another key challenge of transparency is the monitoring and enforcement of environmental flows and water quality [116].

(3) Water Valuation

The principle of water allocation for high economic value use is included in the National Strategy on Water Resources [123], but in practice, water allocation and reallocation is, typically, not based on economic values; although, water pricing [129] and charges [130] have been identified as instruments for water policy. Instead, the administrative allocation of water in Vietnam is principally based on biophysical information [118]. Further, while the concept of environmental water values is part of the national law [125], few studies have been undertaken (for example, Nam and Son [131]; Do and Bennett [132]; Vo and Khai [133]) to support water decision-making [134].

(4) Compensation and Mitigation Mechanisms

Compensation for water-related damages is included in the LWR 2012 and LEP 2014, with the latter comprising two revisions to respond to concerns over compensation [125]. In particular, Article 164 regulates that the head of an organization that incurs an environmental violation is held responsible. Article 162 regulates that the statute of limitations of violation begins when the damage is detected rather than when the violation occurs. Nevertheless, compensation challenges remain including the absence of the capacity and technical knowledge to quantify losses.

(5) Reform Oversight and Champions

MONRE is mandated to oversee overall water resource management but faces challenges in fulfilling its role. This is because water policy development and implementation are fragmented across national ministries with limited coordination between MONRE, other national ministries, and provincial governments [126]. To illustrate, commissions for environmental protection were established in three major river basins (Cau, Nhue-Day and Dong Nai-Sai Gon) in 2007–2009, but they have struggled to achieve environmental targets [118]. Another challenge is cooperation between national and local levels. Dual supervisory roles of central agencies and provincial people's committees, in addition to unclear reporting mechanisms, have also limited the efficiency of water governance [119,126].

Reform oversight is challenged by inadequate transboundary cooperation mechanisms. The Mekong River Commission was formed in 1995 by an agreement between Laos, Thailand, Cambodia and Vietnam, but two upstream countries of China and Myanmar are not members. As a result, decisions on building upstream hydropower plants have not fully taken into account downstream costs in relation to livelihoods and the environment [135].

(6) Capacity to Deliver

Investment in the water sector has increased over the period 2006–2015 and Vietnam has invested more than US$6.4 million in 140 water programs with funding mainly from the state budget and international donors [119]. Yet, the annual investment requirement for water supply and sanitation alone is about US$2.7 billion, while the actual investment currently is less than 40% of this amount. Better planning and allocation, however, should help respond to the funding shortage by improving the efficiency and quality of public spending and also by attracting greater private financing [119].

A lack of staff and insufficient budget for environmental expenditure poses risks in relation to environmental water protection [134]. Inadequate water monitoring systems and limited modeling and management tools also hamper water planning and management [119]. In addition, while there is capacity in water sciences within, for example, the Vietnam Academy of Hydraulic Works, the Institute of Hydraulic Works Planning, and the Institute of Water Resources, key gaps in the social sciences remain.

(7) Risk and Resilient Decision-Making

Water risks are identified and regulated by the LEP, but a limited capacity within relevant water agencies means that some of these risks, especially in terms of water quality, are not effectively managed or mitigated. Without adequate risk assessment, the ad hoc construction of levees in the Mekong Delta began in the early 2000s and has resulted in increased flooding in the areas outside the levees, reduced soil fertility, degraded wetlands and also decreased rice productivity [136]. An adaptive response to this risk has been the 2017 Government Resolution 120, which is intended to promote flood-based livelihoods and adaptation to climate change [137].

4. Discussion

The four applications show that the WGRF is easy-to-apply, flexible and can be used in multiple contexts, and at both a basin and national scale. Collectively, the four cases show that, even in the absence of quantitative analysis and modeling, the strategic considerations provide a means to scope current water governance and to identify both the strengths and weaknesses of existing water reform processes.

The strategic considerations in the WGRF serve multiple purposes. First, as in the case of the Murray–Darling Basin, Australia, they provide a means to evaluate the successes and failures of reform, highlight power asymmetries and, importantly, provide constructive support for adaptive management of the reform process. Second, as applied to the Rufiji Basin, Tanzania, the WGRF scopes the existing water governance structure and, thereby, identifies future possible opportunities for water reforms. Thus, the framework can be used to devise a more effective water reform agenda, ex ante and not just during the reform process or to rectify past mistakes. Third, as in the case in the Colorado Basin, the WGRF frames current and past actions to support dialogues in relation to transboundary institutional reforms. In particular, the framework identifies the necessity for reform oversight and how barriers to water reform might be overcome in the Colorado Basin. Fourth, for Vietnam, the framework shows it can provide an evaluation of water governance strengths and weaknesses at a national level and not just a basin scale. Further, in terms of Vietnam, the WGRF shows how framing of the water institutions, and their roles, provides the opportunities for better coordination, decision-making and improved water outcomes, especially in relation to water quality.

While the WGRF can be used as a stand-alone approach to water reform, we highly recommend that it be part of an overall water policy cycle that can be applied to problem identification and also within each policy sequence that comprises: (1) formulation; (2) adoption; (3) implementation; and (4) monitoring and evaluation. We show that the WGRF is a valuable descriptive tool of water governance and reform, but its principal contribution is to support adaptive decision making and resilient public policy. This is because successful water reform in relation to "wicked problems" [13] cannot be decided ex ante without regard to learning, new information, changes in circumstances of stakeholders or a range of other unknown or unknowable factors. It is in this context of socio-ecological systems that the WGRF facilitates evidence-informed actions and integrative decisions that support desired water outcomes. Thus, we contend that WGRF should be core to any water governance reform process and is flexible enough to be applied at both a local, basin and national scale.

5. Conclusions

The world stands at a critical threshold in terms of how water is extracted and consumed, by whom, when and where. Consequently, decision-makers face key challenges in terms of how to balance water supply with demand without compromising the long-term sustainability of riparian ecosystems or aquifers. This requires water conservation and water reallocation as part of an on-going governance reform process. While there are already a number of water management (such as Integrated Water Resources Management) and governance guidelines (such as OECD Water Governance Principles), we contend that none provide the "sweet spot" in terms of ease of use, flexibility to multiple scales and contexts, or is integrative in relation to the reform research agenda, especially in relation to inequities in water allocation.

In response to the needs of decision-makers in relation to water allocation and water outcomes, we developed the water governance reform framework (WGRF). It is a strategic framework that allows both stakeholders and decision makers to review seven key considerations: (1) well-defined and publicly available reform objectives; (2) transparency in decision-making and public access to available data; (3) water valuation of uses and non-uses to assess trade-offs and winners and losers; (4) compensation for the marginalized or mitigation for persons who are disadvantaged by reform; (5) reform oversight and "champions"; (6) capacity to deliver; and (7) resilient decision-making. In four very different applications spanning five countries, we show how the WGRF can be readily applied to provide valuable insights about water governance and the water reform process, even in the absence of quantitative analysis or modeling. We contend that these applications show that the WGRF is fit for purpose and adds important integrative features to existing governance principles. In our view, if the WGRF is employed within a broader water policy cycle, it will help deliver both improved water outcomes and more effective water reforms.

Glossary (from Grafton [13]):

Water Scarcity: A measure of water use to water availability. A commonly used measure is the ratio of the annual water extracted in a given location to the annual renewable fresh water available.

Water Security: According to the United Nations it is 'The capacity of a population to safeguard sustainable access to adequate quantities of acceptable quality water for sustaining livelihoods, human well-being, and socio-economic development, for ensuring protection against water-borne pollution, and water-related disasters, and for preserving ecosystems in a climate of peace and political stability.'

Water Stress: A measure of per capita annual renewable fresh water availability. Water availability of less than 1000 m^3 is considered to be high water stress and less than 500 m^3 is defined as extreme water stress.

Author Contributions: Authorship is alphabetical after the first author. Conceptualization (R.Q.G.); Applications (R.Q.G., D.G., A.M. and T.N.D.); Writing-Review & Editing (R.Q.G., D.G., A.M. and T.N.D.).

Funding: D.G. received partial funding from Social Sciences and Humanities Council of Canada. Grant Number: 430–2014-00785. A.M. received partial supported by the Australian Centre for International Agricultural Research under project number FSC2013-006.

Acknowledgments: The authors are most grateful to Arjen Hoekstra for the invitation to submit an article to the special 10th anniversary issue of Water.

Conflicts of Interest: The authors declare no conflict of interest. The funders had no role in the design of the study; in the collection, analyses, or interpretation of data; in the writing of the manuscript, and in the decision to publish the results.

References

1. Mekonnen, M.M.; Hoekstra, A.Y. Four billion people facing severe water scarcity. *Sci. Adv.* **2016**, *2*, e1500323. [CrossRef]
2. Vörösmarty, C.J.; McIntyre, P.B.; Gessner, M.O.; Dudgeon, D.; Prusevich, A.; Green, P.; Glidden, S.; Bunn, S.E.; Sullivan, C.A.; Liermann, C.R. Global threats to human water security and river biodiversity. *Nature* **2010**, *467*, 555–561. [CrossRef] [PubMed]
3. Wada, Y.; Bierkens, M.F.P. Sustainability of global water use: Past reconstruction and future projections. *Environ. Res. Lett.* **2014**, *9*, 104003. [CrossRef]
4. Wada, Y.; Gleeson, T.; Esnault, L. Wedge approach to water stress. *Nat. Geosci.* **2014**, *7*, 615–617. [CrossRef]
5. Nechifor, V.; Winning, M. Global Economic and Food Security Impacts of Demand-Driven Water Scarcity—Alternative Water Management Options for a Thirsty World. *Water* **2018**, *10*, 1442. [CrossRef]
6. Aeschbach-Hertig, W.; Gleeson, T. Regional strategies for the accelerating global problem of groundwater depletion. *Nat. Geosci.* **2012**, *5*, 853–861. [CrossRef]
7. Wada, Y.; van Beek, L.; Bierkens, M.F. Nonsustainable groundwater sustaining irrigation: A global assessment. *Water Resour. Res.* **2012**, *48*. [CrossRef]
8. Grafton, R.Q.; Williams, J.; Jiang, Q. Possible pathways and tensions in the food and water nexus. *Earth's Future* **2017**, *5*, 449–462. [CrossRef]
9. Grafton, R.Q.; McLindin, M.; Hussey, K.; Wyrwoll, P.; Wichelns, D.; Ringler, C.; Garrick, D.; Pittock, J.; Wheeler, S.; Orr, S. Responding to Global Challenges in Food, Energy, Environment and Water: Risks and Options Assessment for Decision-Making. *Asia Pac. Policy Stud.* **2016**, *3*, 275–299. [CrossRef]
10. Menga, F. Domestic and international dimensions of transboundary water politics. *Water Altern.* **2016**, *9*, 704–723.
11. Conker, A.; Hussein, H. Hydraulic Mission at Home, Hydraulic Mission abroad? Examining Turkey's Regional 'Pax-Aquarum' and Its Limits. *Sustainability* **2019**, *11*, 228. [CrossRef]
12. World Commission on Dams. *Dams and Development: A New Framework for Decision-Making*; Earthscan: London, UK, 2000.
13. Quentin Grafton, R. Responding to the 'Wicked Problem' of Water Insecurity. *Water Resour. Manag.* **2017**, *31*, 3023–3041. [CrossRef]
14. Kaushal, S.; Gold, A.; Mayer, P. Land Use, Climate, and Water Resources—Global Stages of Interaction. *Water* **2017**, *9*, 815. [CrossRef] [PubMed]
15. Sultana, F. Water justice: Why it matters and how to achieve it. *Water Int.* **2018**, *43*, 483–493. [CrossRef]
16. Dong, G.; Shen, J.; Jia, Y.; Sun, F. Comprehensive Evaluation of Water Resource Security: Case Study from Luoyang City, China. *Water* **2018**, *10*, 1106. [CrossRef]
17. Zwarteveen, M.Z.; Boelens, R. Defining, researching and struggling for water justice: Some conceptual building blocks for research and action. *Water Int.* **2014**, *39*, 143–158. [CrossRef]
18. Garrick, D.E. *Water Allocation in Rivers under Pressure: Water Trading, Transaction Costs and Transboundary Governance in the Western US and Australia*; Edward Elgar Publishing: Cheltenham, UK, 2015.

19. Rahaman, M.M.; Varis, O. Integrated water resources management: Evolution, prospects and future challenges. *Sustain. Sci. Pract. Policy* **2005**, *1*, 15–21. [CrossRef]
20. OECD. *OECD Principles on Water Governance*; OECD: Paris, France, 2015.
21. Pegram, G.; Yuanyuan, L.; Le Quesne, T.; Speed, R.; Jianqiang, L.; Fuxin, S. *River Basin Planning Principles: Procedures and Approaches for Strategic Basin Planning*; Asian Development Bank, GIWP, UNESCO, and WWF-UK: Paris, France, 2013.
22. Ostrom, E. *Understanding Institutional Diversity*; Princeton University Press: Princeton, NJ, USA, 2005.
23. Meinzen-Dick, R. Beyond panaceas in water institutions. *Proc. Natl. Acad. Sci. USA* **2007**, *104*, 15200. [CrossRef] [PubMed]
24. Powell, N.; Kløcker Larsen, R.; de Bruin, A.; Powell, S.; Elrick-Barr, C. Water Security in Times of Climate Change and Intractability: Reconciling Conflict by Transforming Security Concerns into Equity Concerns. *Water* **2017**, *9*, 934. [CrossRef]
25. Cherlet, J. Tracing the Emergence and Deployment of the 'Integrated Water Resources Management' Paradigm. In Proceedings of the 12th EASA Biennial Conference, Ghent, Belgium, 10–13 July 2012.
26. Allouche, J. The birth and spread of IWRM-A case study of global policy diffusion and translation. *Water Altern.* **2016**, *9*, 412–433.
27. Bogardi, J.J.; Dudgeon, D.; Lawford, R.; Flinkerbusch, E.; Meyn, A.; Pahl-Wostl, C.; Vielhauer, K.; Vörösmarty, C. Water security for a planet under pressure: Interconnected challenges of a changing world call for sustainable solutions. *Curr. Opin. Environ. Sustain.* **2012**, *4*, 35–43. [CrossRef]
28. Cook, C.; Bakker, K. Water security: Debating an emerging paradigm. *Glob. Environ. Chang.* **2012**, *22*, 94–102. [CrossRef]
29. Allouche, J.; Middleton, C.; Gyawali, D. Technical veil, hidden politics: Interrogating the power linkages behind the nexus. *Water Altern.* **2015**, *8*, 610–626.
30. Zeitoun, M.; Lankford, B.; Krueger, T.; Forsyth, T.; Carter, R.; Hoekstra, A.Y.; Taylor, R.; Varis, O.; Cleaver, F.; Boelens, R.; et al. Reductionist and integrative research approaches to complex water security policy challenges. *Glob. Environ. Chang.* **2016**, *39*, 143–154. [CrossRef]
31. OECD. *OECD Water Governance Indicator Framework*; OECD: Paris, France, 2018.
32. Zeitoun, M.; Warner, J. Hydro-hegemony—A framework for analysis of transboundary water conflicts. *Water Policy* **2006**, *8*, 435–460. [CrossRef]
33. Hussein, H.; Grandi, M. Dynamic political contexts and power asymmetries: The cases of the Blue Nile and the Yarmouk Rivers. *Int. Environ. Agreem. Politicslaw Econ.* **2017**, *17*, 795–814. [CrossRef]
34. Zeitoun, M.; Allan, J.A. Applying hegemony and power theory to transboundary water analysis. *Water Policy* **2008**, *10*, 3–12. [CrossRef]
35. Grafton, R.Q.; Reinhardt, W. An assessment framework for resilient public policy. In *Transformational Change in Environmental and Natural Resource Management*; Young, M., Esau, C., Eds.; Routledge: London, UK, 2016; pp. 52–74.
36. Grafton, R.Q.; Wheeler, S.A. Economics of Water Recovery in the Murray-Darling Basin, Australia. *Annu. Rev. Resour. Econ.* **2018**, *10*, 487–510. [CrossRef]
37. Council of Australian Governments Canberra. *Intergovernmental Agreement on a National Water Initiative*; Council of Australian Governments Canberra: Canberra, Australia, 2004.
38. Australian Government. Basin Plan 2012. Available online: www.legislation.gov.au/Details/F2018C00451 (accessed on 4 December 2018).
39. Howard, J. *A National Plan for Water Security*; Australian Government: Canberra, Australia, 2007.
40. Murray-Darling Basin Authority. *Basin-Wide Environmental Watering Strategy*; Molloy, K., Ed.; Murray-Darling Basin Authority: Canberra, Australia, 2014.
41. Stewardson, M.J.; Guarino, F. Basin-scale environmental water delivery in the Murray–Darling, Australia: A hydrological perspective. *Freshw. Biol.* **2018**, *63*, 969–985. [CrossRef]
42. Argent, R.M. *Australia State of the Environment 2016: Inland Water*; Department of the Environment and Energy: Canberra, Australia, 2017.
43. Grafton, R.Q.; Pittock, J.; Williams, J.; Jiang, Q.; Possingham, H.; Quiggin, J. Water Planning and Hydro-Climatic Change in the Murray-Darling Basin, Australia. *Ambio* **2014**, *43*, 1082–1092. [CrossRef]

44. Kingsford, R.T.; Bino, G.; Porter, J.L. Continental impacts of water development on waterbirds, contrasting two Australian river basins: Global implications for sustainable water use. *Glob. Chang. Biol.* **2017**, *23*, 4958–4969. [CrossRef] [PubMed]
45. Wentworth Group. *Review of Water Reform in the Murray-Darling Basin*; Wentworth Group: Sydney, Australia, 2017.
46. Grafton, R.Q. Policy Review of Water Reform in the Murray-Darling Basin, Australia: The Do's and Do Nots. *Aust. J. Agric. Resour. Econ.* **2019**. [CrossRef]
47. Productivity Commision. Murray-Darling Basin Plan: Five-year Assessment. Available online: www.pc.gov.au/inquiries/current/basin-plan/draft (accessed on 6 December 2018).
48. Bureau of Meteorology. Water Data Online. Available online: www.bom.gov.au/waterdata/ (accessed on 4 December 2018).
49. Akter, S.; Grafton, R.Q.; Merritt, W.S. Integrated hydro-ecological and economic modeling of environmental flows: Macquarie Marshes, Australia. *Agric. Water Manag.* **2014**, *145*, 98–109. [CrossRef]
50. Bennett, J.; Dumsday, R.; Howell, G.; Lloyd, C.; Sturgess, N.; Van Raalte, L. The economic value of improved environmental health in Victorian rivers. *Australas. J. Environ. Manag.* **2008**, *15*, 138–148. [CrossRef]
51. Dyack, B.; Rolfe, J.; Harvey, J.; O'Connell, D.; Abel, N.J. *Valuing Recreation in the Murray: An Assessment of the Non-Market Recreational Values at Barmah Forest and the Coorong*; CSIRO: Canberra, Australia, 2007.
52. Morrison, M.; Bennett, J.; Blamey, R. Valuing improved wetland quality using choice modeling. *Water Resour. Res.* **1999**, *35*, 2805–2814. [CrossRef]
53. Morrison, M.; Hatton Macdonald, D.A. *Economic Valuation of Environmental Benefits in the Murray-Darling Basin*; Murray-Darling Basin Authority: Canberra, Australia, 2010.
54. Hatton MacDonald, D.; Morrison, M.D.; Rose, J.M.; Boyle, K.J. Valuing a multistate river: The case of the River Murray. *Aust. J. Agric. Resour. Econ.* **2011**, *55*, 374–392. [CrossRef]
55. Rolfe, J.; Windle, J. Valuing the Protection of Aboriginal Cultural Heritage Sites. *Econ. Rec.* **2003**, *79*, S85–S95. [CrossRef]
56. Weir, J.K. *Murray River Country: An Ecological Dialogue with Traditional Owners*; Aboriginal Studies Press: Canberra, Australia, 2009.
57. Grafton, R.Q.; Chu, H.L.; Stewardson, M.; Kompas, T. Optimal dynamic water allocation: Irrigation extractions and environmental tradeoffs in the Murray River, Australia. *Water Resour. Res.* **2011**, *47*. [CrossRef]
58. Settre, C.; Connor, J.; Wheeler, S.A. Reviewing the Treatment of Uncertainty in Hydro-economic Modeling of the Murray–Darling Basin, Australia. *Water Econ. Policy* **2017**, *3*. [CrossRef]
59. Bell, D. Witness Statement 14 June 2018 to the Murray-Darling Basin Royal Commission. Available online: www.mdbrc.sa.gov.au/sites/g/files/net3846/f/david_bell_final_witness_statement.pdf?v=1529025595 (accessed on 4 December 2018).
60. Young, M. Witness Statement to Murray-Darling Basin Royal Commission 24 September 2018. Available online: https://www.mdbrc.sa.gov.au/sites/g/files/net3846/f/mdbrc-witness-statement-professor-mike-young.pdf?v=1537772711 (accessed on 4 December 2018).
61. Wentworth Group in Association with; Grafton, R.Q.; Kowalick, I.; Miller, C.; Stubbs, T.; Verity, F.; Walker, K. *Sustainable Diversions in the Murray-Darling Basin: An Analysis of the Options for Achieving a Sustainable Diversion Limit in the Murray-Darling Basin*; Wentworth Group: Sydney, Australia, 2010.
62. Marshall, V.; Kirby, M.D. *Overturning Aqua Nullius: Securing Aboriginal Water Rights*; Aboriginal Studies Press: Canberra, Australia, 2017.
63. National Water Commission (NWC). *Australia's Water Blueprint: National Reform Assessment 2014*; NWC: Canberra, Australia, 2014.
64. Horne, J.; Grafton, R.Q. The Australian Water Markets Story: Transformation Incrementally. In *Successful Public Policy: Lessons from Australia and New Zealand*; Luetjens, J., Mintrom, M., 't Hart, P., Eds.; ANU Press: Canberra, Australia, 2018.
65. Baldwin, B. Strong Water Reform Continues to Progress. Media Release, 13 May 2015. Available online: www.environment.gov.au/minister/baldwin/2015/mr20150513.html (accessed on 6 December 2015).
66. Commission, M.-D.B.R. Transcript of Proceedings 30 October 2018 by Richard Beasley. Available online: www.mdbrc.sa.gov.au/sites/g/files/net3846/f/30_october_2018_hearing_transcript_final_0.pdf?v=15420942622018 (accessed on 4 December 2018).

67. Wheeler, S.A.; Connor, J.; Grafton, R.Q.; Crase, L.; Quiggan, J. Submission to the Productivity Commission's Murray-Darling Basin Five-Year Assessment Issues Paper. Available online: www.pc.gov.au/__data/assets/pdf_file/0008/227483/sub040-basin-plan.pdf (accessed on 5 December 2018).
68. Matthews, K. Water Management in Australia—Time for a Rethink. In *The 2018 Peter Cullen Lecture 8 November 2018*; Peter Cullen Trust: Bruce, Australia, 2018.
69. Grafton, R.Q.; Horne, J. Water markets in the Murray-Darling basin. *Agric. Water Manag.* **2014**, *145*, 61–71. [CrossRef]
70. Grafton, R.Q.; Pittock, J.; Davis, R.; Williams, J.; Fu, G.; Warburton, M.; Udall, B.; McKenzie, R.; Yu, X.; Che, N.; et al. Global insights into water resources, climate change and governance. In *Water: Security, Economics and Governance*; Grafton, R.Q., Pittock, J., Tait, M., White, C., Eds.; Tilde Publishing and Distribution: Prahran, Australia, 2013; pp. 550–562.
71. CSIRO. *Water Availability in the Murray-Darling Basin. A Report to the Australian Government from the CSIRO Murray-Darling Basin Sustainable Yields Project*; CSIRO: Canberra, Australia, 2008.
72. Alexandra, J. Risks, uncertainty and climate confusion in the Murray–Darling Basin reforms. *Water Econ. Policy* **2017**, *3*, 1650038. [CrossRef]
73. Pittock, J.; Williams, J.; Grafton, R.Q. The Murray-Darling Basin plan fails to deal adequately with climate change. *Water* **2015**, *42*, 28–34.
74. Adamson, D.; Loch, A. Possible negative feedbacks from 'gold-plating'irrigation infrastructure. *Agric. Water Manag.* **2014**, *145*, 134–144. [CrossRef]
75. Chu, L.; Grafton, R.Q.; Stewardson, M. Resilience, Decision-making and Environmental Water Releases. *Earth's Future* **2018**, *6*, 777–792. [CrossRef]
76. University of Dar es Salaam. The Rufiji Basin. Available online: http://ira.udsm.ac.tz/rufijibasin/basin.php (accessed on 3 December 2018).
77. WREM International. *Rufiji Basin IWRMD Plan: Final Report, Volume I: Rufiji IWRMD Plan. Irrigation, M.o.W.a*; Report Prepared for the United Republic of Tanzania, Ministry of Water; WREM International Inc.: Atlanta, GA, USA, 2015.
78. Sokile, C.S.; Kashaigili, J.J.; Kadigi, R.M.J. Towards an integrated water resource management in Tanzania: The role of appropriate institutional framework in Rufiji Basin. *Phys. Chem. Earth* **2003**, *28*, 1015–1023. [CrossRef]
79. Van Koppen, B.; Sokile, C.S.; Hatibu, N.; Lankford, B.A.; Mahoo, H.; Yanda, P.Z. *Formal Water Rights in Rural Tanzania: Deepening the Dichotomy?* International Water Management Institute: Colombo, Sri Lanka, 2004.
80. Kabogo, J.E.; Anderson, E.P.; Hyera, P.; Kajanja, G. Facilitating public participation in water resources management: Reflections from Tanzania. *Ecol. Soc.* **2017**, *22*. [CrossRef]
81. The United Republic of Tanzania. *National Water Policy*; Development, Ministry of Water and Livestock Development, Ed. Available online: www.egov.go.tz/egov_uploads/documents/Water_Policy_en.pdf (accessed on 3 December 2018).
82. Manero, A. *Water Distribution within Smallholder Irrigation Schemes in Tanzania and Its Implications for Economic Inequality*; Australian National University: Canberra, Australia, 2018.
83. Muginya, D.M. *Participation of Local Government Authorities and the Performance of Integrated Water Resources Management in Tanzania: A Case Study of Little Ruaha River, Iringa District*; Mzumbe University: Iringa, Tanzania, 2013.
84. Van Koppen, B.; Sokile, C.S.; Hatibu, N.; Lankford, B.A.; Mahoo, H.; Yanda, P.Z. Formal water rights in rural Tanzania: Deepening the dichotomy. *Int. Water Manag. Inst. Work. Pap.* **2004**, *71*.
85. Van Koppen, B.; van der Zaag, P.; Manzungu, E.; Tapela, B. Roman water law in rural Africa: The unfinished business of colonial dispossession. *Water Int.* **2014**, *39*, 49–62. [CrossRef]
86. Sokile, C.S.; van Koppen, B. Local water rights and local water user entities: The unsung heroines of water resource management in Tanzania. *Phys. Chem. Earth* **2004**, *29*, 1349–1356. [CrossRef]
87. Kashaigili, J.J.; Rajabu, K.; Masolwa, P. Freshwater management and climate change adaptation: Experiences from the Great Ruaha River catchment in Tanzania. *Clim. Dev.* **2009**, *1*, 220–228. [CrossRef]
88. Kadigi, R.M.J.; Mdoe, N.S.; Lankford, B.A.; Morardet, S. The value of water for irrigated paddy and hydropower generation in the Great Ruaha, Tanzania. In Proceedings of the East Africa Integrated River Basin Management Coference, Morogoro, Tanzania, 7–9 March 2005; Sokoine University of Agriculture: Morogoro, Tanzania, 2005; pp. 265–278.

89. Lankford, B.; van Koppen, B.; Franks, T.; Mahoo, H. Entrenched views or insufficient science?: Contested causes and solutions of water allocation; insights from the Great Ruaha River Basin, Tanzania. *Agric. Water Manag.* **2004**, *69*, 135–153. [CrossRef]
90. Mdemu, M.V.; Francis, T. Productivity of water in large rice (paddy) irrigation schemes in the upper catchment of the Great Ruaha River Basin, Tanzania. In *Water Resources Planning, Development and Management*; Wurbs, R., Ed.; InTech: London, UK, 2013. [CrossRef]
91. Stommel, C.; Hofer, H.; East, M.L. The effect of reduced water availability in the Great Ruaha River on the vulnerable common hippopotamus in the Ruaha National Park, Tanzania. *PLoS ONE* **2016**, *11*, e0157145. [CrossRef] [PubMed]
92. Yang, Y.C.E.; Wi, S. Informing regional water-energy-food nexus with system analysis and interactive visualization—A case study in the Great Ruaha River of Tanzania. *Agric. Water Manag.* **2018**, *196*, 75–86. [CrossRef]
93. Coulson, A. Small-scale and large-scale agriculture: Tanzanian experiences. In *Looking Back, Looking Ahead—Land, Agriculture and Society in East Africa*; Ståhl, M., Ed.; The Nordic Africa Institute: Milton Keynes, UK, 2015; pp. 44–73.
94. Mdee, A.; Harrison, E.; Mdee, C.; Mdee, E.; Bahati, E. The Politics of Small-Scale Irrigation in Tanzania: Making Sense of Failed Expectations . In *Working Paper 107—Future Agricultures*; Future Agricultures Consortium: Brighton, UK. Available online: https://assets.publishing.service.gov.uk/media/57a089f6e5274a27b200034d/WP107.pdf (accessed on 3 December 2018).
95. Bjornlund, H.; Zuo, A.; Wheeler, S.A.; Parry, K.; Pittock, J.; Mdemu, M.; Moyo, M. The dynamics of the relationship between household decision-making and farm household income in small-scale irrigation schemes in southern Africa. *Agric. Water Manag.* **2019**, *213*, 135–145. [CrossRef]
96. Manero, A. Income inequality within smallholder irrigation schemes in Sub-Saharan Africa. *Int. J. Water Resour. Dev.* **2017**, *33*, 770–787. [CrossRef]
97. Van Koppen, B.; Sokile, C.; Lankford, B.; Hatibu, N.; Mahoo, H.; Yanda, P. Water rights and water fees in rural Tanzania. In *Irrigation Water Pricing: The Gap between Theory and Practice*; Molle, F., Berkoff, J., Eds.; Colombia International Water Management Institute: Wallingford, UK, 2007; pp. 143–163.
98. Kashenge-Killenga, S.; Meliyo, J.; Urassa, G.; Kongo, V. Extent of Salt-Affected Soils and Their Effects in Irrigated and Lowland Rain-Fed Rice Growing Areas of Southwestern Tanzania. In *Climate Change and Multi-Dimensional Sustainability in African Agriculture: Climate Change and Sustainability in Agriculture*; Lal, R., Kraybill, D., Hansen, D.O., Singh, B.R., Mosogoya, T., Eik, L.O., Eds.; Springer International Publishing: Berlin/Heidelberg, Germany, 2016.
99. Batakanwa, F.; Mahoo, H.; Kahimba, F. Influence of irrigation water quality on soil salinization in semi-arid areas: A case study of Makutopora, Dodoma-Tanzania. *Int. J. Sci. Eng. Res.* **2015**, *6*, 1435–1446, ISSN 2229-5518.
100. Kashenge-Killenga, S.; Tongoona, P.; Derera, J.; Kanyeka, Z. Farmers' perception of salt affected soils and rice varieties preferences in the northeastern Tanzania and their implications in breeding. *Int. J. Dev. Sustain.* **2014**, *3*, 1257–1271.
101. Elisante, E.; Muzuka, A.N. Occurrence of nitrate in Tanzanian groundwater aquifers: A review. *Appl. Water Sci.* **2017**, *7*, 71–87. [CrossRef]
102. Nkotagu, H. Origins of high nitrate in groundwater in Tanzania. *J. Afr. Earth Sci.* **1996**, *22*, 471–478. [CrossRef]
103. Mihale, M.J. Nitrogen and Phosphorus Dynamics in the Waters of the Great Ruaha River, Tanzania. *J. Water Resour. Ocean Sci.* **2015**, *4*, 59–71. [CrossRef]
104. Kashaigili, J.J. Assessment of Groundwater Availability and Its Current and Potential Use and Impacts in Tanzania; International Water Management Institute. Available online: http://gw-africa.iwmi.org/Data/Sites/24/media/pdf/Country_Report-Tanzania.pdf (accessed on 4 December 2018).
105. US Bureau of Reclamation. Colorado River Basin Study. US Department of Interior. Available online: www.usbr.gov/lc/region/programs/crbstudy.html (accessed on 10 December 2018).
106. Fleck, J. *Water Is for Fighting over and Other Myths about Water in the West*; Island Press: Washington, DC, USA, 2016.
107. Udall, B.; Overpeck, J. The twenty-first century Colorado River hot drought and implications for the future. *Water Resour. Res.* **2017**, *53*, 2404–2418. [CrossRef]

108. Wheeler, K.G.; Robinson, C.J.; Bark, R.H. Modelling to bridge many boundaries: The Colorado and Murray-Darling River basins. *Reg. Environ. Chang.* **2018**, *18*, 1–13. [CrossRef]
109. Garrick, D.; Jacobs, K.; Garfin, G. Models, Assumptions, and Stakeholders: Planning for Water Supply Variability in the Colorado River Basin1. *JAWRA J. Am. Water Resour. Assoc.* **2008**, *44*, 381–398. [CrossRef]
110. Squillace, M. Water Transfers for a Changing Climate. *Nat. Resour. J.* **2013**, *53*, 55–116. [CrossRef]
111. Weismann, M.L. CBT Water Market Indicator: Prices Continue to Approach $30,000/UNIT. Available online: https://journalofwater.com/jow/cbt-water-market-indicator-prices-continue-to-approach-30000-unit/ (accessed on 10 December 2018).
112. Grafton, R.Q.; Libecap, G.D.; Edwards, E.C.; O'Brien, R.J.; Landry, C. Comparative assessment of water markets: Insights from the Murray–Darling Basin of Australia and the Western USA. *Water Policy* **2011**, *14*, 175–193. [CrossRef]
113. Batker, D.; Christin, Z.; Colley, C.; Graf, W.; Jones, K.B.; Loomis, J.; Pittman, J. Nature's Value in the Colorado River Basin; Earth Economics. Available online: https://wrrc.arizona.edu/sites/wrrc.arizona.edu/files/Earth%20Economics%20Colorado%20River%20Basin%20ESV%20FINAL.pdf (accessed on 10 December 2018).
114. Robison, J.; Cosens, B.; Jackson, S.; Leonard, K.; McCool, D. Indigenous water justice. *Lewis Clark Law Rev.* **2018**, *22*, 873–953. [CrossRef]
115. Sullivan, A.; White, D.D.; Hanemann, M. Designing collaborative governance: Insights from the drought contingency planning process for the lower Colorado River basin. *Environ. Sci. Policy* **2019**, *91*, 39–49. [CrossRef]
116. 2030 Water Resources Group. Vietnam Hydro-Economic Framework for Assessing Water Sector Challenges; World Bank Group Water. Available online: https://www.2030wrg.org/wp-content/uploads/2018/02/ENG_Vietnam-Hydro-Economic-Analysis_August-2017.pdf (accessed on 20 December 2018).
117. IGES. *WEPA Outlook on Water Environmental Management in Asia 2018*; Institute for Global Environmental Strategies, Ministry of Environment of Japan: Hayama, Japan, 2018.
118. Do, T.N. Vietnam's Challenges to Go with the Flow: Tackling the Country's Stark Water Trade-Offs. Available online: https://www.policyforum.net/vietnams-challenges-go-flow/ (accessed on 20 December 2018).
119. World Bank. *Vietnam: Toward a Safe, Clean, and Resilient Water System*; World Bank Group: Washington, DC, USA, 2019.
120. Vietnam MONRE. *Report on Task Completion in the First Nine Months of 2018 and Plan for the Rest of the Year*; Ministry of Natural Resources and Environment: Hanoi, Vietnam, 2018. (In Vietnamese)
121. USDCITA. Vietnam—Environmental and Pollution Control Equipment and Services. Available online: www.export.gov/apex/article2?id=Vietnam-Environmental-and-Pollution-Control-Equipment-and-Services (accessed on 30 November 2018).
122. Australian Government. Vietnam Climate Change Delivery Strategy 2011–2016; AusAID, Canberra. Available online: https://dfat.gov.au/about-us/publications/Documents/vietnam-climate-change-strategy-2011-16.pdf (accessed on 30 November 2018).
123. (Government of Vietnam) GOV. *National Strategy on Water Resource Management by 2020*; Government of Vietnam: Hanoi, Vietnam, 2006. (In Vietnamese)
124. (Government of Vietnam) GOV. *Law on Water Resources*; Government of Vietnam: Hanoi, Vietnam, 2012. (In Vietnamese)
125. (Government of Vietnam) GOV. *Law on Environmental Protection*; Government of Vietnam: Hanoi, Vietnam, 2014. (In Vietnamese)
126. Wyrwoll, P.R.; Grafton, R.Q.; Daniell, K.A.; Chu, H.L.; Ringler, C.; Lien, L.T.H.; Khoi, D.K.; Do, T.N.; Tuan, N.D.A. Decision-Making for Systemic Water Risks: Insights From a Participatory Risk Assessment Process in Vietnam. *Earth's Future* **2018**, *6*, 543–564. [CrossRef]
127. Molle, F.; Thai, H.C. Implementing integrated river basin management in the Red River Basin, Vietnam: A solution looking for a problem? *Water Policy* **2011**, *13*, 518–534. [CrossRef]
128. Du, T.L.T.; Bui, D.D.; Quach, Z.T.; Robins, L. Water Governance in a Changing Era: Perspective on Viet Nam. In *Water Governance Dynamics in the Mekong Region*; Blake, D.J.H., Robins, L., Eds.; Strategic Information Research Development Centre: Jaya, Malaysia, 2016; pp. 241–278.
129. (Government of Vietnam) GOV. *Law on Hydraulic Works*; Government of Vietnam: Hanoi, Vietnam, 2017. (In Vietnamese)

130. (Government of Vietnam) GOV. *Decree 82/2017/ND-CP on Guidance for Methods and Calculation of Charges of Issuing Water Exploitation Permits*; Government of Vietnam: Hanoi, Vietnam, 2017. (In Vietnamese)
131. Nam, P.K.; Son, T.V.H. *Tapping into Demand: A New Approach to Water Supply in Vietnam*; Economy and Environment Program for Southeast Asia (EEPSEA): Laguna, Philippines, 2005.
132. Do, T.N.; Bennett, J. Estimating wetland biodiversity values: A choice modelling application in Vietnam's Mekong River Delta. *Environ. Dev. Econ.* **2009**, *14*, 163–186. [CrossRef]
133. Vo, D.T.; Huynh, K.V. Estimating residents' willingness to pay for groundwater protection in the Vietnamese Mekong Delta. *Appl. Water Sci.* **2017**, *7*, 421–431. [CrossRef]
134. Do, T.N. *Economic Instruments for Environmental Management: International Experience and Applications in Viet Nam*; Justice Publishing House: Hanoi, Vietnam, 2011. (In Vietnamese)
135. HDR; DHI. *Study on the Impacts of Mainstream Hydropower on the Mekong River*; Ministry of Natural Resources and Environment: Hanoi, Vietnam, 2015.
136. Do, T.N. *Impacts of Alternative Dyke Management Strategies on Wetland Values in Viet Nam's Mekong River Delta*; Australian National University: Canberra, Australia, 2008.
137. (Government of Vietnam) GOV. *Resolution 120/2017/ND-CP on Mekong Delta's Sustainable Development and Climate Resilience*; Government of Vietnam: Hanoi, Vietnam, 2017. (In Vietnamese)

water

MDPI

Article

A Philosophical Justification for a Novel Analysis-Supported, Stakeholder-Driven Participatory Process for Water Resources Planning and Decision Making

David R. Purkey [1,*], Marisa Isabel Escobar Arias [2], Vishal K. Mehta [2], Laura Forni [2], Nicholas J. Depsky [3], David N. Yates [4] and Walter N. Stevenson [5]

1 Latin America Center, Stockholm Environment Institute, Bogota 110231, Colombia
2 U.S. Water Group, Stockholm Environment Institute, Davis, CA 95616, USA; marisa.escobar@sei.org (M.I.E.A.); vishal.mehta@sei.org (V.K.M.); laura.forni@sei.org (L.F.)
3 Energy and Resources Group, University of California, Berkeley, CA 94720, USA; nicholas.depsky@gmail.com
4 Research Applications Program, National Center for Atmospheric Research, Boulder, CO 80305, USA; yates@ucar.edu
5 Department of Classical Studies, University of Richmond, Richmond, VA 23173, USA; wstevens@richmond.edu
* Correspondence: david.purkey@sei.org; Tel.: +57-314-282-6360

Received: 29 May 2018; Accepted: 20 July 2018; Published: 31 July 2018

Abstract: Two trends currently shape water resources planning and decision making: reliance on participatory stakeholder processes to evaluate water management options; and growing recognition that deterministic approaches to the evaluation of options may not be appropriate. These trends pose questions regarding the proper role of information, analysis, and expertise in the inherently social and political process of negotiating agreements and implementing interventions in the water sector. The question of how one might discover the best option in the face of deep uncertainty is compelling. The question of whether the best option even exists to be discovered is more vexing. While such existential questions are not common in the water management community, they are not new to political theory. This paper explores early classical writing related to issues of knowledge and governance as captured in the work of Plato and Aristotle; and then attempts to place a novel, analysis-supported, stakeholder-driven water resources planning and decision making practice within this philosophical discourse, making reference to current decision theory. Examples from the Andes and California, where this practice has been used to structure participation by key stakeholders in water management planning and decision-making, argue that when a sufficiently diverse group of stakeholders is engaged in the decision making process expecting the discovery of the perfect option may not be warranted. Simply discovering a consensus option may be more realistic. The argument touches upon the diversity of preferences, model credibility and the visualization of model output required to explore the implications of various management options across a broad range of inherently unknowable future conditions.

Keywords: scientific analysis; decision-support; classical views on knowledge and authority; designing participatory processes involving stakeholders

1. Introduction

The use of computer models to support the evaluation of water resources management options has grown steadily since the first hydrologic [1] and water resource computer modeling tools [2]

were described in the 1960's and 1970's. In laying out their blueprint for a physically-based, digitally simulated hydrological response model, Freeze and Harlan echoed the developers of the Stanford Watershed Model [3] in claiming that *the ability to accurately predict behavior is a severe test of the adequacy of knowledge in any subject* [Emphasis added]. As part of a review and evaluation of multi-objective programming techniques in water resources Cohon and Marks paraphrase the assertion in Loucks and Dorfman [4] that these models *should be able to predict the outcome of the decision-making process* [Emphasis added]. These early pioneers in the field of water modeling were sanguine about the power of early computers to provide this predictive power, with Freeze and Harlan even conceding that their blueprint was more of an *artist's conception*. Nonetheless, the expectation was clear, as data quality and computing power improved; it would become increasingly possible to predict outcomes related to hydrology and water management and to use these predictions to identify the best possible course of action.

This expectation related to the power of rigorous analysis, common to modern science, is consistent with a line of thinking that came to prominence in Europe during the 18th Century Enlightenment. The Enlightenment ideal, which argued for a society based on reason instead of faith and doctrine, placing rational analysis in a favored position in terms of identifying the best course for social and political evolution, emerged following more than a century of thought on the proper balance between reason and faith. In 1615, Galileo wrote in a letter to the Grand Duchess Christina of Tuscany that *nothing ... which necessary demonstrations prove to us, ought to be called in question ... upon the testimony of biblical passages which may have some different meaning beneath their words* [5]. In his work *Pensées*, published in 1670, Pascal opined that *the supreme function of reason is to show man that some things are beyond reason* [6]. In 1693, following the 1687 publication of his landmark Principia, Newton wrote in a letter to Richard Bently that *gravity may put the planets into motion, but without the divine Power, it could never put them into such a circulating motion as they have about the Sun* [7].

By the 18th Century, however, this search for a holistic balance between reason and faith faded in the face of the predominance of reason and analysis. The Scottish philosopher David Hume [8] went so far as to suggest that *there is no question of importance whose decision is not comprised in the science of man; and there is none, which can be decided with certainty, before we become acquainted with that science.* Hume would likely have included the complex water management decisions confronting society today amongst his *questions of importance*. In the decades since the publication of the early work on hydrologic and water resources computer modeling, the water management community has worked diligently to capitalize on improved data availability and computing power. This has been done in order to assess whether a particular bit of infrastructure should be built (e.g., [9]), operating rule, encoded (e.g., [10]), policy, promulgated (e.g., [11]), or behavior, incentivized (e.g., [12]), based on the application of ever more sophisticated models and analysis.

Yet in the face of this growing sophistication within the water resources practice, and the continued aspiration that better data and computing power can uniquely identify the rational water management choice, two important trends conspire to complicate matters. The first relates to the increasing reliance on participatory stakeholder processes as a forum for evaluating water management options and selecting the appropriate course of action. This trend was launched in the United States by the passage of the National Environmental Policy Act, signed into law by President Nixon in 1970, which increased the influence of stakeholders in environmental policymaking [13]. Similar laws exist elsewhere in the world. The second trend relates to the growing recognition that earlier deterministic analytical approaches to this evaluation of options may no longer be appropriate in the face of future uncertainty. For example, the weakness of the assumption of hydrologic stationarity in the face of climate change, an assumption that has underpinned most water modeling work carried out since the 1970s, has been well documented [14].

The convergence of these two trends poses questions as to the proper role of data, information, analysis, and expertise in the inherently complex social and political process of negotiating water resources management agreements and implementing water resources management interventions.

The authors have encountered this convergence while working in Bolivia on the Pilot Program on Climate Resilience (PPCR) and in California working on Integrated Regional Water Management (IRWM) planning. In response, it has proven useful to look back before the Enlightenment, to the origins of the western philosophical tradition in ancient Greece, and to re-discover that still important conversations on the proper role of knowledge and expertise within systems of governance and authority took place over 2000 years ago, conversations notable for the absence of consideration of faith. This paper argues that these conversations offer insights regarding the proper design of analysis supported, stakeholder driven, participatory water resources management planning, and decision-making processes.

2. A Conversation during an Athenian Drought

An article about the evolving mythical and metaphorical power of the weather in literature [15] states the obvious, we all talk about the weather, and we always have. As such, it is not impossible to imagine that two leading minds of the western philosophical tradition would have discussed the drought conditions effecting Athens around 360 BC. At that time, Plato, 67 and the leader of his own academy, and Aristotle, 24 and studying with him, were engaged in a dialogue about the proper role of the state, and coming to somewhat different conclusions. Looking back on this dialogue from the perspective of the Renaissance, Rafael, in his fresco the School of Athens (Figure 1), depicted Plato gesturing towards the sky where presumably some divine truth about society could be discovered while Aristotle gestures towards the ground where mortals walk. Assuming Rafael captured them discussing the Athenian drought of 360 BC, what might we expect them to be saying to each other?

Figure 1. View of the central portion of the Raphael's *School of Athens* painted between 1509 and 1511 in the Vatican's Apostolic Place, depicting a supposed conversation between Plato (left) and Aristotle (right). (Stanza della Segnatura, Palazzi Pontifici, Vatican).

Before imagining such a hypothetical exchange, some context. Writings by a contemporary of Plato and Aristotle, Demosthenes, from 361 BC, observes that his *land not only produces no crops, but that year, as you all know, the water even dried up in wells, so that not a vegetable grew in the garden* (Polykles 61) and in 357 BC, he found that *there was a universal shortage of grain* (Leptines 33), chronicling that around 360 BC Athens, with its Mediterranean climate, was indeed gripped by a severe drought. Archeological evidence [16] suggests that by 360 BC Athens had become increasingly reliant on private rainwater catchment cisterns. This water management strategy stood in stark contrast to earlier expenditures by pre-democracy autocrats, even "tyrants", in more centralized water management arrangements, most notably the 7.5 km long aqueduct constructed during the reign of Peisistratos, also known as the Tyrant of Athens, who ruled from 546 BC to 527 BC. The Peisistratos Aqueduct conveyed water from springs in the foothills of the Hymottos Mountains to a point near the Acropolis and was comprised of both tunnels hewn from the solid rock and intricate terra cotta pipelines. It is one of antiquity's early examples of hydraulic engineering [17]. Against this backdrop, a possible conversation in Plato's Academy:

Plato: Things are getting worse, the wells are drying up, crops are failing, and conflicts over access to the cisterns still holding water are increasing. Something needs to be done.

Aristotle: Might I suspect that you favor some bold, ambitious plan?

Plato: Of course. You recall how Peisistratos responded to earlier water shortages by constructing his aqueduct that filled the fountains of the Acropolis. Why are we not considering similar responses today?

Aristotle: Because he was only able to achieve such a system by imposing severe taxes on his subjects. Surely you are not proposing that we sacrifice our Athenian democracy simply to increase water supplies. You must know that the empowered citizens of today's Athens are loathe to fund such autocratic endeavors.

Plato: But such ambitious responses need not be autocratic. Even though the citizens are fixed in the shadows[1], so to speak, ignorant of the real options to manage current water shortages, I am confident that someone driven by the quest for the pure drought management option[2] could identify the correct response that would truly benefit the polis.

Aristotle: It sounds like you would consider the role of Officer of the Fountains to be one fit for your ideal Philosopher-King[3].

Plato: I would, and that seems to be consistent with the way our democracy is heading. I have seen your notes for your pending treatise on the Athenian Democracy where you document how the position of Officer of the Fountains is no longer filled by lot[4].

Aristotle: Indeed, but as much as we cannot let a small group of recalcitrant citizens, based on self-interest alone, reject any collective action to respond to the drought, we must also avoid the temptation to cede all authority to a small group of the elite. You know my views on the importance of a strong middle-class[5]. That group must ultimately arbitrate the merits of any response to the current drought.

Plato: That sounds to me like a recipe for paralysis. Surely something as important as the management of water during these dry times warrants ceding control to those with the knowledge and experience to discover the correct course of action[6].

Aristotle: I am not sure that the Officer of the Fountains alone could identify such a correct course of action when each citizen brings to the decision his or her own desires and passions. Better to engage the polis in the process of deliberating on what to do. Such a process might not yield the universally perfect drought response, but it might yield something that could practically be implemented[7]. In addition, there is the issue of uncertainty surrounding this decision. How do we know if the drought will persist or the rains return?

Plato: Uncertainty is indeed tricky. While some would accuse me of being an absolutist in my pursuit for the truth, be it in terms of the proper drought response or any other issue confronting the polis, I am cognizant of my unfortunate friend Socrates who recognized the wisdom of acknowledging what he did not know[8].

Aristotle: It seems then that perhaps you should be willing to ascribe similar wisdom regarding the unknown and uncertain to the Officer of the Fountain. Adhering to an absolute version of the true drought response is not reasonable[9]. Can you defer to the deliberations of the middle class?

Plato: I cannot. The issue of how to navigate the current drought is simply too important to the survival of the polis to be left to the middle class. I adhere to my conviction that in this case the Officer of the Fountain must indeed by granted the authority of a Philosopher-King. Given the gravity of the situation and import of our response, it would seem that we are not of a like mind as to how best to respond.

Aristotle: So it seems. Let us hope it rains soon, and if it does not let us hope that our political deliberation, in whatever from it takes, leads to a decision. Doing nothing does not seem like a viable option.

[1]In the Allegory of the Cave (Republic, VII 514 a, 2 to 517 a, 7) Plato wrote of individuals chained in a cave, convinced that the shadows on the wall in front of them are reality. He goes on to speculate how someone freed from the cave would come to see both (i) the real world, *he would be able to view the things themselves, the beings, instead of the dim reflections* and (ii) his or her responsibility to those remaining in the cave because *If he again recalled his first dwelling, and the "knowing" that passes as the norm there, and the people with whom he once was chained, don't you think he would consider himself lucky and, by contrast, feel sorry for them?* [2]In his Theory of Forms, Plato (Republic 477a-478e) asserts that there is a difference between knowledge (*epistémé*) and *doxa*, translated as opinion. *Knowledge is infallible—you cannot know what is false. Opinion, however, can be mistaken. So opinion cannot be knowledge.* As Plato applies the Theory of Forms to all things and ideas, he would argue that knowledge about the perfect drought response existed to be discovered. [3]Plato (Republic V.473c11-d6) asserts that those with the ability to distinguish between knowledge and opinion should be endowed with authority because *unless philosophers become kings or those whom we now call kings and rulers philosophize and there is a conjunction of political power and philosophy there can be no cessation of evils.* While there is some debate as to whether Plato felt that all decision should be under the purview of his Philosopher-King, or simply the constitutional decisions, he does make reference to laws and institutions, which are related to the implementation of a constitution, stating that *knowledge of what justice is, which requires knowledge of the definition of its essence (form), will in some cases improve judgments about how far possible or actual laws or institutions are just* (Republic VI.484cd, VII.520cd). Major water management decisions would seem to be one such case. [4]From Aristotle (Athenaion Politeia, 43.1) we learn that among the leadership positions on the Athenian Democracy, *the Officer of Fountains was one of the few that were elected by vote whereas most other officers were chosen by lot; so important was this position within the governance system of classical Athens.* As such, the citizens of the Athenian Democracy seemed to have already identified the need for a "professional" class of water managers. [5]Aristotle (Politics 1296b-1297a 4.12) argues that a large and strong middle class is essential to the stability of a democracy as a stabilizing force against the conflict between the autocratic elite and the disgruntled masses. He writes that the *lawgiver in his constitution must always take in the middle class* and that *everywhere it is the arbitrator that is most trusted, and the man in the middle is an arbitrator.* The assumption here is that Plato would not agree that this middle class possessed the sort of training that he ascribed to his Philosopher-King. [6]Plato (Republic, Book 6, lines 487d-488e), describing the state of affairs aboard a ship at sea, writes that sailors *have no idea that the true navigator must study the seasons of the year, the sky, the stars, the winds and all the other subjects appropriate to his profession if he is to be really fit to control a ship.* The inference is that the true captain is in a better position to steer the "Ship of State". [7]Aristotle who wrote that sensation, reason and desire all contribute to action, stressed the difference between scientific knowledge and practical wisdom. Of the latter he wrote (Nicomachean Ethic, Book VI) it is *concerned with things human and things about which it is possible to deliberate; for we say this is above all the work of the man of practical wisdom, to deliberate well. Nor is practical wisdom concerned with universals only-it must also recognize the particulars; for it is practical, and practice is concerned with particulars.* This suggests that unlike Plato, Aristotle would not expect that a perfect, context independent, drought response to be available to be discovered. [8]In his recounting of the trial of his mentor (Apology 21d), Plato suggested that in describing one of his accusers Socrates said *I am wiser than this man; for neither of us really knows anything fine and good, but this man thinks he knows something when he does not, whereas I, as I do not know anything, do not think I do either. I seem, then, in just this little thing to be wiser than this man at any rate, that what I do not know I do not think I know either*. [9]Aristotle (Nicomachean Ethic, Book I) acknowledged uncertainty in the setting of public policy by stating that *we must therefore be content if, in dealing with subjects and starting from premises thus uncertain, we succeed in presenting a broad outline of the truth: when our subjects and our premises are merely generalities, it is enough if we arrive at generally valid conclusions for it is the mark of an educated mind to expect that amount of exactness in each kind which the nature of the particular subject admits.*

History records that in 346 BC, and again in 333 BC, Kephisodoros of Hagnous and Pytheas of Alopeke, respectively, were named in decrees commending their efforts as Officers of the Fountains to restore the public water system of Athens. The archeological evidence also points to increased investment in public water supply in the third quarter of the 4th Century BC [16]. As such, it seems that the citizens of Athens did take collective action in response to the drought of 360 BC. We can only imagine the conversations between the Officers of the Fountains and the citizens of Athens that took place in the Agora in more than a decade between 360 and 346 BC, leading to this apparent change in water policy. One can surmise, however, that they may have been contentious.

3. Implications for Water Management for the 21st Century

This hypothetical dialogue is offered in order to suggest that, as in antiquity, the current efforts by a group of water management professionals to convince, through analysis, the wider body politic that a particular course of action is in its collective best interest, is rooted in a longstanding philosophical debate. Thanks to the early pioneers in the field of water modeling, today's water management professionals have access to powerful tools as they seek to discover the best water management option in the face of uncertainty related to climate change, demography, economic development, and regulatory reform. New approaches to decision making under uncertainty based on rigorous analysis [18,19] have been proposed to assist in this effort. However, the fundamental question of whether the "perfect" option, the Platonic form, even exists to be discovered remains, motivating new questions as to the proper role of data and analysis in the creation of knowledge (epistemology from Plato's reference to *episteme*) that will guide society towards more sustainable patterns of water use and conservation as pressure on this vital resource grows.

This paper will argue that the proper response to these questions lies not in rejecting either the view of Plato, with his focus on the quest for the perfect form, or that of Aristotle, with his recognition of the importance of human desires in shaping decisions, but in their integration. Based on that argument, this paper reports on an attempt to translate recent academic work on decision making under uncertainty into a structured, analysis-supported, stakeholder-driven participatory process focused on river basin planning and management that includes the sustained participation of the full spectrum of stakeholders in a river basin and not simply interactions between the decision makers themselves. The process has been designed to facilitate negotiations amongst water managers and stakeholders holding distinct opinions as to the definition of a successful outcome, analogous to Plato's Philosopher-Kings" and Aristotle's "middle-class", respectively, leading, hopefully, to broad and stable agreements.

In presenting this process, the authors acknowledge that there is a rich literature pertaining to the use of participatory processes in water management, including literature on the use of models to inform these processes. Vionov et al. [20], presented the current state of the practice of modeling with stakeholders, identifying seven components of potential stakeholder participation in the environmental modeling process. Basco-Carrera et al. [21] proposed a framework for distinguishing between participatory and collaborative modeling in water resources management. Halbe et al. [22] propose a framework for initiating, designing, and institutionalizing participatory modeling in water resources management. The work presented here might reasonably be considered a single case within the general frameworks that these papers propose. What distinguishes this approach is the bottom up manner in which it has been developed and improved based on our experience as water modelers, trained to aspire to the Platonic role of Philosopher-King, yet working with stakeholders as members of the Aristotelian middle class, in a number of diverse settings over a number of years. While the approach may conform with more general frameworks, it is distinguished by its intentional design references to philosophical discourse.

It could be argued that the initial seeds of this approach were sown during the same period of time as the origins of water modeling. A recent review of the seminal psychological research related to decision making [23] referenced experiments conducted in the 1970s where individual subjects refused to apply accepted axioms from rational choice theory while playing games of chance, even after these axioms were fully explained to them [24]. Extrapolating these findings into the broader realm of decision-making, Slovic suggested that:

> *An analysis which fails to appreciate the concern for regret and ambiguity is likely to violate the decision maker's preferences. Therefore we must work to devise methods for incorporating such psychological variables into the decision analysis, despite the aesthetic and practical complications that will arise when utilities and preferences are context-dependent.* [Emphasis added]

This conclusion, backed up by volumes of subsequent research in the fields of psychology and behavioral economics, suggests that no single "right" answer will be independently discovered by every player confronting the same game of chance, as suggested by the rational choice model. Gintis [25], demonstrated that counter to rational choice theory, when facing individual decisions game players are variably more averse to loss than attracted to gain, with some individuals placing greater value on what they possess than on what they could acquire. In public goods games, where freeriding is the rational choice, Lopes [26], reported that participants demonstrated notions of fairness, cooperation, reciprocity, and social pressure that led to non-rational decisions. Finally, Henrich et al. [27] found that the degree of pre-existing social cohesion and market integration within a group of players of a public good game largely explained the degree of cooperation expressed.

Slovic's conclusion that the distinct preferences that individuals bring to a decision-making challenge will uniquely shape their notion of the correct course of action, has been repeatedly confirmed. The conclusion strongly evokes Aristotle's suggestion that in addition to reason, sensation and desire also contribute to action. The challenge is to create an intentional space for this sensation and desire in the water management decision-making process, which heretofore has been structured more along technical and analytical lines. Moreover, the suggestion that the degree of social cohesion contributes to higher levels of cooperation implies that a process that actually contributes to the creations of such cohesion can better harmonize the distinct sensations and desires of individual stakeholders. The practice described in this paper has been structured so that the distinct preferences held by individual stakeholders in a particular river basin remain the central focus, with each stakeholder taking responsibility for articulating and defending his or her own interests or values, as the group collectively seeks to define a mutually acceptable course of action.

This focus is consistent with the conclusions of the National Research Council [28], which in a useful review of models of decision-making, draws a distinction between the more common "predict-then-act" framework and an alternative "analysis with deliberation" approach. As its name implies, the first framework imagines a sequential progression of gathering information, using this information to construct knowledge, leading to decisions. This is how Plato imagined the path towards knowledge in his Theory of Forms. The second approach imagines a more iterative process, which begins with the stakeholders and decision makers participating in a joint problem formulation based on their particular preferences that guides data gathering and analytical design. This analysis will produce intermediate results that are collectively reviewed by these same participants leading to new cycles of data gathering and analysis, and ultimately to a decision. Aristotle's model of the deliberating middle class corresponds well with this approach.

As part of a research program on the "analysis with deliberation" approach, researchers at the RAND Corporation proposed a decision-making model called Robust Decision Making (RDM) that attempts to formalize this approach (Figure 2). A novel problem formulation framework referred to as XLRM initiates the iterative process. In the XLRM framework, a decision making challenge is divided into four component parts. The first corresponds to uncertain e<u>x</u>ogenous (X) factors that are outside the control of stakeholders and decision makers but which have the potential to influence outcomes (e.g., climate change, population growth). The second component relates to the <u>l</u>evers (L) or options that decision makers can implement in order to improve outcomes (e.g., new infrastructure, new regulations), with the option of maintaining the current system configuration (Business as Usual) always being considered. The R in the framework corresponds to the analytical tools deployed to <u>r</u>elate the articulated uncertainties to the identified management options so that values of stakeholder specific <u>m</u>etrics (M) of performance used to evaluate and compare potential outcomes can be produced. These different performance metrics can be equated to the different human sensations and desires that Aristotle referred to in the Nocomachean Ethic and the context-dependent preferences that Slovic discovered during his research on the validity of rational choice theory.

It is worth noting, however, that the iterative process shown in Figure 2 and described in Lempert et al [18] suggests that the "analysis with deliberation" approach will ultimately lead to the identification of the "Robust Strategy", which is one that will perform "well" over a range of possible future conditions. This strategy is presumably discovered by moving through a series of steps focused on case generation, scenario discovery, and tradeoff analysis leading to the identification of the robust approach. There are, however, several potential pathways that the analysis can follow after the XLRM participatory scoping has been completed which do not lead directly to the robust strategy. These include:

1. Returning to the XLRM scoping exercise after determining that the performance of the identified strategies is too poor.
2. Returning to the XLRM scoping exercise after determining that the vulnerabilities of the identified strategies are too high.
3. Returning to the XLRM scoping exercise after determining that the tradeoffs among the identified strategies are too extreme.

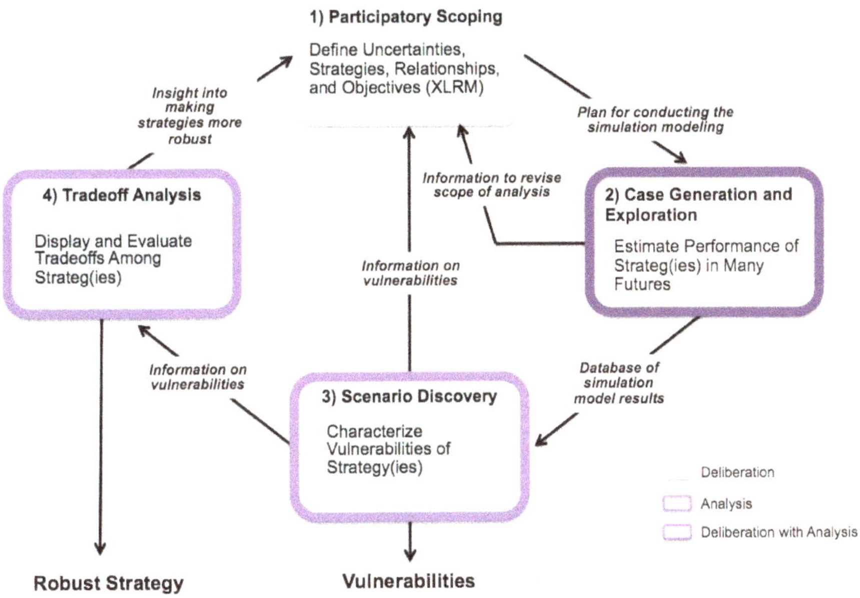

Figure 2. The Robust Decision Making framework, an example of the Analysis with Deliberation Approach to decision making under uncertainty. [18].

The figure is not explicit, however, in defining what constitutes a too poor performance, a too high vulnerability, or a too extreme tradeoff. In fact, in many applications of RDM in the water sector [29,30], efforts have been made to inform the definition of these thresholds through some algorithmic procedure. When taken to the extreme, this would render the process defined in Figure 2 a more complex version of predict-then-act, in effect reducing efforts to reconcile the different desires and preferences held by the stakeholders to simply add another deterministic step in an analytical procedure.

The participatory process described in this paper, referred to as the Robust Decision Support (RDS) practice attempts to operationalize the process described in Figure 2 while avoiding the inclination towards the Platonic inherent in overly algorithmic assignment of thresholds of performance, vulnerability, and tradeoff. Here is it important to be clear that the proposed RDS practice fully adopts the analytical framework within RDM (Figure 2), with its rigorous combination of uncertainties and strategies within model runs designed to produce values of a range of performance metrics. What is novel is the manner in which this analytical framework is embedded within a broadly subscribed participatory process, where the participating stakeholders, not algorithms, drive the decision to return to the participatory scoping step. These decisions are made based on the desires of individual or groups of stakeholders involved in a particular river basin planning exercise, desires that Slovic showed may have little relation to rational choice. While RDS should be considered a derivative of

RDM, it is distinct in how it understands the motivations of individual stakeholders within a decision making process, and the importance it places on these distinct motivations.

While developing the RDS practice in the context of water management planning and decision making, the authors have spent five years working in collaboration with water management stakeholders and decision makers in Latin America, the United States, Southeast Asia, and Africa in order to test, refine, and formalize this analysis-supported participatory process. In developing this practice, the research team sought to recognize the unique preferences of each stakeholder engaged a decision-making process. The result is a formal process for accompanying stakeholders and decision makers through a series of modules, as shown in Figure 3.

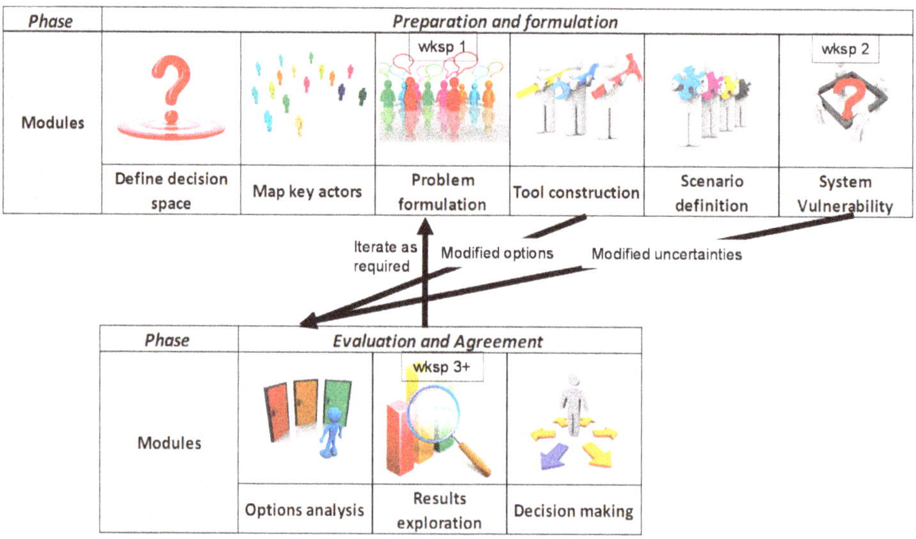

Figure 3. Steps within the Robust Decision Support (RDS) practice designed to lead stakeholders representing different water management constituencies though a process of highly participatory negotiations beginning with a problem formulation and culminating with a joint commitment to implement a consensus program of action. Potential iterative loops back to the initial problem formulation can follow model results exploration. Adjustments in the assumed future uncertainties will require a new system vulnerability assessment, while adjustments to the management options under consideration can be accommodated without generating a new vulnerability baseline. The timing of stakeholder workshops within the process is indicated.

More complete descriptions of each of modules, in Figure 3, which are implemented in two phase Preparation and Formulation followed by Evaluation and Agreement, are included below, including some justification for their inclusion in the RDS practice in terms of managing stakeholder dynamics.

1. Define decision space. River basin management decisions do not typically take place in a vacuum, rather they emerge from a legacy of prior discussions and decisions. In this step, a thorough review of past reports and plans, as well as interviews with key decision makers, are used to document what broadly defines the decision space, focusing on the legal, regulatory, political, or financial factors that motivate the decision-making process. If there are no such motivating factors present, it may be hard to initiate the RDS practice.
2. Map key actors. Once the motivating factors for a decision are understood, the next step involves administering a survey designed to identify which stakeholders should be invited and encouraged to participate in the RDS process, and to define the sorts of information, experiences,

and perspectives they will bring to the process. The results are used to produce a map of social networks that highlight potential conflicts and coalitions, as well as a plan to encourage the contribution of information and insights to the process.

3. Problem formulation. Once the key stakeholders are convened, a workshop is held to develop a first version of the XLRM matrix framing the decision making challenge. This workshop begins with a session designed to articulate key planning uncertainties, many of which are not contentious. All interest groups, for example, can usually agree that climate change is an uncertainty that has the potential to impact outcomes related to any individual stakeholder desire. This exercise serves to create common purpose amongst disparate stakeholders. The next session focuses on the particular strategies that each stakeholder favors in order to improve outcomes. This discussion can be quite contentious as many stakeholders strongly oppose strategies being offered by others, so no judgement can be cast on any particular strategy suggested at this point. This exercise serves to encourage respect amongst the stakeholders. The final session is the most important, as it focuses on the definition of the distinct metrics of performance that each stakeholder will use to evaluate the outcome of each strategy identified, their own preferred strategy as well as those offered by others. These metrics should be independent of any particular strategy, leaving open the possibility that a strategy proposed by one stakeholder might actually improve outcomes in terms of the metrics defined by another. This is the basis of tradeoff analysis and compromise.
4. Tool construction. Once the problem formulation is complete, work can begin to assemble the analytical tools (R) required to capture the articulated uncertainties (X), represent the identified management options (L), and produce the desired metrics of performance (M). In this step, it is important to assure that the analytical tool responds fully to the stakeholder-driven problem formulation. Failure to do so will degrade the creditability of the tool itself while success will further the commitment to knowledge co-creation amongst the participants. This is critical in order to avoid the model as a "black box" outcome that can lead to all too common and unproductive model critiques that divert attention from the real task of balancing the distinct values held by each stakeholder.
5. Scenario definition. Once an acceptable analytical tool is developed, a set of scenarios based on the articulated planning uncertainties must be defined in order to construct an ensemble of model runs spanning possible future conditions. This entails defining plausible future ranges for each uncertainty, and here the stakeholders must be involved. The goal is to define endmembers that the stakeholders feel would both stress the system and be easily handled, and to then populate the intervening space with a series of discrete intermediate assumptions. The ensemble of scenarios is constructed by implementing a full combinatorial sorting of each discrete condition associated with each articulated uncertainty.
6. System vulnerability. Once an ensemble of scenarios is constructed it is run supposing that current management regimes are maintained. This Business as Usual case is critical to the RDS practice as it allows for an assessment of the baseline vulnerability of the current system in the face of the articulated uncertainties. In fact, the second stakeholder workshop in the RDS practice involves exploring the modeled values for the stakeholder-defined metrics of performance for each member of the scenario ensemble under current management in order to co-create an assessment of the potential vulnerability of the existing system. The discussion of stakeholder preferences in the absence of any analysis of the performance of any particular management options is critical in order to avoid the case where each constituency locks on to the strategy that will produce the best outcome with respect to their particular metric of performance, typically the one they suggested. This workshop also includes the definition of performance thresholds for each stakeholder preference corresponding to the minimum acceptable and maximum aspiration levels.

7. Option analysis. Only once the vulnerability assessment is complete, and the sideboards of "could live with" and "would love to have" thresholds are defined for each stakeholder's individual desires, is the ensemble run again to include representations of the current formulation of the stakeholder-proposed management options.
8. Results exploration. When only the Business as Usual case is run, the model ensemble contains a single run for each scenario constructed, each run producing model output associated with each stakeholder-defined metric of performance. The model output database produced from option analysis doubles in size with each proposed management strategy considered. As such, the use of innovative, interactive data visualization tools to explore the outcome space defined by the desired metrics of performance for each combination of articulated uncertainties and identified management options is critical to the success of the RDS practice. This exploration is carried out in close collaboration with key stakeholders in order to promote the creation of shared knowledge and insights about the system and potential outcomes that allow the discussion to focus on the particular values held by each participant and not of the merits of the analytical tools themselves.
9. Decision support. Based on the shared insights developed through the participatory exploration of the ensemble model output database, the performance of specific management options can be evaluated relative to the Business as Usual base case and to each other. Using the model results related to metrics of performance of particular interest to each stakeholder, the participants can decide to either reformulate the problem (refine uncertainties, and/or modify existing or propose new management options), leading to the reformulation of the XLRM matrix and the initiation of a new analytical cycle. Eventually, if the process is successful, broad acceptance of a preferred set of options is achieved. Experience with the RDS practice suggests that, following the vulnerability assessment step, it is useful to first configure the set of model runs to represent in isolation the distinct management options suggested by each stakeholder. The results typically suggest that while the strategy will improve outcomes in terms of the particular metrics of performance proposed by that stakeholder, it will have negligible or negative impacts on the metrics of performance proposed by the other. This typically leads to a negotiation focused on defining integrated programs of action that combine key features of several of the distinct strategies originally offered by the participants. The subsequent evaluation of the ensemble output associated with these integrated programs typically leads to the definition of a preferred program of action.

The entire process, from the initial definition of the decision space to the identification of the preferred integrated program of actions can take anywhere from 9 to 18 months to complete and requires stakeholder participation in up to five all-day workshops. The RDS practice should not be viewed as a quick effort that will miraculously lead to the identification of the perfect management option. The goal is not to create a common vision of what constitutes the true best outcome, rather, the RDS practice is a sustained effort to create a stable agreement between disparate parties, each of whom expresses and advocates for his or her own desired outcome within a process that fosters social cohesion.

We cannot know exactly what conversations took place in the Athenian Agora between the height of the 360 BC drought and the recognition of the Kephisodoros of Hagnous in 346 BC for his efforts to restore the fountains of Athens. Nonetheless it is clear that the process leading to the restoration of the fountains took years to complete. The great contribution of the early pioneers in the development of hydrologic and water resources model is that they provided the tools required to accelerate the complex social and political process leading to improved policy setting and decision-making around water resources. A decision that took more than a decade to reach in Ancient Athens can now be completed in a little more than a year.

4. Implementing the RDS Practice in the Andes

Water managers across the tropical Andean Region must plan for an uncertain future prompted by climate change [31]. These changes are already being felt in high elevations. The glaciated area of Peru's Cordillera Blanca—which represents 35% of all glaciers in the country—have retreated from 728 km^2 in 1960 to 536 km^2 in 2003 [32]. In Bolivia's Cordillera Real, the glaciated area shrank 43% by volume between 1963 and 2006 [33]. Colombia has lost more than 50% of its glaciers, with only six glacier capped mountains remaining as of 2012 [34]. Also vulnerable to climatic changes are the Andean páramos (high altitude moorlands) and bofedales (high altitude marshes), which with their high infiltration and soil moisture storage capacities provide crucial water production for downstream users, particularly during low flow periods. The unique biodiversity of páramos and bofedales underscores their fragility, with climate change compromising their ability to supply water [35], as well as reducing their total area [36]. For the 40 million people that depend directly on Andean ecosystems (glaciers, páramos, and bofedales) for their water resources [37], such changes can have far-reaching consequences, creating water management challenges in terms of meeting urban and agricultural demands, capturing hydropower generation potential, and preserving important ecosystems.

One Andean region where these consequences are already being felt is the La Paz/El Alto Region of Bolivia (Figure 4). These neighboring cities possess very different histories and geographies that have translated in to very different levels of water service provision. The historical city of La Paz lies in a canyon to which several glaciated watersheds drain providing the city with reasonable levels of water supply, albeit one at risk to future retreat and disappearance of these glaciers. El Alto situated on the western edge of La Paz lies atop a 4000 meter high altiplano that drains to Lake Titicaca. Only one watershed draining the Cordillera Real has been tapped to provide water to El Alto, complemented by a groundwater pumping plant in the southern part of the city. The water quality in the Tuni watershed is also threatened by contamination associated with relic mining activity. While the two cities are actually served by the same water utility, the residents of El Alto typically rely heavily on water procured from private tanker trucks that ply the streets of the city. Based on its own analysis, the utility developed ambitious plans for increased water sharing between La Paz and El Alto and expansion of the El Alto capture zone far to the north. These plans encountered extreme skepticism and strong resistance from both the City of La Paz and existing irrigators in the targeted watersheds.

In 2012, in order to build trust and resolve the conflict between the stakeholders in La Paz and in rural areas to the north of El Alto, the Ministry of the Environment and Water convened a focal group of key stakeholders and decision makers to plan for new water management investments in the region. Over the course of a year, these stakeholders participated in an RDS process, beginning with the problem formulation workshop that produced the XLRM matrix shown in Table 1. Two of the articulated uncertainties merit further explanation. To a large extent, the current government in Bolivia draws its political power from indigenous communities, which are more present in rural areas and in El Alto than in historic La Paz. There was an interest in exploring the implications of a possible shift away from the current policy related to historical water rights in La Paz and/or the lower priority assigned to agricultural water use as against potable supply. Also, the uncertainty related to further expansion of irrigated areas had a particular local character as it was tied strongly to a global quinoa boom that was underway when the problem formulation was completed. In total, the definition of the five dimension of planning uncertainty resulted in the definition of an ensemble of 192 different future scenarios.

Using a widely available water resources systems model [38] as an analytical engine, a model of the pertinent region of the Cordillera Real was constructed and calibrated to simulate glacier evolution, bofedal dynamics, and rainfall-runoff hydrology in existing and potential source watersheds. The current urban and agricultural demand and supply systems were also represented, along with several strategies to expand the urban water supply system to several watersheds to the north along the Cordillera Real and to invest in watershed protection. To explore the six identified strategies across the full range of articulated uncertainties required an ensemble of 1152 WEAP model runs (192 future scenarios × 6 management strategies), each producing a time series output for each of the metrics of performance. The challenge of exploring the large database of model output associated with such an ensemble is to develop a shared conceptual model amongst the RDS participants regarding the interpretation of results associated with each unique combination of scenario and strategy. It is in the development of this shared conceptual model that advanced data exploration and visualization tools proved particularly useful. Forni et al. [2016] reports on how advanced visualization was used in Bolivia to guide the RDS participants in the construction of a shared conceptual model and in the extraction of shared insights and knowledge from the ensemble of model output.

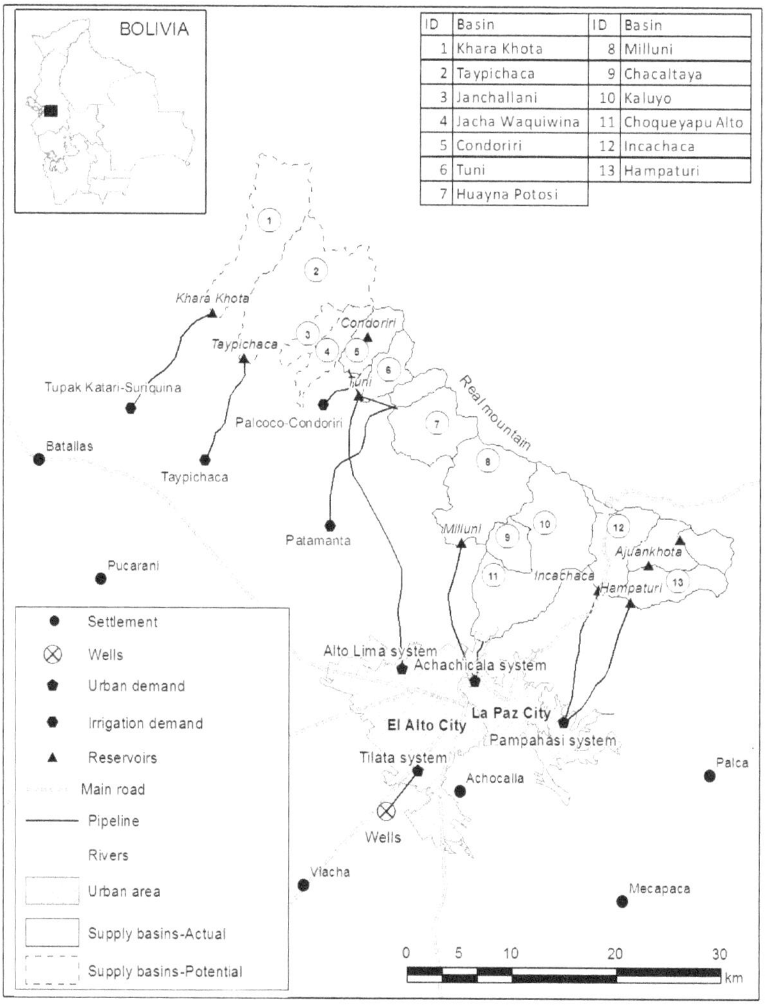

Figure 4. The La Paz/El Alto Region of Bolivia depicting current sources of water supply for La Paz and El Alto (solid outlines), along with potential new source watersheds (dashed outlines), existing reservoirs (triangles), conveyance facilities (solid lines), and points of urban (hexagons) and agricultural (pentagons) demand. The wellfield (circular cross hatch) providing water to El Alto is also shown. (Data source: Ministry of Environment and Water).

Table 1. Initial XLRM problem formulation matrix developed by a focal group of stakeholders representing various constituencies convened by the Bolivian Ministry of the Environment and Water during a one-day problem formulation workshop held in La Paz, Bolivia.

X (Exogenous Factors/Uncertainties)	L (Levers/Management Strategies)
Climate change and variability Population growth Increased per capita demand Changes in water allocation priority Expanded agricultural production.	Current system 3 new urban system expansion plans Conservation of bofedales Reduced urban distribution losses Reduced agricultural distribution losses
R (Relationships/Model)	**M (Metrics of Performance)**
Cordillera Real Model (in WEAP)	Urban water demand satisfaction Agricultural water demand satisfaction Total system losses Reservoir storage levels

At this point it is sufficient to say that once this shared model was available it became increasingly easy to add additional layers of complexity to the data visualizations, without overwhelming the participants in the process. In spite of the complexity of the system, the RDS participants were able to grasp the implications of the information presented, co-creating knowledge about the performance of the system under various future scenarios and management strategies. At the end of the 9-month process a consensus was reached that the capture zone for the El Alto water supply system should be extended north to include new watersheds, but only in combination with investments to improve the performance of existing irrigation systems in these watershed. This decision was subsequently refined through engineering design and costing and submitted to the PPCR for funding. By 2018, the realization of these new investments was nearing completion.

5. Implementing the RDS Practice in California

California has one of the most complex engineered water resource systems in the world. More than a century of water resources development has deeply impacted the environment. Most of California's historical wetlands are gone, water quality is impaired, habitats for salmonid fish are severely reduced, and close to 90% of riparian woodlands have been lost [39]. Much of this development occurred over the 20th century, from a combination of local, state, and federal infrastructure projects that invariably sourced water from a distance after local sources failed to keep up with demand. Since the 1970's, environmental demands have come to the forefront, and state and federal government's role has shifted a bit, from supporting infrastructure development, to protecting environmental needs [40]. Nonetheless, today, close to 80% of human water use in California is for irrigated agriculture, supporting a 40+ billion dollar industry and making California, especially the Central Valley, a globally important agricultural producer [41].

In California, the RDS practice has been applied to support local efforts to complete an Integrated Regional Water Management (IRWM) planning process in order to propose specific projects for financial support from the State of California. In Yuba County (Figure 5), at the downstream end of the Yuba River basin, which originates upstream in the Sierra Nevada, a stakeholder group representing municipal water supply, irrigation, hydropower, flood management, and environmental interests worked together over a period of 18 months to define the uncertainties, objectives and management options required to create an XLRM problem formulation matrix and to explore the potential performance of various management options contained within the Yuba County Integrated Regional Water Management Plan. To support the analysis a model was calibrated to represent the hydrology of the Yuba River Basin and the North and South Sub-Basins of the Yuba County Groundwater Aquifer. Within this hydrologic context, the operations of the major hydropower, flood control, and water supply infrastructure in the basin were simulated along with the allocation of water to meet municipal, irrigation, and ecological water needs.

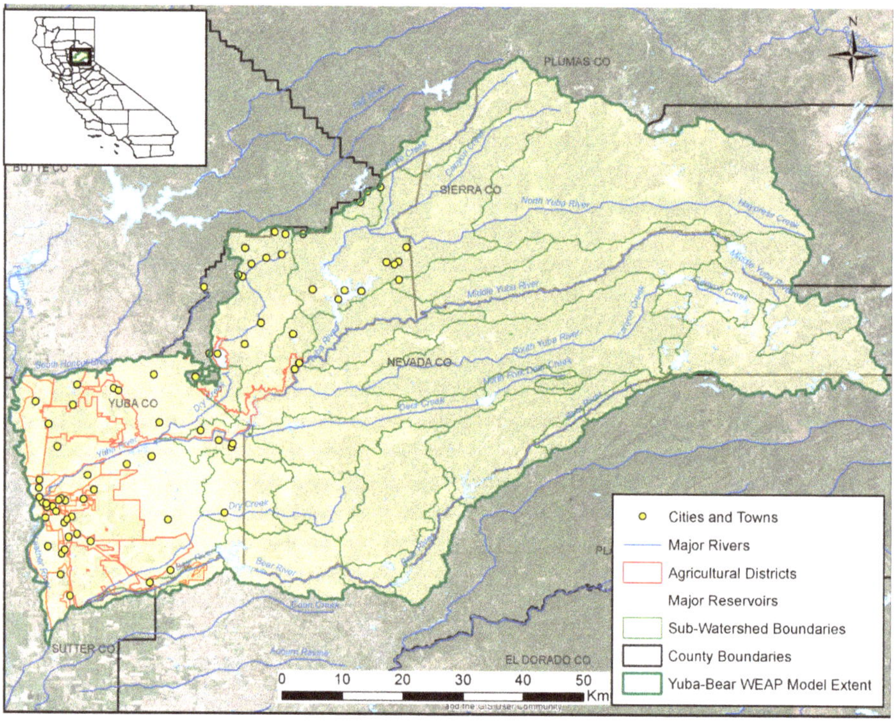

Figure 5. The Yuba River Watershed in Northern California, depicting the source water region in the Sierra Nevada as well as the water management infrastructure located within Yuba County in the western portion of the watershed. Red dashed lines show the position of irrigation districts with access to some portion of the flow in the Yuba River Watershed, yellow dots show the location of population centers where drinking water is supplied by a public utility. (Data source: California Spatial Information Library).

The final XLRM matrix, co-created over several rounds of ensemble analysis, is shown in Table 2. It contains two dimensions of planning uncertainty, climate change, and regulatory reform associated with current efforts to update the Bay-Delta Water Quality Control Plan that contemplate requiring the tributaries in the Sacramento and San Joaquin River Basin to constantly contribute a percentage of their full natural flow to ensure the health of the Bay Delta system. Four climate projections of critical interest were formulated based on the concerns articulated by the stakeholders. Two corresponded to end members extracted from the IPCC database (hot-dry and warm-wet) and two were constructed to contain climate attributes of particular concern to stakeholders (repeating extended drought and dry-fall/wet spring). Two Delta regulation futures were also formulated, the first corresponding to the current regulatory regime and the second representing a 50% of full natural flow contribution to Delta health, meaning that in the final round of analysis, a total of eight future scenarios were ultimately explored.

Table 2. Final XLRM problem formulation matrix developed by a focal group of stakeholders representing various constituencies convened by the Yuba County Water Agency in Marysville, California. The initial problem formulation contained two additional dimensions of uncertainty associated with hydropower relicensing and land use change, but these were eliminated from consideration due to the discovery of their relatively small impact, relative to climate change and the eventual Delta regulatory regime, on the vulnerability of the system.

X (Exogenous Factors/Uncertainties)	L (Levers/Management Strategies)
Climate change Delta regulations	Current system 12 Yuba IRWMP projects
R (Relationships/Model)	**M (Metrics of Performance)**
Yuba Model (in WEAP)	Ecological Water supply Hydropower Flood safety

As in the case of Bolivia, a widely available water resource systems model was used as the primary analytical engine to produce values for the metrics of performance and innovative data visualization techniques [42] were used to create a shared conceptual model of the results of the ensemble analysis. The RDS analysis showed that the Yuba system is highly vulnerable in a drought scenario, combined with a new 50% of full natural flow Delta regulatory regime. The analysis of strategies, drawn from the IRWM plan, showed that most of the individual strategies proposed did not broadly reduce the regional vulnerabilities of the system as it relates to the metrics of performance posited by the full spectrum of stakeholders. In particular, urban conservation projects, while helping local constituencies, would not significantly decrease system level vulnerabilities because of the very small proportion of total water used by the non-agricultural sector. In this setting standalone urban water management actions, while helping to improve overall water use awareness, cannot produce broad benefits. However, an ambitious combination of multiple discrete strategies—a portfolio of infrastructure and river restoration action—could create positive cross-sectoral regional impact. Based on this exercise in negotiation, the participants in the Yuba RDS process were able to prioritize and combine discrete actions into a package for which they have sought financial support from the State of California. While there is no guarantee that this application will prove successful, it does represent the sort of integrated package of broadly supported actions that the IRWM planning process was intended to promote, but which has proved elusive over California's decade-long experiment with Integrated Regional Water Management Planning [43].

6. Discussion and Potential Limitations

Beginning with the work in Bolivia, and continuing through the work in California, the research team has continued to refine the RDS practice based on feedback from participants. This has produced the current design of the RDS practice shown in Figure 3. One current feature of the practice, which was implemented in California but which was not undertaken in Bolivia, is the discrete step of assessing the vulnerability of the current system, and defining threshold values for the user specific performance metrics associated with the "would love to see" and "could live with" criteria prior to actually presenting any results associated with the performance of a particular strategy. In Bolivia, the results for the base case were simply presented along with the results associated with the strategies. It has become clear that taking the time to develop a baseline system vulnerability assessment and to define performance thresholds before evaluating the implications of strategies is an effective way to diffuse potential conflicts that typically surround strategies. This is particularly true for strategies related to new infrastructure, as it focuses the discussion where it should be, on the distinct set of values that each stakeholder brings to negotiation, values which must be broadly addressed by any proposed water management intervention, including new infrastructure investments.

Another feature of later RDS efforts has been subsequent sets of ensemble model runs that combine features of individual strategy options into integration programs of action. In Bolivia, the exploration of performance focused only on the proposed strategies in isolation. Fortunately the stakeholders were able to extract from this analysis the broad outlines of an integrated program that became the kernel of eventual engineering design and costing work. This effort was complicated somewhat, however, by issues related to timing and scale that could have been resolved as part of the RDS work had additional rounds of analysis been completed. In the Yuba case, and in all other current RDS efforts, these additional rounds have proved very useful in creating a consensus around not only the timing of specific actions but also around more coordinated operating rules associated with these interventions.

Nonetheless, it is safe to say that in all venues where the RDS practice has been deployed the experience has been positively received. Table 3 contains some written impressions of the process from both the Bolivia and California cases. In general, the experience suggests that the co-creation of knowledge is an effective way to diffuse distrust between various stakeholder constituencies that often characterize water resource planning and decision-making process. This co-creation process begins with the joint problem formulation, with its open statement of stakeholder-specific values and preferences, continues with the joint vetting and validation of the modeling tools that will be used to inform the process, and culminates in an open discussion of mutually acceptable tradeoffs and compromises. The entire process is facilitated by the ability to dynamically explore the implications of various management options using powerful data visualization tools, which is both empowering for the RDS participants, and, based on feedback provided following workshops, enjoyable. In addition to helping the participants deliberate on the relative merits of each individual strategy proposal, the RDS process itself seems to build the social cohesion that can lead to increased levels of cooperation.

Table 3. Some observations on the utility of the RDS practice from a survey of participants following the completion of the exercise in both La Paz/El Alto, Bolivia and Yuba County, California.

Survey Questions	Responses from Bolivia Case Study	Responses from Yuba Case Study
Was the exercise useful?	Yes	Yes
How was it useful in extracting valuable information?	- Direct comparison of the considered alternatives. - Recollection and organization of basin data. - Discovery of new research areas and monitoring improvements needed for reliable future databases. - Democratization of information by reaching a wider audience, not just modelers.	The visualization platform was invaluable to our group's efforts to evaluate the complex quantitative data produced during the RDS ensemble analysis. It helped move us from being overwhelmed to comprehension and gave important insights regarding future conditions and the efficacy of various water management projects.
How was the visualization useful in your future management design?	- To establish the magnitude of the operation that needed to be done in order to guarantee water supply for the cities of El Alto and La Paz in the next 30 years. - The study demonstrated that small interventions in the basin would not bring about the desired levels in the short term and were vulnerable to climate change.	The visualization platform was transformative to our water planning process. We initially started down the well-worn path of traditional water management planning. Following established guidelines, we seemed bound to write yet another formulaic plan that would define desired outcomes for our region and then propose, rank, and elicit funding for the subset of projects that appeared most likely to help achieve our desired outcomes. The visualization techniques allowed us to move beyond conventional wisdom and recognize that the projects included in our plan would likely not get us to where we needed to be as a region. Based on this experience, our group plans to fund a project that would internalize these techniques to our formal planning process.

That said, there are potential drawbacks associated with the RDS practice that need to be considered and compensated for, if possible.

1. The process is time-consuming, with experience suggesting that the entire process can take anywhere between 9 and 18 months to complete, depending on data availability and system complexity. Participants need to be aware of the time commitment from the outset.
2. In addition to being time consuming, the RDS practice requires the sustained participation of the stakeholders in the process. Each workshop is designed to continue the construction of the shared conceptual model that allows the complex data visualizations to be understood and useful. The process is severely hampered by the representatives of the various stakeholder constituencies dropping in or out. This risk should be made clear at the outset.
3. There is no guarantee that a consensus preferred program of action that emerges from the RDS process will be optimal in the classic sense of the word as the strategies that are considered are limited by the imagination and creativity of the participants. That said, if the participants each maintain their own value-specific definitions of what constitutes a successful outcome, and agree to a common preferred program of action, there is a high likelihood that it will at least avoid being Pareto sub-optimal.
4. Only human beings can directly participate in the RDS process, so the environment will always be a silent party to the negotiations. If the interests of the environment are not being actively defended in the RDS process, there is a risk that the consensus preferred program of actions will not be environmentally sustainable. This is why the active participation of environmental organizations in the RDS process is vital.
5. There is always the possibility that some stakeholder or stakeholder constituency will emerge at the end of the process, claiming that they were not involved in the development of the consensus preferred program of action and that they are not in favor of its implementation. Avoiding this eventuality is the reason why the decisions space definition and key actor mapping steps must be taken seriously.
6. There always exists the possibility that the participating stakeholders will not be able or define a consensus preferred program of action. While this has yet to occur in any of the roughly ten RDS exercises the research team has implemented in California, Latin America, Africa, and Asia, were it to occur there might be a temptation to characterize the decision space as a wicked problem [44]. Wicked problems generally require authoritarian responses, although it could be argued that a failed RDS process would aid in discovering that requirement, making the eventual implementation of a decision by fiat more politically acceptable.

Recognizing these potential limitations, and making plans to address them early in the RDS process will greatly improve the usefulness and potential success of the investment of time and talent required to actually implement what is a powerful negotiation technique. It must be acknowledged, however, that the RDS practice has not been applied in a context of institutionalized power imbalance, such as a caste system. It would be interesting to see if the RDS practice would prove successful in such a setting.

7. Conclusions

While there is a long heritage of participatory process design in the water resources planning and management space as well as other public policy arenas [45–47], the experiences reported upon from Bolivia and California suggest that the RDS practice should be considered to be a particularly effective approach. One potential explanation for its effectiveness might be found by looking again at the Allegory of the Cave referenced in the hypothetical dialogue between Plato and Aristotle. Plato actually concludes the allegory by posing the following question.

Now if once again, along with those who had remained shackled there, the freed person had to engage in the business of asserting and maintaining opinions about the shadows ... would he not be exposed to ridicule down there?

It is almost as if Plato recognized the impossible task of his Philosopher-King in a democratic context where decisions cannot easily be imposed by fiat. Even if the expert believed that these opinions about the shadows held by the enchained, Aristotle's desires and sensations, were wrong, he or she would have no way to change them. Slovic discovered this too when even after explaining the axioms of rational choice theory to the participants in his study, he could not get them to uniformly apply them. The great insight of Aristotle was to recognize the importance of practical wisdom associated with deliberation within a particular decision-making or policy-setting context. The RDS practice attempts to capitalize on this insight, not by allowing the shadows alone to dictate, but by bringing the monumental contributions of the early pioneers in the field of water modeling to bear in an attempt to reduce the fuzziness of the flickering shadows. While many other factors beyond good models and effective negotiations determine success, some problems are indeed wicked, thanks to the early water modeling pioneers, and the countless contributions of the water management community over the decades; these tools allow us to do more when confronting a drought or any other water management challenge than simply pray for rain.

Author Contributions: Conceptualization, D.R.P., D.N.Y. and W.N.S.; Formal analysis, L.F. and N.J.D.; Funding acquisition, D.R.P.; Methodology, M.I.E.A. and V.K.M.; Project administration, M.I.E.A. and V.K.M.; Software, N.J.D.; Visualization, L.F.; Writing-original draft, D.R.P.; Writing-review & editing, M.I.E.A., V.K.M., D.N.Y. and W.N.S.

Acknowledgments: The work carried out in Bolivia was supported first by the Challenge on Water and Food of the CGIAR system and later by the Pilot Program on Climate Resistance within the Inter-American Development Bank. The work carried out in the Yuba Basin was supported by the California Water Foundation, with a small matching grant from the California Department of Water Resources. Other RDS work carried out in California, Latin America, Africa, and Asia, but nor reported upon in this paper, was supported by the NASA ROSES program, the United States Agency for International Development, the International Development Research Council, the World Bank and the Swedish International Development Agency. While the work reported on here reflect the views of the authors alone, the support of all of these organizations is deeply appreciated. In addition, three anonymous reviewers provided very helpful feedback on the initial version of this article. Particular thanks to one reviewer for helping to construct a more appropriate title and to another for reminding the authors that pre-Enlightenment thinkers had much to do with the emergence of reason as the standard for governance. His or her suggestion that the more holistic view of the balance between reason and faith adopted by these 17th century thinkers might actually be more appropriate for modern times merits consideration. Finally, the authors would like to thank the participants in all of the RDS processes organized over these past six years. Their enthusiastic engagement helped make the case that RDS could be an effective approach to water management decision-making and their helpful suggestions and critiques contributed mightily to its improvement over time. Our deepest thanks. Readers interested in obtaining the data associated with this paper can contact David Purkey at david.purkey@sei.org or at #57 3142826360.

Conflicts of Interest: The authors declare not conflicts of interest.

References

1. Freeze, R.; Harlan, R. Blueprint for a physically-based, digitally-simulated, hydrologic response model. *J. Hydrol.* **1969**, *9*, 237–258. [CrossRef]
2. Cohon, J.; Marks, D. A review and evaluation of multi-objective programing techniques. *Water Resour. Res.* **1975**, *11*, 208–220. [CrossRef]
3. Crawford, N.; Linsley, R. *Digital Simulation in Hydrology Stanford Watershed Model 4*; Technical Report No. 39; Stanford University: Palo Alto, CA, USA, 1966.
4. Loucks, D.; Dorfman, P. An evaluation of some linear decision rules in chance-constrained models for reservoir planning and operation. *Water Resour. Res.* **1975**, *11*, 777–782. [CrossRef]
5. Shea, W.; Davie, M. *Galileo: Selected Writings*; Oxford World Classics: Oxford, UK, 2012.
6. Levi, H. *Blaise Pascal: Pensées and Other Writings*; Oxford World Classics: Oxford, UK, 1995.
7. Janiak, A. Correspondence with Richard Bentley. In *Isaac Newton: Philosophical Writings*; Cambridge University Press: Cambridge, UK, 2007.

8. Hume, D. *A Treatise of Human Nature*; Oxford University Press: London, UK, 1739.
9. Scholten, L.; Schuwirth, N.; Reichart, P.; Lienert, J. Tackling uncertainty in multi-criteria decision analysis—An application to water supply infrastructure planning. *Eur. J. Oper. Res.* **2015**, *242*, 243–260. [CrossRef]
10. Najil, A.; Haghigi, A.; Vali Semani, H. Simultaneous optimization of operating rules and rule curves for multi-reservoir systems using a self-adaptive simulation-GA model. *J. Water Resour. Plan. Manag.* **2016**, *142*. [CrossRef]
11. Haasnoot, M.; van Deursen, W.; Guillaume, J.; Kwakkel, J.; van Beek, E.; Middelkoop, H. Fit for purpose? Building and evaluating a fast, integrated model for exploring water policy pathways. *Environ. Model. Softw.* **2014**, *60*, 99–120. [CrossRef]
12. Akhbari, M.; Grigg, N. Managing water resource conflicts: Modelling behavior in a decision tool. *Water Resour. Manag.* **2015**, *29*, 5201–5216. [CrossRef]
13. Dreyfus, D.; Ingram, H. The National Environmental Policy Act: A review of intent and practice. *Nat. Resour. J.* **1976**, *16*, 243–262.
14. Milly, P.; Betancourt, J.; Falkenmark, M.; Hirsch, R.; Kundzewicz, Z.; Lettenmaier, D.; Stouffer, R. Stationarity is dead: Wither water management. *Science* **2008**, *319*, 573–574. [CrossRef] [PubMed]
15. Schultz, K. *Writers in the Storm: How Weather Went from Symbol to Sciences and Back Again*; The New Yorker: New York, NY, USA, 2015.
16. Camp, J. Drought and famine in the 4th Century B.C. Hesperia Supplements: Studies in Athenian Architecture. *Sculpt. Topogr.* **1982**, *20*, 9–17.
17. Koutsoyiannis, D.; Zarkadoulas, N.; Angelakis, A.; Tchobanoglous, G. Urban water management in Ancient Greece: Legacies and lessons. *J. Water Resour. Plan. Manag.* **2008**, *134*, 45–54. [CrossRef]
18. Lempert, R.; Popper, S.; Bankes, S. *Shaping the Next One Hundred Years: New Methods for Quantitative*; Long-Term Policy Analysis; Rand Corporation: Santa Monica, CA, USA, 2003.
19. Brown, C.; Ghile, Y.; Laverty, M.; Li, K. Decision scaling: Linking bottom-up vulnerability analysis with climate projections in the water sector. *Water Resour. Res.* **2012**, *48*. [CrossRef]
20. Vionov, A.; Kolgani, N.; McCall, M.; Glynn, P.; Kragt, M.; Ostermann, F.; Pierce, S.; Pamu, P. Modelling with Stakeholders—Next Generation. *Environ. Model. Softw.* **2016**, *77*, 196–220. [CrossRef]
21. Basco-Carrera, L.; Warren, A.; van Beek, E.; Jonoski, A.; Giardino, A. Collaborative modelling or participatory modelling? A framework for water resources management. *Environ. Model. Softw.* **2017**, *91*, 95–110. [CrossRef]
22. Halbe, J.; Pahl-Wostl, C.; Adamowsky, J. A methodological framework to support the initiation, design and institutionalization of participatory modeling processes in water resources management. *J. Hydrol.* **2018**, *556*, 701–716. [CrossRef]
23. Kahneman, D. *Thinking Fast and Slow*; Farrar, Straus and Giroux: New York, NY, USA, 2011.
24. Slovic, P.; Tversky, A. Who accepts Savage's Axiom? *Behav. Sci.* **1974**, *19*, 368–373. [CrossRef]
25. Gintis, H. Beyond *Homo economicus*: Evidence from experimental economics. *Ecol. Econ.* **2000**, *35*, 311–322. [CrossRef]
26. Lopes, H. From self-interest motives to justice motives: The challenges of some experimental results. *Am. J. Econ. Sociol.* **2008**, *67*, 287–314. [CrossRef]
27. Henrich, J.; Boyd, R.; Bowles, S.; Camerer, C.; Fehr, E. In search of homo economicus: Behavioral experiments in 15 small-scale societies. *Am. Econ. Rev.* **2001**, *91*, 73–78. [CrossRef]
28. National Research Council. *Understanding Risk: Informing Decisions in a Democratic Society*; National Academy Press: Washington, DC, USA, 1996.
29. Groves, D.; Yates, D.; Tebaldi, C. Developing and applying uncertain global climate change projections for regional water management planning. *Water Resour. Res.* **2008**, *44*. [CrossRef]
30. Lempert, R.; Groves, D. Identifying and evaluating robust adaptive policy responses to climate change for water management agencies in the American west. *Technol. Forecast. Soc. Chang.* **2010**, *77*, 960–974. [CrossRef]
31. Vuille, M.; Bradley, R. Mean annual tempera ture trends and their vertical structure in the tropical Andes. *Geophys. Res. Lett.* **2000**, *27*, 3885–3888. [CrossRef]

32. Condom, T.; Escobar, M.; Purkey, D.; Pouget, C.; Suarez, W.; Ramos, C.; Apaestegui, J.; Tacsi, A.; Gomez, J. Simulating the implications of glaciers' retreat for water management: a case study in the Rio Santa basin, Peru. *Water Int.* **2012**, *37*, 442–459. [CrossRef]
33. Escobar-Arias, M.; Lima, N.; Purkey, D. Modelación hidrológica y escenarios de cambio climático en cuencas de suministro de agua de las ciudades La Paz y El Alto, Bolivia. *Aqua-LAC* **2013**, *5*, 23–34.
34. Ceballo, L. *Glaciares de Colombia Más que Montañas con Hielo*; Instituto de Hidrología; Meteorología y Estudios Ambientales: Bogotá, Colombia, 2012.
35. Buytaert, W.; Vuille, M.; Dewulf, A.; Urrutia, R.; Karmalkar, A.; Celleri, R. Uncertainties in climate change projections and regional downscaling in the tropical Andes: Implications for water resources management. *Hydrol. Earth Syst. Sci.* **2010**, *14*, 1247–1258. [CrossRef]
36. Buytaert, W.; Cuesta-Camacho, F.; Tobón, C. Potential impacts of climate change on the environmental services of humid tropical alpine regions. *Glob. Ecol. Biogeogr.* **2011**, *20*, 19–33. [CrossRef]
37. Josse, C.; Cuesta, F.; Navarro, G.; Barrena, V.; Becerra, M.; Cabrera, E.; Chacón-Moreno, E.; Ferreira, W.; Peralvo, M.; Saito, J. *Physical Geography and Ecosystems in the Tropical Andes*; Inter-American Institutefor Global Change Research: Montevideo, Uruguay, 2010.
38. Yates, D.; Sieber, J.; Purkey, D.; Huber-Lee, A. WEAP21-A demand-, priority- and preference-driven water planning model: Model Characteristics. *Water Int.* **2005**, *30*, 487–500. [CrossRef]
39. Carle, D. *Introduction to Water in California: Updated with a New Preface*; University of California Press: Oakland, CA, USA, 2004.
40. Hanak, E.; Lund, J.; Dinar, A.; Gray, B.; Howitt, R.; Mount, J.; Moyle, P.; Thompson, B. *Managing California's Water: From Conflict to Reconciliation*; Public Policy Institute of California: San Francisco, CA, USA, 2011.
41. USDA. *California Agricultural Statistics: Agricultural Overview, Crop Year 2013*; USDA Pacific Regional Field Office: Sacrament, CA, USA, 2015.
42. Forni, L.; Galaitsi, S.; Mehta, V.; Escobar, M.; Purkey, D.; Lima, N.; Depsky, N. Exploring scientific information for policy making under deep uncertainty. *Environ. Model. Softw.* **2016**, *86*, 232–247. [CrossRef]
43. Lubell, M.; Lippert, L. Integrated regional water management: A study of collaboration or water politics-as-usual in California, USA. *Int. Rev. Adm. Sci.* **2011**, *77*, 76–100. [CrossRef]
44. Rittel, H.; Webber, M. Dilemmas in a general theory of planning. *Policy Sci.* **1973**, *4*, 155–169. [CrossRef]
45. Pahl-Wostl, C.; Sendzimir, J.; Jeffrey, P.; Aerts, J.; Berkamp, G.; Cross, K. Managing change toward adaptive water management through social learning. *Ecol. Soc.* **2007**, *12*, 30. [CrossRef]
46. Bouwen, R.; Taillieu, T. Multi-party collaboration as social learning for interdependence: developing relational knowing for sustainable natural resource management. *J. Community Appl. Soc. Psychol.* **2004**, *14*, 137–153. [CrossRef]
47. Reed, M.S.; Graves, A.; Norman, D.; Posthumus, H.; Hubacek, K.; Morris, J.; Prell, C.; Quinn, C.; Stringer, L. Who's in and why? A typology of stakeholder analysis methods for natural resource management. *J. Environ. Manag.* **2009**, *90*, 1933–1949. [CrossRef] [PubMed]

water

MDPI

Review

Improving Irrigation Water Use Efficiency: A Review of Advances, Challenges and Opportunities in the Australian Context

Richard Koech [1,*] and Philip Langat [2]

1 School of Health, Medical and Applied Sciences, Central Queensland University, Bundaberg Campus, University Drive, Bundaberg, QLD 4670, Australia

2 School of Environmental and Rural Science, University of New England, Armidale, NSW 2351, Australia; plangat@myune.edu.au

* Correspondence: richardkoech@hotmail.com; Tel.: +61-74150-7191

Received: 2 November 2018; Accepted: 28 November 2018; Published: 2 December 2018

Abstract: The demand for fresh water is on the increase, and the irrigation industry in Australia is looking to a future with less water. Irrigation consumes the bulk of the water extracted from various sources, and hence the efficiency of its use is of outmost importance. This paper reviewed the advancements made towards improving irrigation water use efficiency (WUE), with a focus on irrigation in Australia but with some examples from other countries. The challenges encountered, as well as the opportunities available, are also discussed. The review showed that improvements in irrigation infrastructure through modernisation and automation have led to water savings. The concept of real-time control and optimisation in irrigation is in its developmental stages but has already demonstrated potential for water savings. The future is likely to see increased use of remote sensing techniques as well as wireless communication systems and more versatile sensors to improve WUE. In many cases, water saved as a result of using efficient technologies ends up being reused to expand the area of land under irrigation, sometimes resulting in a net increase in the total water consumption at the basin scale. Hence, to achieve net water savings, water-efficient technologies and practices need to be used in combination with other measures such as incentives for conservation and appropriate regulations that limit water allocation and use. Factors that affect the trends in the irrigation WUE include engineering and technological innovations, advancements in plant and pasture science, environmental factors, and socio-economic considerations. Challenges that might be encountered include lack of public support, especially when the methods used are not cost-effective, and reluctance of irrigations to adopt new technologies.

Keywords: water use; irrigation efficiency; remote sensing; emerging technologies; real-time control; surface irrigation

1. Introduction

Irrigation is an essential agricultural practice for food, pasture and fibre production in semiarid and arid areas. In many countries including Australia, efficient water use and management are today's major concerns. The bulk of the irrigation water is sourced from rivers and dams and conveyed via open channels or pipelines to irrigated farms for storage before use or direct application to root zones. Irrigators who use groundwater often have storage tanks on their properties. At the farm level, irrigation systems or methods commonly in use may be broadly classified as sprinkler, surface, and drip or trickle systems. In sprinkler systems (e.g., solid sets, centre pivots and travelling irrigators), water is delivered in form of sprays using overhead sprinklers. In drip or trickle systems, water is delivered in small amounts via small nozzles installed in pipes or tapes, which can either be above the ground or underground. The sprinkler and drip/trickle systems are also referred to as pressurised systems, as they operate under low pressure which often involves some form of pumping. In surface systems (e.g., furrow and basin/border), water is conveyed over the field surface by the gravitational force. The furrow system is the most common method for the irrigation of row crops in Australia and in the world.

Globally, it is estimated that about 70% of fresh water abstracted is used to irrigate 25% of the world's croplands (399 million ha) which supply 45% of global food [1]. Water used for industrial and domestic purposes account for approximately 20% and 10% of the total global water usage, respectively. In Australia, for instance, in the year 2016–2017, 9.1 million mega litres were used to irrigate 2.2 million ha [2]. The demand for fresh water resources is on the increase, and the trend is likely to continue with the increasing population that comes with increased demand for food and fibre, and the predicted negative impacts of climate change. There is also increased awareness of the need to provide sufficient water to serve other ecological services. There appears to be consensus that irrigated agriculture in general is up against a future with less water.

This, therefore, calls for increased effectiveness in the utilisation of the scarce water resources, a concept that is technically called water use efficiency (WUE) or simply irrigation efficiency. From an engineering standpoint, WUE is often defined using a volumetric or hydrological approach, simply as the proportion of the water supplied through irrigation that is productively or beneficially used by the plant (Equations (1) and (2)). This definition is predominantly used when referring to field-scale irrigation water management. However, it should be noted that WUE may also be assessed at the catchment or basin scale [3].

The two most commonly used efficiency measures of an irrigation system are (i) application efficiency (AE) and (ii) requirement efficiency (RE), which can be written as:

$$\mathrm{AE} = \frac{\text{volume of water stored in the root zone}}{\text{total volume of water applied}} \tag{1}$$

$$\mathrm{RE} = \frac{\text{volume of water stored in the root zone}}{\text{water deficit prior to irrigation}} \tag{2}$$

The efficiency performance measures, AE and RE, are only applicable at the field scale. However, losses of water also occur in conveyance and distribution channels prior to delivery to the field. If the water is stored in dams prior to usage, then further losses may occur as a result of evaporation and seepage. Performance measures used in these cases include conveyance, distribution and storage efficiencies.

On the other hand, the efficiency of irrigation water use may also be seen in a plant physiological sense, and in particular as a comparison of the yield or economic return of an irrigated crop or pasture to the total amount of water transpired by the crop or pasture. In fact, in recent literature (e.g., [4]), this is commonly referred to as irrigation water productivity and not WUE. In the cotton industry in Australia, this is sometimes referred to as irrigation water use index (IWUI) and relates cotton production only to the amount of irrigation water used [5].

In Australia, surface irrigation is the main irrigation method used, and in 2013–2014, it accounted for 59% of the total irrigated land [2]. However, the system in general is associated with high labour requirement and low WUE. This explains why modernisation and automation projects (discussed later in this paper) have tended to focus on this irrigation system. Conversely, the pressurised irrigation methods (sprinkler and drip) are generally less labour-intensive and have significantly higher WUE.

With the advancement of technology, thanks largely to the many years of investment and research and development in agriculture, there are new and emerging opportunities for further improving the WUE in irrigated agriculture. Examples of these include use of remotely sensed data (from drones or satellites), communication networks and the availability of cheap sensors.

It is clear from the above discussion that in order to improve the irrigation WUE, losses that occur along the conveyance and distribution channels must be minimised, and the timing and the quantity of water applied (or irrigation scheduling) must be optimised. Improvement of the irrigation WUE may lead to water savings which may be used to irrigate more land, which is particularly relevant where water is the limiting factor of production. The purpose of this paper was to review the advancements that have been made to improve the irrigation WUE, document the challenges encountered as well as exploring opportunities for further development. Although the bulk of the review is on Australia's irrigated agriculture, examples from other countries are also used, and it is anticipated that the findings will inform researchers and policy makers in general. The paper starts by looking at the nexus between irrigation modernisation and automation in Australia, particularly focusing on irrigation distribution channels and on-farm development. The review then discusses the role of irrigation scheduling in improving the WUE, and the concept of real-time control and optimisation that is still under development. The emerging and potential opportunities for improved WUE through remote sensing techniques, and sensors and communication networks are discussed. In the final section, we discuss the challenges to the achievement of higher WUE, with focus on water consumption at the basin scale and factors affecting trends in WUE which are broadly categorised as: engineering and technological; environmental; advancements in plant and pasture science; and socio-economic.

2. Irrigation Modernisation and Automation

The terms irrigation modernisation and automation have become common in irrigation literature in the recent decades, and in some cases have been used interchangeably. In the context of this paper, modernisation is the process of replacing ageing irrigation infrastructure and methods, often with new or "modern" equipment and technologies that have been developed in the recent past. The aims of irrigation modernisation include water saving and improved water delivery, and reduced operating and labour costs, which leads to sustainable agricultural production and enhanced livelihoods of farmers [6]. On the other hand, automation of irrigation systems is the use of equipment that allows the irrigation process to proceed with minimum human involvement, except for periodic inspections and routine maintenance. In Australia, modernisation and automation have been undertaken in the water distribution networks at the farm level as well. The nexus between automation and modernisation, and WUE will be explored in this paper.

2.1. Irrigation Water Distribution Systems

Irrigation has been practiced for many decades in Australia, especially in the Murray Darling Basin (MDB) (Figure 1), which consumes about two-thirds of the water abstracted for irrigation of crops and pasture [2]. The bulk of the water in the basin is delivered to irrigators by private and farmer-owned companies. Murray Irrigation in the MDB, which supplies water to 2300 farms with a total area of 748,000 ha is the largest in the country. Other large companies include Murrumbidgee Irrigation and Coleambally Irrigation Co-Operative Limited, which supply water to the Murrumbidgee Irrigation Area and the Coleambally Irrigation District, respectively.

In the last few years, irrigation water in the MDB has been largely conveyed via open channels. In addition, a significant portion of the infrastructure supporting irrigation, with some built in the early years of the last century, have started to age and become inefficient. From the WUE perspective, irrigation water conveyed via open channels is of interest because of the associated losses. Hence, the Federal and State and Territory Governments developed the Murray Darling Basin Plan and embarked on a program of modernising and automating this critical infrastructure, as discussed below.

Seepage losses, particularly in earthen open channels, may consume up to approximately 14% of the total water supplied to an irrigation scheme. Evaporation losses, especially in large open channels, may also be considerable [7], especially in arid parts of Australia. Therefore, one of the priorities of the modernisation plan of the MDB was to reduce these losses using a variety of methods including lining the canals with clay or rubber, repair of earthen and concrete channels, installation of gravity pipelines in place of open channels, and upgrade of on-farm irrigation infrastructure (discussed in greater detail in the next subsection). One example of such a project undertaken in the MDB from 2011 to 2015 is the Trangie-Nevertire Irrigation Scheme, which returned about 40% of the original water entitlements, significantly reduced water losses, and improved the water delivery efficiency from about 65% to 93% [8].

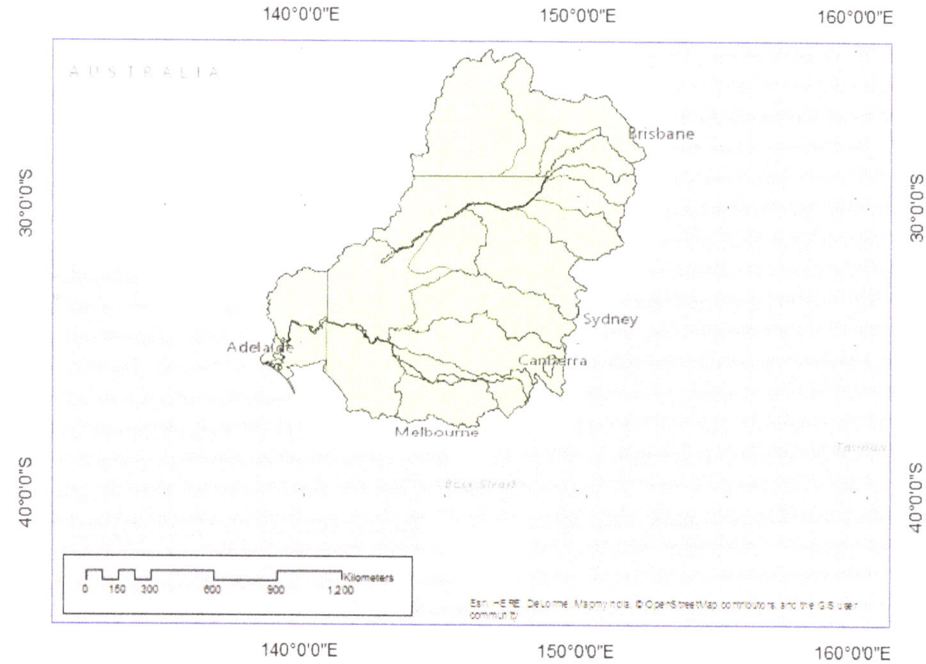

Figure 1. The Murray Darling Basin, which consumes about 70% of the water abstracted for irrigation in Australia.

A critical step towards improving the WUE is the ability to accurately measure the amount of water supplied to irrigators. This is in line with the saying that goes: "You cannot manage what you cannot measure." However, previous research undertaken in Australia showed that inaccurate flow measurement techniques, for instance using the Dethridge wheels, led to the supply of irrigation water in excess of entitlement volumes. A study by Goulburn Water [9] showed that large Dethridge wheels operated with inaccuracies of between −18% to +3%. Hence, regulatory requirements were put in place to ensure that irrigation water meters operate at an acceptable level of performance [10]. The requirements include pattern approval by the National Measurement Institute (NMI) and the ability of the meters to perform within maximum limits of error of ±5% in field conditions. The irrigation modernisation program involved the replacement of Dethridge wheels with water meters that were compliant with these regulations.

The above examples, therefore, demonstrate the opportunities for improving the WUE before the water is delivered to the farm and provide an indication of the magnitude of savings that have occurred in specific projects. There are still many irrigation enterprises that rely on open and unlined channels for their water supply, although statistics on their proportion are not immediately available. Some of the strategies used to modernise and automate flow of water in irrigation canals are also used on-farm to control the flow of water to different portions of the field. These include automatic regulators or gates and telemetry systems. These are discussed in the next subsection.

2.2. On-Farm Irrigation Development

In the recent past, a number of on-farm research and development projects aimed at improving the WUE have been undertaken. The funding scheme discussed above mainly focused on large irrigation schemes, but the Australian Government also initiated the On-farm Irrigation Efficiency Program to help individual irrigators improve their irrigation infrastructure or change irrigation practices (e.g., convert to more efficient irrigation methods) in order to save water. The most significant developments, especially in the surface system, appear to be automatic gates or outlets, water metering and use of telemetry systems. Some of the on-farm modernisation and automation projects have also been funded by the irrigators themselves.

Gates are structures placed in irrigation channels or bay/basin outlets to control flow of water, and be may be controlled by a mechanical timer or electric solenoid. Some gates also have the capacity to measure the flow rate. There has been an increase in the use of telemetry systems in irrigation to allow for remote measurement and control of various parameters. The mode of communication used by these telemetry systems include radio, telephone, infrared, satellite and internet. In Australia, there are commercially available on-farm irrigation telemetry systems manufactured by local companies, such as AWMA and Rubicon that utilise the Supervisory Control and Data Acquisition (SCADA) platform.

These on-farm automation developments may appear to be focussed on reducing the irrigation labour requirement especially in the surface systems, but they also play a role in improving the WUE. This is because, with automated systems, there is less chance of human error that may lead to water loss. A good example of this is that in manual surface irrigation systems, inflow is turned on and off at the completion of the irrigation by an operator. A delay of cutting off the flow will therefore lead to water losses. Improvement of application efficiencies of well-managed automated-bay-irrigated fields in Northern Victoria was demonstrated by Smith et al. [11].

2.3. Irrigation Modernisation in Developed and Developing Countries

The above discussion on irrigation modernisation and automation has largely focused on Australia. To provide context, this subsection will briefly examine the scenario in the rest of the world, particularly in Spain and the Unites States, two countries which are also major irrigating economies in Europe and North America, respectively.

According to Plusquellec [6], developed and some emerging countries generally possess conditions favourable for automation and modernisation of especially large-scale irrigation systems, for instance:

- Organised manner of supply of water to users;
- Irrigation systems that are generally well-maintained, hence less costly to upgrade;
- Strong policy and regulatory environment, and the willingness and capacity to enforce laws related to water use;
- Availability of technical expertise and equipment; and
- Well-developed infrastructure such as roads.

Background information about irrigation modernisation and automation in Spain can be obtained from a number of articles, for instance González-Cebollada [12] and Lecina et al. [13]. Significant Spanish Government reforms and modernisation to manage demand of water began in 2002, and as was the case in Australia, the projects were largely taxpayer-funded, and the objectives included revitalisation of the irrigation sector and water conservation. The programs included improvement of the irrigation water distribution network and promotion of water-efficient methods such as the drip system. Similarly, the modernisation of irrigation systems in the United States aimed to improve the water delivery system using methods such as modification of check structures, use of reticulation systems, improved measurement and control, and the use of SCADA systems [14]. However, in a survey conducted by the United States Department of Agriculture (USDA) a few years ago, it was estimated that at least half of the United States cropland was still irrigated with less efficient irrigation methods [15].

3. Irrigation Scheduling

Irrigation scheduling is the process of determining how much water to apply and when to irrigate, and thus has a direct effect on WUE: the application of more water than that is necessary for optimal plant consumption reduces the irrigation WUE. Irrigation scheduling requires an understanding of the pattern of plant water use, which is affected by factors such as weather, growth stage and canopy wetness. The meteorologic component varies seasonally, daily and diurnally.

Irrigation may be scheduled based on the plant water status, which may be measured directly using a pressure bomb, or indirectly by monitoring the flow of the stem sap. Other indirect methods include the measurement of soil moisture content using probes and estimation of crop evapotranspiration (ET). A summary of the main irrigation scheduling approaches is presented in Jones [16]. In Australia, the most common tools used for irrigation scheduling are soil probes and tensiometers [17]. The major drawback of using these soil-moisture-based tools for scheduling is that they give point-based measurements, while the soil characteristics are known to vary spatially and temporally [18].

Farmers who do not use any scheduling tool often rely on their experiences to schedule their irrigations. However, previous studies have shown that such farmers who rely on the "rule-of-thumb" may be losing water [19]. Emerging methods of measuring or estimating crop water status for irrigation scheduling purposes are discussed later in this paper.

With the advancement in technology including the internet, a number of computer-based irrigation scheduling systems have been developed to help farmers in their decision-making process. Typical examples of these used in Australia include WaterSense, WaterTrack Rapid and IrriSatSMS [20]. However, despite the proven benefits of improved WUE using these technologies, their adoption is still limited because of reasons ranging from complexity to cost [17]. In the recent years, cheaper and more versatile sensors have become available (for instance Figure 2).

(**a**) (**b**)

Figure 2. An example of a device with multiple sensors (**a**), with the ability to measure soil properties (**b**) (moisture, conductivity and temperature), air temperature and relative humidity. The device has a 2G/3G connectivity, and its web platform has an open Application Programming Interface (API) for ease of integration with other systems.

It is clear from the above discussion that scheduling of irrigation is easier if the irrigation system is automated, and with features such as accurate metering and sensors.

4. Real-Time Control and Optimisation

Due to factors such as differences in soil composition and weather, the infiltration characteristics at the field scale will vary both spatially and temporally. For most conventional irrigation systems that seek to apply water uniformly, this will mean that the on-farm WUE will be equally variable across the field [21]. The variability in WUE is more pronounced in surface irrigation systems (e.g., furrow) whereby irrigation water is conveyed over the soil surface. Hence, in the recent times, the concept of real-time control and optimisation, which was traditionally used in other branches of engineering, has gained prominence in irrigation water management.

In the context of irrigation, real-time control implies measurements taken during an irrigation event (e.g., advance of water in a furrow system) are processed for the modification of the same irrigation event. This is at variance with conventional management systems which typically rely on previous or historical measurements, which are affected by the temporal nature of infiltration characteristics. Real-time control is feasible when the control process is automated so that the feedback can be implemented rapidly. On the other hand, optimisation is the process of manipulating various variables of an irrigation system with the aim of achieving the best possible outcome. This has traditionally been achieved through trial and error or irrigator experience; however, owing to the advancement in computing technology in the recent past, the use of simulation models has been on the increase [22].

Surface irrigation systems that can be controlled and optimised in real time are sometimes referred to as smart irrigation systems (Figure 3). They are regarded as improvements from purely automated systems, which are mostly designed to reduce irrigation labour requirement through automation of some tasks. The traditional or conventional irrigation systems are associated with high labour requirement and low WUE (Figure 3).

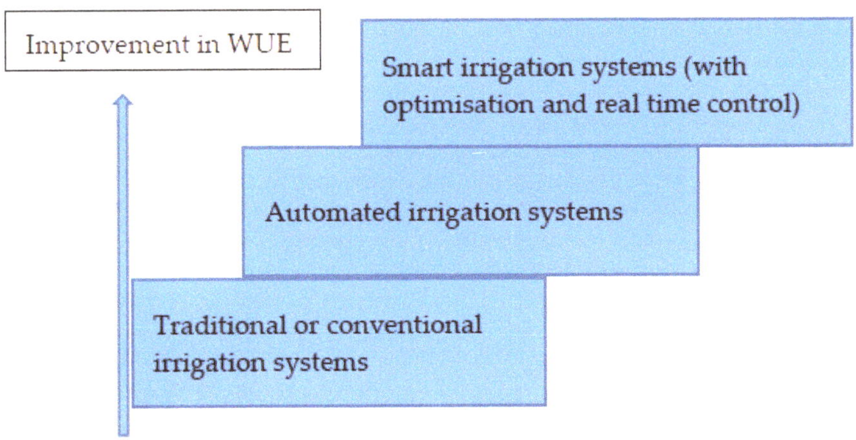

Figure 3. Advances in irrigation technologies.

In surface irrigation systems in particular, adaptive real-time control has been proposed for the management of temporal infiltration variability [23–25]. A real-time optimisation system for furrow irrigation was tested in a field of commercially grown cotton in Queensland, Australia, demonstrated potential for an improvement in WUE and a reduction in labour requirement [26]. The system involved measurement of the inflow rate, sensing of the advance of water along the furrow, a computing system with a simulation model, and a telemetry system to facilitate communications between different components. A commercial prototype of this system was produced, and trials in a commercially irrigated field showed that it is able to control irrigation events by cut-off time to achieve the maximum application efficiency [27].

Hence, it is clear that real-time control and optimisation in the Australian irrigation industry is still at its infancy, particularly in surface irrigation. However, based on the amount of research and the progress made so far (for instance the commissioning of prototype systems), it is conceivable that, in the future, it will play a bigger role in irrigation water management and improvement of WUE.

5. Emerging and Potential Opportunities for WUE

Investments in research and development projects in the recent decades and advancement in technology, in general, have yielded new or emerging opportunities for increased WUE in irrigated agriculture. This has come in the form of new and advanced equipment and techniques, as well as cheaper and relatively accurate alternatives.

5.1. Remote Sensing

As discussed above, in irrigated agriculture, improvement of WUE is achieved by optimising the timing and quantity of irrigation applications. The scheduling methods described, whether plant-, soil- or meteorologic-based (evapotranspiration), are normally used on the ground. These methods are generally expensive, time-consuming and cannot be easily automated [16], and also mostly location-specific and not suitable for use in large areas. The option of remote sensing, which is not at all a new concept in agriculture, has, in the recent years, been an active area of irrigation water management research due to its advantages in systematic measurements across space and time, ability to cover large areas and capability to be integrated into models and with Geographic Information Systems (GIS).

New approaches using remotely sensed data to estimate the crop or plant water status and hence schedule irrigations are emerging. The first is satellite imagery which has been applied in many agricultural applications, for example yield and disease monitoring. In the last few decades, methods using algorithms to derive vegetation indices from satellite imagery in combination with ground-based measurements to estimate evapotranspiration (ET) over large areas have emerged [21,28]. Use of Landsat thermal infrared (TIR, https://lta.cr.usgs.gov/L8) imagery to derive spatial variability information of ET at the field scale and uniformity of water consumption for the purposes of improving WUE [29,30] is a recent step towards improved irrigation water management.

Use of remote sensing in irrigation water use monitoring, evaluation and management are underutilized due to issues of spatial and temporal resolution, quality of results and one-time/one-place syndrome among others. However, the current Lansat-8 satellite series comes with a 30 m spatial resolution and can be used to assess actual crop evapotranspiration and crop water use at the field and farm scale. There are a number of commercial satellites now available that may be used for agricultural purposes, for instance Sentinel-2 (https://sentinel.esa.int/web/sentinel/missions/sentinel-2) and Planet (https://www.planet.com/markets/monitoring-for-precision-agriculture/).

Another remote sensing approach to determining the crop water status, which is still in research phase, is the use of thermal and multispectral imagery collected using unmanned aerial vehicles (UAV) or drones. Research has shown that the plant canopy temperature is correlated to the plant water status, and hence can be used for irrigation water management [31]. Applications using reflectance of near and mid-infrared regions of the electromagnetic spectrum to assess water status in cereal crops, fruit trees, grapevine, and pasture are described in Cozzolino [31].

The main advantage of remote sensing is the ability to estimate the crop water status over spatial scales, which cannot be possibly realised with the conventional methods such as soil probes or plant-based techniques. It is also expected that, with the increased uptake of drone technology, their prices will decrease and therefore become more accessible to many farmers. However, increased effort is also needed to connect irrigators and remote sensors to maximise economies of scale. Key opportunities and advances to watch include future collection of very high resolution (<10 m) data through hyperspectral sensors such as the current commercial IKONOS and Quickbird satellites, rapid data access availability of data from multiple sensors with a wide array of spatial, spectral, and radiometric features and remote sensing multi-data synthesis through streaming technology.

5.2. Sensor and Communication Networks

Sensors are equipment used to collect a range of data such as soil moisture and weather in order to improve agricultural management. Typical examples of sensors used in irrigation water management include soil moisture probes and weather stations. Traditionally, equipment for monitoring crop or soil water status were connected using cables and often required manual reading and the data used to schedule future irrigations. Apart from the inaccuracies that come with using historical data for future water management, such manual processes are time-consuming and often expensive. The use of wireless sensor technologies to improve WUE in irrigated agriculture is on the increase.

More often, a series of wireless sensors are used to monitor various parameters in the field, for example soil moisture and weather data. This is especially driven by the fact that the recent advancement in technology and competition has led to the availability of cheap sensors. A wireless sensor network consists of a number of individual sensors (sensor nodes), a sink node or hub to receive and process data from the sensor nodes, and a communication technology [32]. The sensor networks may also have actuators that can be used to automate the irrigation system.

The wireless communication technologies that are used for agricultural purposes including water management are discussed in many texts, for instance Rehman et al. [32]. The current communication technologies used in agriculture are ZigBee, Bluetooth, WiFi, GPRS/3G/4G, Long Range Radio (LoRa) and SigFox [33]. The ZigBee technology is commonly preferred in irrigation water management because of the range, low cost, energy efficiency and reliability [32,34].

The use of wireless sensors that measure soil moisture, temperature and humidity and relay the data over the 3G internet network is described in Reference [35]. The automation of such crucial data collection means that the irrigation system can be controlled in real time, thereby achieving higher WUE.

It appears from the literature reviewed that research into water loss through leakage has concentrated on urban water supply distribution networks. The techniques used to detect leakages in urban water distribution networks can be applied in irrigation settings. Pressure sensors connected to a wireless sensor network can play a vital role in in detecting leakages, and thus facilitate faster repair and prevention of further losses [36]. There is also a potential for using smart water technology to detect losses in water pipelines [37].

There are substantial ongoing research activities involving sensors and communication networks which are likely to lead to improved products and services in the future. It is also likely that communication networks will be used in a more integrated manner to achieve multiple objectives, for example irrigation and urban water supply using smart water meters.

6. Irrigation Water Productivity

This paper has largely been written from an engineering perspective, which predominantly defines WUE as the ratio of the irrigation water beneficially used by the plant or pasture to the water supplied through irrigation. This section will briefly discuss irrigation water productivity, with a focus on plant genetics and agronomic practices used to achieve higher yields using less water.

Through plant breeding, scientists have managed to develop high-yielding crop varieties. This implies that, all other factors kept constant, with the same amount available water, farmers can achieve a higher irrigation water productivity. Detailed information on how the plant WUE can be improved through molecular genetics is described in Ruggiero et al. [38]. The research provides an overview of the manipulation of genes that strongly impacts on WUE such as these that control root traits and stomatal development. Some genetically modified varieties are also resistant to pests and diseases, leading to higher yields. A study conducted within the cotton industry in Australia found that the water use productivity had increased by 40% over a period of ten years as a result of yield increases achieved by developments in plant breeding, the use of genetically modified varieties, and improved crop and water management systems [5].

Deficit irrigation, which is the application of less water than that is required by the plant or pasture, is a strategy that is often used when water is limiting. In a trial conducted in a dairy region of Victoria, Australia, where pasture is often irrigated, Rogers et al. [39] demonstrated that lucerne under deficit irrigation can fully recover once full irrigation is restored, and thus ideal forage can be grown under water limiting conditions. Tejero et al. [40] in a trial undertaken in a citrus orchard in Spain concluded that deficit irrigation strategies have the potential to improve WUE. Du [41] proposed the adoption of deficit irrigation strategies in areas of China where conventional irrigation is no longer sustainable because of water shortages.

7. Water Consumption at the Basin Scale and Trends in WUE

7.1. WUE and Water Consumption at the Basin Scale

The need for water users to achieve greater WUE is often seen as a prerequisite for saving water for the benefit of other users as well as the environment. However, literature reviewed suggests that a higher WUE does not necessarily equal to net water saving, particularly at the basin scale.

When seen from the dimension of a water basin, what may be assessed as a loss in one perspective (e.g., deep drainage losses that may occur in surface systems), may be a gain in another way (e.g., recharge of groundwater resources). Some research has shown that significant improvement in delivery and on-farm WUE may in fact lead to a decline in groundwater resources [21] or reduce water for environment and downstream users [3]. Therefore, although improvement of on-farm irrigation WUE may lead to water savings on the farm, it will not necessarily be beneficial on a catchment or basin scale [15].

An overall increase of water consumption at the basin scale may occur if water savings ultimately leads to an expansion of the irrigated area [42]. This was demonstrated by research conducted in Morocco which saw the overall water consumption rise as a result of subsidised drip irrigation kits promoted as a means of increasing productivity and saving water [43]. In this example, although the drip system is generally regarded as water-efficient, farmers were found to shift to more water-intensive crops and generally used the "saved" water to expand the acreage under irrigation. This view has been corroborated by other studies, for instance a FAO-funded research project undertaken in North Africa and the Middle East region [44]. The study found that at the field scale, water saving may appear to be substantial, but at the basin scale, the total water consumption may actually increase while the crop water productivity gains for the most important crops may be modest at best. In a study undertaken in India, the widespread adoption of water efficient methods such as the sprinkler and drip systems were found to have the capacity to substantially reduce overextraction of groundwater resources, but half of the water saved was reused to expand the area under irrigation [45].

Figure 4 is used as an example to demonstrate the simultaneous increasing uptake of water-efficient technologies (sprinkler and drip) and the increasing total area under irrigation (especially between 1994 and 2013) in the Unites States. This suggests the potential reuse of any water savings to expand the area under irrigation. The graph in Figure 4 also shows the corresponding decrease in surface irrigation systems that are generally regarded as inefficient.

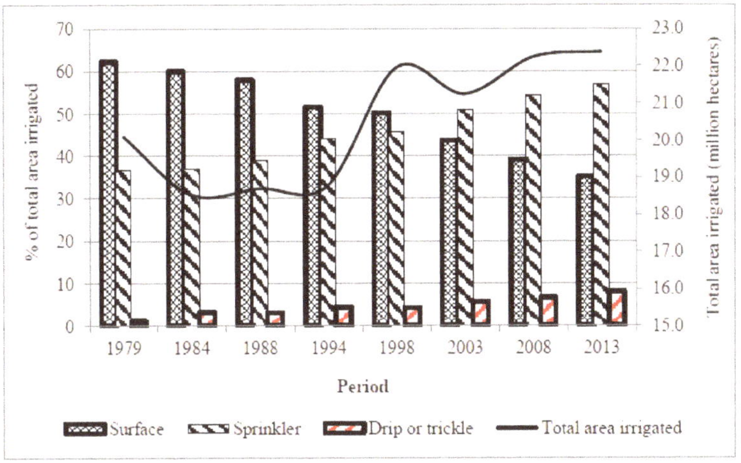

Figure 4. Area of land in the Unites States irrigated using different methods. (Plotted from data obtained from: USDA 1990, Table 4; USDA 1999, Table 4; USDA 2006, Table 4.6.1; USDA 2009, Table 4; USDA 2014, Table 28) [46].

An analysis of the MDB in Australia showed that the environment may become the unintended casualty (receive less water on average) of the increases in WUE driven by the adoption of water-efficient technologies [47] with most of the saved water being reused. The reuse of the saved water seems to be corroborated by the trend of the total irrigation water use in Australia between 2002 and 2017 (Figure 5). The graph shows that the total irrigation water use between 2002 and 2006 was above 10,000 gigalitres (GL) but reduced to a low of just above 6000 GL in the four-year period: 2007–2008 and 2010–2011. The reduced irrigation water use in the period 2006–2011 was as a result of a severe drought that drastically reduced the availability of water for irrigation. In the period 2012–2014, the water use increased back to a similar level to the early part of the available data (approximately 11,000 GL), effectively signalling no net water saving. There was a decrease of irrigation water use in 2014–2016, but increased slightly in 2016–2017 to just over 9000 GL. The trends appear to be largely dependent on weather patterns.

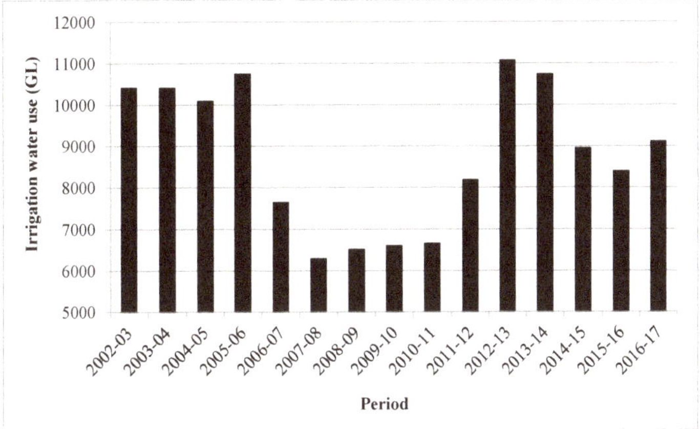

Figure 5. Total irrigation water use in Australia. (Plotted from data obtained from: Australian Bureau of Statistics (ABS), 2004–2018.)

While Figure 5 shows the trend of irrigation water use in the whole of Australia, Figure 6 is specific to the MDB, which consumes the bulk of the water used for irrigation in the country as previously discussed. In addition, Figure 5 shows the trend of the irrigated land in the basin and correlation of the irrigation water use with the area irrigated, meaning when farmers have access to more water, they irrigate more land (and vice versa). Therefore, it is likely that water saved as a result of the water-efficient technologies and practices adopted is reused as suggested by the studies quoted above.

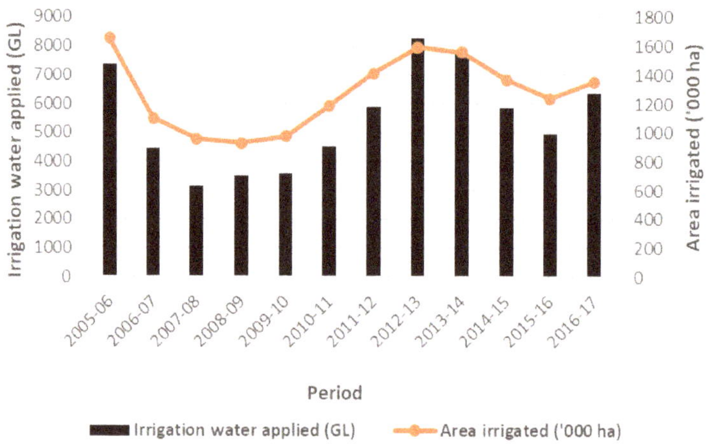

Figure 6. Irrigation water use and area irrigated ('1000 ha) in the Murray Darling Basin, Australia. (Plotted from data obtained from: ABS 2004–2018.)

However, there are strategies that can be used to attain a good balance between improved irrigation efficiency and environmental conservation, including groundwater recharge. A typical example is the water saving initiatives funded by the Australian Government, with the understanding that the water saved is released for environmental use [6]. The regulatory return of the saved water to the environment therefore mitigates the "rebound effect" phenomenon, which suggests that the increase in efficiency of use of a resource may lead to an increase in the rate of consumption of that resource [48].

Many other studies undertaken in different parts of the world have also linked widespread adoption of water-efficient technologies to overall increase in water consumption mostly due to expansion of land under irrigation, and not a decrease as intended. Nonetheless, as shown by a study undertaken in Spain, the water-efficient technologies have come with other side benefits such as reduced use of fertilisers and better accounting of water use [49].

In the literature reviewed for this study, an almost unanimous view that emerges is that overall reduction of water at the basin scale cannot simply be attained through the promotion or subsidies provided for water-efficient technologies. These technologies will thus need to be used in tandem with other measures such as incentives for conservation [45] and regulations to limit water allocation [44], among others.

Another interesting dimension is the nexus between the irrigation methods deemed to be generally more water-efficient, energy consumption and greenhouse gas emissions. For instance, modelling has demonstrated that although pressurised irrigation systems such as sprinkler and drip methods are generally more efficient and productive, they are more energy-consuming (compared to conventional systems such as furrow irrigation), resulting in the production of additional greenhouse gas emissions [50]. The energy costs in many countries (in the case of irrigation electricity for pumping water) has been rising steadily. This is thus likely to impact on the adoption of the water-efficient but high-energy consuming irrigation methods.

7.2. Factors Affecting Trends in WUE

From the above discussion, it is clear that the trends in the WUE of irrigated agriculture are affected by a range of factors which may be broadly categorized as shown in Figure 7. Engineering and technological factors include improvement of water distribution networks and on-farm irrigation development, irrigation scheduling, real-time control and optimisation, remote sensing and sensor and communication networks. These factors improve irrigation WUE mainly by reducing water losses. In the recent past, a variety of hardware and software gadgets has become commercially available and is used to enhance irrigation WUE. Advancements in plant genetics have led to the development of high-yielding and disease-resistant varieties with higher WUE. There has been greater environmental awareness, leading to some governments around the world funding water-saving initiatives with the understanding that the water saved is released as environmental flows. Socio-economic factors are also important drivers of WUE. This will be covered in this section, with a focus on technology adoption and the decision-making processes of irrigation water users.

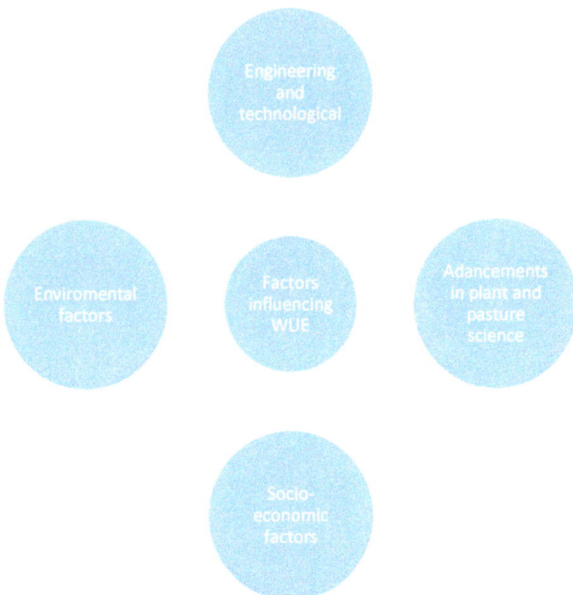

Figure 7. Factors influencing water use efficiency.

When farmers are faced with limited availability of irrigation water, they have to make challenging decisions on how best to operate. This is a common problem in Australia, where to a large extent, water is the limiting factor of production, but land is virtually unlimited. It is common to see farmers irrigate part of their land and cultivate the rest under rain-fed conditions. A study undertaken in Southern Spain found that the majority farmers growing irrigated intensive olive groves used deficit irrigation in order to maximise the value of limited available water [51].

Some researchers have observed that water saving initiatives have mainly focussed on engineering solutions such as reduction of seepage losses and suggested that further improvement in on-farm WUE could be achieved by the adoption of new irrigation technologies [52]. However, it must be noted that technology adoption is a complex sociological phenomenon, and its success will largely depend on the willingness of water users to change their attitudes. Irrigation water users, like other members of the community in general, are predisposed to continue with the farming practices they are most familiar with, for fear of the unknown. In most cases, water users have access to information on new technologies, but getting people to change their attitudes and adopt new practices or technologies is a slow process. Research undertaken in the United States suggested that the requirement to learn a new set of skills may act as a deterrent to irrigators investing in new technologies or adopting new practices [15].

In both Australia and the United States, the cost of changing to new technologies and practices has been cited as a significant factor causing non-adoption [15,52]. This includes the capital required as well as the associated on-going costs. A typical example is when farmers decide to change from surface irrigation methods to pressurised systems such as centre pivots and lateral move machines which are generally associated with higher WUE. These pressurised systems not only require substantially higher capital costs to install, but come with higher energy consumption and are therefore expensive to run [5].

A European-Union-funded research project undertaken in Italy and Portugal concluded that lack of adequate knowledge and incentives may prevent farmers from exploiting the full potential of available technologies to optimise WUE [53]. The study thus recommended that there should be a continuous knowledge exchange among scientific experts, farmers and other stakeholders, and appropriate support be provided to encourage environmental conservation.

As discussed earlier, one of the main approaches used in Australia to attain increased WUE is the upgrade of irrigation infrastructure and provision of subsidies for on-farm improvements. However, some studies [3] have shown that sometimes these investments are not cost-effective, especially when compared to other alternatives such as water trading which is aimed at efficiently allocating water across competitive use. Therefore, when the economic benefits of these taxpayer-funded initiatives are not apparent, public support is not guaranteed.

It should be noted, however, that WUE cannot be improved infinitely. In the case of an irrigation system that has a lower irrigation performance to start with, it would be easier to notice an increase in WUE when improvements to the system are undertaken. However, for a system that is already operating at or near the optimum level, the WUE will increase (if at all) at a much slower rate. A study conducted in the Guadalquivir River Basin in Spain found that the impact on the WUE of technological innovations, such as deficit irrigation, new crop varieties and other water-saving technologies, had decreased considerably after a number of years [54]. As previously noted, plant breeding has been used for a number of decades now to develop plant varieties with higher WUE. However, this improvement is not expected to continue at the same rate as before [55]. This implies that the world cannot solely rely on the past and the present technologies to improve WUE, but must continue to undertake research to generate newer technologies that can be used to further improve the WUE.

8. Conclusions

The purpose of this paper was to review the steps that have been taken to improve the use of the scarce water resources, with a focus on irrigation in Australia but with examples from other countries. We also looked at the challenges that have been encountered and explored opportunities that may lead to improved WUE in the future.

The Australian Federal, State and Territory Governments have facilitated the modernisation and automation of irrigation infrastructure such as distribution channels and improvements on on-farm irrigation hardware. This has led to improved WUE, and some of the water saved has been used for environmental purposes. Improved irrigation scheduling and the real-time control and optimisation have also demonstrated potential for further water savings. The use of remote sensing and communication sensor networks are emerging in the irrigation industry and are expected to contribute to improve WUE. However, challenges lie in the way of enhanced WUE. These include lack of public support, especially when the methods used are not cost-effective, and reluctance of irrigations to adopt new technologies.

The review has demonstrated that the adoption of water-efficient technologies has delivered water savings at the field scale, with some of the savings being released as environmental flows. However, the net water saving at the basin scale is not always achievable. In fact, some studies have demonstrated that a net increase in water consumption, largely due to the reuse of the saved water to expand the area of land under irrigation. Hence, an overall reduction of water consumption at the basin scale is likely to be achieved when water-efficient technologies are used in combination with other measures, such as provision of incentives for water conservation and regulations to limit water allocation.

Author Contributions: R.K. conceptualised this review study and produced the first draft of the manuscript. P.L. reviewed the draft, rewrote some sections and formatted the paper.

Funding: This research received no external funding.

Acknowledgments: We sincerely acknowledge and appreciate the comments and suggestions we received from the anonymous reviewers and editor of this journal.

Conflicts of Interest: The authors declare no conflicts of interest.

References

1. Thenkabail, P.S.; Hanjra, M.A.; Dheeravath, V.; Gumma, M. Global Croplands and Their Water Use from Remote Sensing and Nonremote Sensing Perspectives. In *Advances in Environmental Remote Sensing-Sensors, Algorithms, and Applications*; Weng, Q., Ed.; CRC Press: Boca Raton, FL, USA, 2011.
2. ABS. *Water Use on Australian Farms*; Australian Bureau of Statistics (ABS): Canberra, Australia, 2018.
3. Qureshi, M.E.; Grafton, R.Q.; Kirby, M.; Hanjra, M.A. Understanding irrigation water use efficiency at different scales for better policy reform: A case study of the Murray-Darling Basin, Australia. *Water Policy* **2011**, *13*, 1–17. [CrossRef]
4. Han, X.; Wei, Z.; Zhang, B.; Han, C.; Song, J. Effects of crop planting structure adjustment on water use efficiency in the irrigation area of Hei River Basin. *Water* **2018**, *10*, 1305. [CrossRef]
5. Roth, G.; Harris, G.; Gillies, M.; Montgomery, J.; Wigginton, D. Water-use efficiency and productivity trends in Australian irrigated cotton: A review. *Crop Past. Sci.* **2013**, *64*, 1033–1048. [CrossRef]
6. Plusquellec, H. Modernization of large-scale irrigation systems: Is it an achievable objective or a lost cause. *Irrig. Drain.* **2009**, *58*, 104–120. [CrossRef]
7. Moghazi, H.; Ismail, E.-S. A study of losses from field channels under arid region conditions. *Irrig. Sci.* **1997**, *17*, 105–110. [CrossRef]
8. Murray-Darling Basin (MDB) Authority. *Towards Healthy, Working Murray-Darling Basin: Basin Plan Annual Report 2015–2016*; Murray-Darling Basin (MDB) Authority: Canberra, Australia, 2016.
9. Goulburn Murray Water. *In-situ REVS Testing of Large Dethridge Meter Outlets in the GMID*; Goulburn Murray Water: Tatura, Australia, 2008.
10. Australian Government. *National Framework for Non-Urban Water Metering—Policy Paper*; Australian Government: Canberra, Australia, 2009.
11. Smith, R.J.; Uddin, M.; Gillies, M.H.; Clurey, P.M. Evaluating the performance of automated bay irrigation. *Irrig. Sci.* **2016**, *34*, 175–185. [CrossRef]
12. González-Cebollada, C. Water and energy consumption after the modernization of irrigation in Spain. *WIT Trans. Built Environ.* **2015**, *168*, 457–465.
13. Lecina, S.; Isidorob, D.; Playánc, E.; Aragüésb, R. Irrigation modernization and water conservation in Spain: The case of Riegos del Alto Aragón. *Agric. Water Manag.* **2010**, *97*, 1663–1675. [CrossRef]
14. Burt, C. Irrigation district modernization in the US and worldwide. In *Irrigation Systems Performance*; Options Méditerranéennes: Série B; Lamaddalena, N., Lebdi, F., Todorovic, M., Bogliotti, C., Eds.; CIHEAM: Paris, France, 2005; Volume 52, pp. 23–29.
15. Schaible, G.; Aillery, M. *Water Conservation in Irrigated Agriculture: Trends and Challenges in the Face of Emerging Demands*; EIB-99; U.S. Department of Agriculture, Economic Research Service, September 2012. Available online: https://www.ers.usda.gov/publications/pub-details/?pubid=44699 (accessed on 1 December 2018).
16. Jones, H.G. Irrigation scheduling: Advantages and pitfalls of plant-based methods. *J. Exp. Bot.* **2004**, *55*, 2427–2436. [CrossRef]

17. Hornbuckle, J.W.; Car, N.J.; Christen, E.W.; Stein, T.-M.; Williamson, B. *IrriSatSMS. Irrigation Water Management by Satellite and SMS-A Utilisation Framework*; CRC for Irrigation Futures and CSIRO: Sydney, NSW, Australia, 2009.
18. Gillies, M.H.; Smith, R. Infiltration parameters from surface irrigation advance and run-off data. *Irrig. Sci.* **2005**, *24*, 25–35. [CrossRef]
19. Keen, B.; Slavich, P. Comparison of irrigation scheduling strategies for achieving water use efficiency in highbush blueberry. *N. Z. J. Crop Hort. Sci.* **2012**, *40*, 3–20. [CrossRef]
20. Car, N.J.; Christen, E.W.; Hornbuckle, J.W.; Moore, G.A. Using a mobile phone Short Messaging Service (SMS) for irrigation scheduling in Australia-Farmers' participation and utility evaluation. *Comput. Electron. Agric.* **2012**, *84*, 132–143. [CrossRef]
21. Ahadi, R.; Samani, Z.; Skaggs, R. Evaluating on-farm irrigation efficiency across the watershed: A case study of New Mexico's Lower Rio Grande Basin. *Agric. Water Manag.* **2013**, *124*, 52–57. [CrossRef]
22. McClymont, D. Development of a Decision Support System for Furrow and Border Irrigation. Ph.D. Thesis, University of Southern Queensland, Toowoomba, QLD, Australia, July 2007.
23. Camacho, E.; Pérez-Lucena, C.; Roldán-Cañas, J.; Alcaide, M. IPE: Model for management and control of furrow irrigation in real time. *J. Irrig. Drain. Eng.* **1997**, *123*, 264–269. [CrossRef]
24. Mailhol, J.-C.; Gonzalez, J.-M. Furrow irrigation model for real-time applications on cracking soils. *J. Irrig. Drain. Eng.* **1993**, *119*, 768–783. [CrossRef]
25. Khatri, K.L.; Smith, R. Real-time prediction of soil infiltration characteristics for the management of furrow irrigation. *Irrig. Sci.* **2006**, *25*, 33–43. [CrossRef]
26. Koech, R.; Smith, R.; Gillies, M. A real-time optimisation system for automation of furrow irrigation. *Irrig. Sci.* **2014**, *32*, 319–327. [CrossRef]
27. Uddin, J.; Smith, R.J.; Gillies, M.H.; Moller, P.; Robson, D. Smart automated furrow irrigation of cotton. *J. Irrig. Drain. Eng.* **2018**, *144*. [CrossRef]
28. Nagler, P.L.; Glenn, E.P.; Nguyen, U.; Scott, R.L.; Doody, T. Estimating riparian and agricultural actual evapotranspiration by reference evapotranspiration and MODIS enhanced vegetation index. *Remote Sens.* **2013**, *5*, 3849–3871. [CrossRef]
29. Anderson, M.C.; Allen, R.G.; Morse, A.; Kustas, W.P. Use of Landsat thermal imagery in monitoring evapotranspiration and managing water resources. *Remote Sens. Environ.* **2012**, *122*, 50–65. [CrossRef]
30. Senay, G.B.; Friedrichs, M.; Singh, R.K.; Velpuri, N.M. Evaluating Landsat 8 evapotranspiration for water use mapping in the Colorado River Basin. *Remote Sens. Environ.* **2016**, *185*, 171–185. [CrossRef]
31. Cozzolino, D. The role of near-infrared sensors to measure water relationships in crops and plants. *Appl. Spectrosc. Rev.* **2017**, *52*, 837–849. [CrossRef]
32. Rehman, A.U.; Abbasi, A.; Islam, N.; Shaikh, Z.A. A review of wireless sensors and networks' applications in agriculture. *Comput. Stand. Interfaces* **2014**, *36*, 263–270. [CrossRef]
33. Jawad, H.M.; Nordin, R.; Gharghan, S.K.; Jawad, A.M.; Ismail, M. Energy-efficient wireless sensor networks for precision agriculture: A review. *Sensors* **2017**, *17*, 1781. [CrossRef]
34. Ojha, T.; Misra, S.; Raghuwanshi, N.S. Wireless sensor networks for agriculture: The state-of-the-art in practice and future challenges. *Comput. Electron. Agric.* **2015**, *118*, 66–84. [CrossRef]
35. McCulloch, J.; McCarthy, P.; Guru, S.M.; Peng, W.; Hugo, D.; Terhorst, A. Wireless sensor network deployment for water use efficiency in irrigation. In Proceedings of the Workshop on Real-World Wireless Sensor Networks, Glasgow, Scotland, 1 April 2008; pp. 46–50.
36. Kumar, S.A.; Ilango, P. The impact of wireless sensor network in the field of precision agriculture: A review. *Wirel. Pers. Commun.* **2018**, *98*, 685–698. [CrossRef]
37. Koech, R.; Gyasi-Agyei, Y.; Randall, T. The evolution of urban water metering and conservation in Australia. *Flow Meas. Instrum.* **2018**, *62*, 19–26. [CrossRef]
38. Ruggiero, A.; Punzo, P.; Landi, S.; Costa, A.; Van-Oosten, M.J.; Grilio, S. Improving plant water use efficiency through molecular genetics. *Horticulturae* **2017**, *3*, 31. [CrossRef]
39. Rogers, M.; Lawson, A.; Kelly, K. Lucerne yield, water productivity and persistence under variable and restricted irrigation strategies. *Crop Past. Sci.* **2016**, *67*, 563–573. [CrossRef]
40. Tejero, I.G.; Zuazo, V.H.D.; Bocanegra, J.A.J.; Fernández, J.L.M. Improved water-use efficiency by deficit-irrigation programmes: Implications for saving water in citrus orchards. *Sci. Hortic.-Amsterdam* **2011**, *128*, 274–282. [CrossRef]

41. Du, T.; Kang, S.; Zhang, J.; Davies, W.J. Deficit irrigation and sustainable water-resource strategies in agriculture for China's food security. *J. Exp. Bot.* **2015**, *66*, 2253–2269. [CrossRef]
42. Berbel, J.; Gutierrez-Marín, C.; Expósito, A. Impacts of irrigation efficiency improvement on water use, water consumption and response to water price at field level. *Agric. Water Manag.* **2018**, *203*, 423–429. [CrossRef]
43. Molle, F.; Tanouti, O. Squaring the circle: Agricultural intensification vs. water conservation in Morocco. *Agric. Water Manag.* **2017**, *192*, 170–179. [CrossRef]
44. Perry, C.; Steduto, P.; Karajeh, F. *Does Improved Irrigation Technology Save Water? A Review of the Evidence, in Discussion Paper on Irrigation and Sustainable Water Resources Management in the Near East and North Africa*; Food and Agriculture Organization of the United Nations: Cairo, Egypt, 2017.
45. Fishman, R.; Devineni, N.; Raman, S. Can improved agricultural water use efficiency save India's groundwater? *Environ. Res. Lett.* **2015**, *10*. [CrossRef]
46. Koech, R.; Smith, R.; Gillies, M. Trends in the use of surface irrigation in Australian irrigated agriculture: An investigation into the role surface irrigation will play in future Australian agriculture. *Water: J. Aust. Water Assoc.* **2015**, *42*, 84.
47. Loch, A.; Adamson, D. Drought and the rebound effect: A Murray-Darling Basin example. *Nat. Hazards* **2015**, *79*, 1429–1449. [CrossRef]
48. Berbel, J.; Gutiérrez-Martín, C.; Rodríguez-Díaz, J.A.; Camacho, E.; Montesinos, P. Literature review on rebound effect of water saving measures and analysis of a Spanish case study. *Water Resour. Manag.* **2015**, *29*, 663–678. [CrossRef]
49. López-Gunn, E.; Mayor, B.; Dumont, A. Implications of the Modernization of Irrigation Systems. In *Water, Agriculture and the Environment in Spain: Can we Square the Circle*; CRC Press: Boca Raton, FL, USA, 2012; pp. 241–253.
50. Ahmad, A.; Khan, S. Water and energy scarcity for agriculture: Is irrigation modernization the answer? *Irrig. Drain.* **2017**, *66*, 34–44. [CrossRef]
51. Expósito, A.; Berbel, J. Microeconomics of deficit irrigation and subjective water response function for intensive olive groves. *Water* **2016**, *8*, 254. [CrossRef]
52. White, D.H.; Beynon, N.; Kingma, O. Identifying opportunities for achieving water savings throughout the Murray-Darling Basin. *Environ. Model. Softw.* **2006**, *21*, 1013–1024. [CrossRef]
53. Levidow, L.; Zaccaria, D.; Maia, R.; Vivas, E.; Todorovic, M.; Scardigno, A. Improving water-efficient irrigation: Prospects and difficulties of innovative practices. *Agric. Water Manag.* **2014**, *146*, 84–94. [CrossRef]
54. Expósito, A.; Berbel, J. Agricultural irrigation water use in a closed basin and the impacts on water productivity: The case of the Guadalquivir river basin (Southern Spain). *Water* **2017**, *9*, 136. [CrossRef]
55. Richards, R.; López-Castañeda, C.; Gomez-Macpherson, H.; Condon, A.G. Improving the efficiency of water use by plant breeding and molecular biology. *Irrig. Sci.* **1993**, *14*, 93–104. [CrossRef]

water

MDPI

Review

A Review of Cost Estimates for Flood Adaptation

Jeroen C. J. H. Aerts

Institute for Environmental Studies (IVM), VU University Amsterdam, De Boelelaan 1087, 1081HV Amsterdam, The Netherlands; jeroen.aerts@vu.nl

Received: 29 September 2018; Accepted: 8 November 2018; Published: 13 November 2018

Abstract: Floods are the most devastating of global natural disasters, and flood adaptation measures are needed to reduce future risk. Researchers have started to evaluate the costs and benefits of flood adaptation, but information regarding the cost of different flood adaptation measures is often not available or is hidden in non-peer-reviewed literature. Recent review studies have explored cost estimates for different aspects of flood adaptation, such as nature-based solutions. This study aims to contribute empirical data regarding the cost of flood adaptation by compiling peer-reviewed literature and research reports. The focus is on construction costs and expenses for operation and maintenance. This paper integrates the unit cost information of six main flood adaptation measure categories: (1) the flood-proofing of buildings, (2) flood protection, (3) beach nourishment and dunes, (4) nature-based solutions for coastal ecosystems, (5) channel management and nature-based solutions for riverine systems, and (6) urban drainage. Cost estimates are corrected for inflation and converted to U.S. dollars (2016). Measures are described, and cost figures for both developed and developing countries are provided. The results of this study can be used as input for economic-assessment studies on flood adaptation measures.

Keywords: flood; cost; adaptation; flood management; cost–benefit

1. Introduction

Floods are the most devastating of global natural disasters; they caused billions of dollars in damage and the loss of life of thousands of people in 2017 [1]. Flood hazards can be categorized into different types: e.g., coastal storm surges, river floods, river- and flash floods, and local inundations caused by extreme precipitation. Climate change and sea-level rise further increase the frequency and severity of flood hazards, while population and economic growth further exacerbate flood exposure in low-lying coastal areas [2]. Urgent action is needed to anticipate future losses, but designing and evaluating long-term adaptation strategies is a complex and challenging process for decision makers [3].

In recent decades, a vast array of studies have been conducted to assess and evaluate options for flood adaption so as to reduce current and future flood risk [4–9]. Such studies provide insights into the effects of sea level and climate change on flood hazard (e.g., depth, extent, and duration) [10] and related socio-economic effects on, for example, exposed populations and economic assets [2]. More recently, researchers have started to evaluate the costs and benefits of flood adaptation by using various future projections. In such analyses, benefits are expressed as reduced flood risk (or "expected annual damage," EAD) achieved by implementing proposed adaptation measures [9,11,12]. For example, a recent study show the future benefits of investing in flood protection are much higher than the cost, assuming different future scenarios [9].

Estimating flood adaptation cost is no sinecure, and detailed cost estimates are usually made on the local level—for example, during the engineering design of individual measures [13]. In most cases, however, only aggregate maintenance and investment costs are available from case studies after implementation of the adaptation measure. Furthermore, cost information is often "hidden" in non-peer-reviewed reports, and efforts to collect such information and check its quality are time-consuming. Despite these challenges, enhanced cost estimates of flood adaptation are urgently required to support the economic analysis of flood adaptation research and decision making [14].

With regard to estimating the unit cost of flood adaptation, progress has been made at various scales over the past 10 years regarding flood management measures. In terms of flood protection measures, Linham et al. [15] estimate flood adaptation costs for global cities, while Jonkman et al. [14] provide an overview of coastal flood protection for three case studies in The Netherlands, Vietnam, and the United States. Aerts et al. [16,17] and Lasage et al. [18] provide detailed cost estimates for flood protection in New York City, Los Angeles, and Ho Chi Minh City, respectively; they also provide cost estimates for the flood-proofing of individual buildings. Recently, novel literature has emerged regarding the role of nature-based solutions and how these could be developed to reduce flood risk. Such measures include mangrove restoration, rehabilitation of coral reefs, and the development of coastal ecosystems in general. Cost estimates for such activities have been reported in extended reviews by, for example, Bayraktarov et al. [19], Lamond et al. [20], and Narayan et al. [21].

While researchers have made significant progress in providing cost estimates for different categories of flood adaptation, no study has yet combined these estimates into one overview. The main goal of this paper is therefore to increase our empirical database on the cost of flood adaptation by compiling existing peer-reviewed literature and additional reports. The purpose is to provide regional to global flood risk assessment studies with an overview of different flood adaptation measures and their unit costs so they can be included in, for example, cost–benefit analyses or other evaluation studies [12].

2. Data and Approach

This study reviews the unit cost information of measures divided over six main flood adaptation categories (flood-proofing buildings, protection, beach nourishment, coastal nature-based solutions, riverine adaptation, and urban drainage) and adds new information. Flood protection measures are provided for both urban and rural areas.

The most relevant and recent overview studies that pertain to these different categories of flood adaptation form the basis of the analysis. In addition to these existing review studies, the following new research aspects are provided: (1) cost estimates for adaptation measures related to urban drainage (e.g., green roofs and pumping capacity), and (2) coastal-defense measures (sluices, groins, breakwaters, and rip-rap). Sources of uncertainty in cost estimates are discussed, as are the major gaps in our knowledge of adaptation costs. Recommendations for future research are also offered.

Research shows that, for the following reasons, it is hard to estimate the unit cost of flood-management measures in terms of finding reliable data:

(1) Cost estimations are mostly made during the design phase of a flood adaptation measure [14]. However, while the aggregate cost of such projects after construction can be found, the underlying cost details and the different cost components, are rarely available online.
(2) The unit costs differ greatly across the literature in terms of what cost components (labor, land purchase, and materials, etc.) are included. This makes the comparison of unit cost prices difficult [13].
(3) Unit cost estimates for the same flood management measures vary across countries and regions depending on local geographic and socio-economic conditions [22]. For example, in many countries, constructing a levee in a rural area is much cheaper than developing a similar structure in an urban area, as labor rates and land prices are often higher in cities. Furthermore, the design requirements (e.g., protection levels) are higher in urban environments due to the larger exposed population and greater economic assets [17].
(4) The unit cost estimates vary over time due to changes in socio-economic fluctuations, which affect labor cost, supply of materials, and land values [12,22].

In view of these challenges, and considering the very diverse set of flood-management measures, the focus in this study is on just two types of unit cost estimates, which are often appraised in detailed engineering studies: (1) construction costs and (2) operation and maintenance costs (O&M). Construction costs are fixed, one-time expenses that include costs for planning, purchasing materials and machinery, land acquisition, construction labor, permits, etc. Operation and maintenance costs include yearly costs needed to operate (e.g., storm surge barriers and sluices), maintain (e.g., cleaning sewer systems), monitor (e.g., restored mangroves), and replace equipment [23].

The approach followed in this paper to select unit cost prices comprises several steps:

(1) A few studies already provide a comprehensive overview of some flood-management categories: flood protection [14,16,17] and nature-based solutions [19,21]. These studies were used as a basis for finding additional information.

(2) Peer-reviewed literature was accessed using the Web of Science search engine and different combinations of general key words (*#flood*, *#cost*, *#management*, *#adaptation*, *#cost-benefit*). Key words representing the different adaptation categories and measures (*protection*, *nature-based solutions*, *levee*, *mangrove*, *beach nourishment*, *coral reefs*, *urban drainage*, etc.) were also used. This resulted in a few hundred papers, of which the unit cost information was manually ordered into project and/or construction- and/or maintenance-related activities. In many papers, though economic aspects of flood management investment were discussed, specific cost estimates were either not provided or cost information was incomplete. Only the papers which contained unit cost prices were selected

(3) Following [19], the unit cost estimates provided in this study—provided that information is available—are distributed over developed and developing countries. In this way, differences in socio-economic conditions that exist across countries and that may affect unit cost prices are addressed, albeit in a simplified manner.

(4) All numbers are converted into comparable units, and cost estimations were converted to U.S. dollars at 2016 price levels, unless otherwise specified. This was done using inflation rates for each country based on the consumer price index (CPI) (https://data.worldbank.org/indicator/fp.cpi.totl). For global average numbers, a flat inflation rate of 4% per year was applied, as in [14]. Local currencies for a particular country were first converted to U.S. dollars by using the XE currency converter (www.xe.com/currencyconverter). At the time of submission of this paper, EUR1 was equal to US$1.202. Unit cost prices were not converted into a "purchasing power parity" unit [24], but future studies can use the estimates in this study to further process the data if required.

(5) Cost estimates are presented in six sub-sections: flood-proofing buildings, flood protection, beach nourishment and dunes, nature-based solutions (coast, channel management, and nature-based solutions for riverine systems), and urban drainage. In a few instances, this classification is somewhat arbitrary, and some measures could have been placed in another category. Each section starts with a short description of the measures followed by a discussion of cost estimates.

(6) In some studies, operation and maintenance costs are specified, whereas other sources use fixed percentages of the total construction costs. If available, both types are addressed, and it is assumed that these are valid for the lifetime of a measure.

A peer-reviewed paper is a first step in a quality check of data, and some studies have conducted quite advanced quality checks and statistical analyses to indicate the uncertainty margins of the cost estimates. For example, a review by Bayraktarov et al. [19] on the cost of nature-based solutions considers a number of studies that were used to perform statistical estimates. However, such review papers are rare, and most reviewed papers describe single case studies, for which cost estimates are based on reports from engineering companies, expert knowledge provided during workshops, or estimates that were communicated via personal communication.

Another issue that pertains to quality and comparability is that it is not always clear what is included in the cost estimates and what sub-categories have been neglected [14]. This problem has been addressed by providing upper and lower limits of cost estimations. However, due to the above-mentioned limitations, the cost estimates that are listed in this paper have high uncertainty parameters; they are probably conservative estimates, as not all cost categories are included in the estimation.

3. Results

3.1. *Flood-Proofing and Elevating Buildings*

Three types of measures are commonly used to flood-proof individual buildings: (1) elevation of (new or existing) buildings, (2) dry flood-proofing, and (3) wet flood-proofing [25] (Appendix A Table A1). The lifespan of both wet and dry flood-proofing is estimated at 20–30 years [26], though Kreibich et al. [27] mention a lifespan of 75 years. In some countries, such as the United States, these measures are linked to building code guidelines provided by the state or municipalities. Building-code requirements apply for buildings in designated flood zones (e.g., the 1/100 flood zone), which are mapped by the government. Furthermore, building codes are often linked to an insurance system in which policy holders get a discount on their flood insurance premiums, when they implement flood-proofing measures. In the United States, the National Flood Insurance Program (NFIP) requires homeowners who have a state-backed mortgage to purchase flood insurance, and the base floor of new structures must be raised above the expected 1/100 flood levels [28].

3.1.1. Elevation and Re-Location

Houses can be elevated to prevent floodwater from entering them. This method is mostly applied to new buildings in flood zones, but it can also be applied to existing buildings, though at higher cost than new buildings. The cost estimates for elevating existing buildings (Table 1) is in the range of ~$19,000–194,000 for buildings in the United States—depending on the type of building and how much it must be elevated [16,29,30]. Factors that determine elevation costs include the following: Condition of the house, electrical and plumbing adjustments, grading, excavation, permits, labor and insurance [31]. Elevating buildings in larger cities such as New York are more expensive due to higher labor costs. The U.S. Army Corps of Engineers (USACE) [29], for example, estimates elevation cost at $194,496/building, which is much higher than the estimate made by Jones et al. [30] (maximum $102,888). Furthermore, Jones et al. [30] estimate the costs of adding "freeboard" (elevating the base floor of a building above the required 1/100 flood levels) for new buildings and for different foundations types. In developing countries such as Bangladesh and Vietnam, the cost for elevating houses in rural areas is estimated at $1287–2574 per house, depending on whether the stilts are made from bamboo or reinforced concrete. This is similar to the results found by Lasage et al. [18] for elevating houses (+2 m) in Ho Chi Minh City, Vietnam ($1544–3088 per house). The USACE [29] estimates the re-location of buildings at $353,537, which includes labor cost, building a new foundation, distance to transport, special permits, house dimensions, and road obstacles. Maintenance costs for elevation were not considered in this study and are probably low (<1%).

Table 1. Cost for elevation (U.S.$) and re-location of buildings.

Country	Building Type	Measure	Cost	Year	$2016/CPI [1]	Reference [2]
United States	Average residential building	Elevation existing building +2ft	$33,239–82,498/building	2009	$37,281–92,531/building	[30] [16] P
United States	Average residential building	Elevation existing building +4ft	$35,464–87,535/building	2009	$39,777–98,180/building	[30] [16] P
United States	Average residential building	Elevation existing +6ft	$37,319–91,732/building	2009	$41,857–102,888/building	[30] [16] P
United States	Average building	Elevation	$19,231–192,000/building	2015	$19,481–194,496/building	[29] [31]
Bangladesh/Vietnam	Rural house, wooden frame	Stilts bamboo, reinforced concrete	$1250–2500/house	2015	$1287–2574/house	[32] P
Vietnam	Residential house	Fill sand +2 m	$1500–3000/building	2014	$1544–3088/building	[18] P
United States	Average building	Re-location	$349,000	2015	$353,537	[29]

[1] Values calculated using the consumer price index (CPI); [2] P = peer-reviewed.

3.1.2. Dry Flood-Proofing

Dry flood-proofing techniques are designed to prevent floodwater from entering a building. Measures include the protection of doors and other openings with permanent or removable flood shields [16] by sealing walls with waterproof coatings, impermeable membranes or supplemental layers of masonry or concrete. Dry flood-proofing has disadvantages: When a house is surrounded by water, the pressure on the walls may cause them to cave in—especially in frame constructions. Both the Federal Emergency Management Agency (FEMA) [25] and Keating et al. [26], therefore, advise to not build dry flood-proof houses when floodwater heights exceed +1 m. Construction costs (Table 2) vary between ~$9000 and $23,000/building in developed countries, and between ~$500 and $10,000 per building in developing countries [18]. Costs depend on the type of measure and the flood depth they are designed to withstand. Wright and Pierce [33] calculate the cost for flood-proofing seven waste-water pumping stations that lie below the expected flood elevation. The average cost for each station is ~$45,000, of which 59% is for "miscellaneous items" (e.g., overhead costs and 20% contingencies). Maintenance costs per year are estimated at 1–3% of the investment cost [26].

Table 2. Cost of dry flood-proofing buildings. The column "measure" shows for which water level measures are designed.

Country	Building Type	Measure	Cost	Year	$2016/CPI [1]	O&M/year $2016	Reference [2]
United States	Residential building	+0.6 m	$8290–13,690	2009	$9298–15,354	n.a.	[16] P
United States	Residential building	+2 m	$12,576–21,126	2009	$14,105–3695	n.a.	[16] P
United Kingdom	Residential building	+0.9 m	$13,000–18,200	2008	$15,299–21,418	1–3%	[26]
Germany	Average building	+1 m	$732/m length	2011	$771/m length	n.a.	[27] P
United States	Waste water pump station	+1 m	$45,571	2016	$45,571	n.a.	[33]
Vietnam	Residential building	+1 m	$500–9361	2013	$569–10,667	n.a.	[18] P
Vietnam	Residential building	+1 m	$516/m^2	2014	$588/m^2	n.a.	[18] P
Bangladesh	Residential building (23 m^2)	n.a.	$679–1300	2010	$773–1481	n.a.	[34]

[1] Values calculated using the 2016 consumer price index (CPI); [2] P = peer-reviewed; n.a. = not available; O&M: operation and maintenance costs.

3.1.3. Wet Flood-Proofing

Wet flood-proofing measures allow floodwater to enter a building but limit the damage to the structure and its contents (Figure 1). This minimizes the risk that the walls of the house will collapse due to hydrostatic pressure from rising floodwaters on the outside. Measures include, for example, building utility installations and high-value areas above flood levels, raising electrical sockets, fitting tiled floors so that the building can quickly be returned to use after the flood, and sealing walls with water-resistant building materials [16]. For an extended table of individual flood-proofing measures, see Appendix B Tables A2 and A3. Construction cost for wet flood-proofing are presented in Table 3 and range between $2412 for residential structures and $34,070 for office buildings. The cost for flood-proofing office buildings is usually higher, as office buildings are relatively large, and the value of building contents is higher than that of residential buildings. Wet flood-proofing the basement of a residential building costs around $35–206/m^2. Maintenance costs are low—estimated at <1% of the total investment cost [26]. The cost of flood-proofing a 1500-L oil tank against buoyancy, buried in the garden, is estimated at $1550 [27].

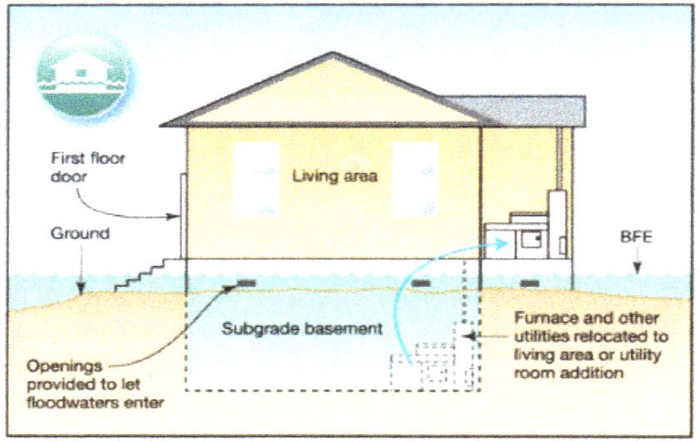

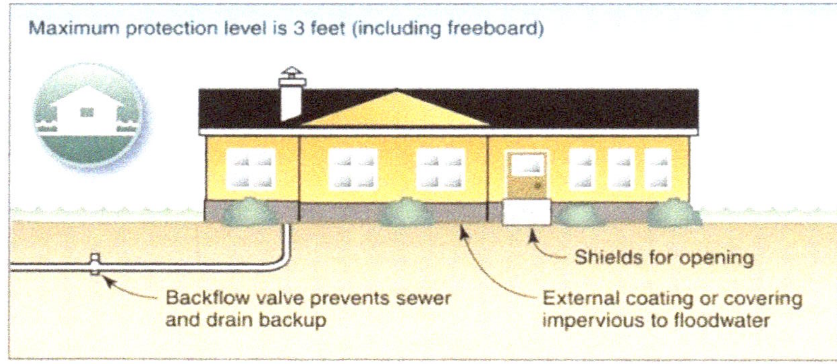

Figure 1. Example of a wet flood-proofed building (**Left**) and a dry flood-proofed building (**Right**) (Source: Federal Emergency Management Agency, FEMA [25]).

Table 3. Cost of wet flood-proofing buildings. The column "measure" shows for which water level measures are designed.

Country	Building Type	Design Requirement	Costs	Year	$2016/CPI [1]	O&M/year $2016	Reference [2]
United States	Basement waterproof	+0.6–2.7 m	$31–184/m^2	2009	$35–206/m^2	n.a.	[16] P
Germany	Basement waterproof	n.a.	$606/m^2	2011	$638/m^2	n.a.	[27] P
United States	Residential building	+0.6 m	$2151–4869	2009	$2412–5461	n.a.	[16] P
United States	Residential building	+2.7 m	$8531–19,307	2009	$9561–21,655	n.a.	[16] P
United Kingdom	Residential building	+0.9 m	$8073–18,369	2008	$9054–20,602	<1%	[26] P
United Kingdom	Office	+0.9 m	$14,937–24,895	2008	$20,442–34,070	<1%	[26] P
Germany	Residential (65 m^2)	n.a.	$22,237	2011	$23,424	n.a.	[27] P
Brazil	Residential	n.a.	$962	2010	$1024	n.a.	[35]
Vietnam	Residential (60 m^2)	+1 m	$248	2014	$273	n.a.	[18] P
Germany	Oil tank	Proofing against buoyancy	$1210/tank	2011	$1550	n.a.	[27] P

[1] Values calculated using the 2016 consumer price index (CPI); [2] P = peer-reviewed; n.a. = not available.

3.2. Flood Protection

Different types of engineered flood-protection measures exist, from simple earth-filled dikes to concrete sea walls and sophisticated storm surge barriers. Table 4 shows different flood-protection methods with their unit costs. Costs are provided both for developing a new structure and for strengthening existing structures.

Storm surge barriers: Storm surge barriers are engineering structures in rivers or estuaries which are designed to protect the high value of economic assets and urban areas from coastal flooding. Storm surge barriers can have movable gates to allow shipping and tidal flows, which are closed during an extreme flooding event. Non-navigable barriers allow only the inflow and outflow of water. Costs vary greatly and are largely determined by the share of the movable parts of the design: Extra gates or a shipping lock can sharply increase costs. Furthermore, geographical and hydrodynamic requirements, such as "span" or "hydraulic head," determine, respectively, the size and the required strength of the design; they may also increase costs [14]. Aerts et al. [16] estimate the cost of movable parts to be between $0.45 billion/km and $3.6 billion/km, depending on the type of gates and head of the barrier (Appendix C Table A4). General costs, including both closure dams and movable parts vary between $0.27 and $3.6 billion/km, where the lower cost range includes barriers with relatively long closure-dam parts. Operation and maintenance costs vary between $0.6 and $22 million/year, depending on the length and number of moveable parts of the barrier system [16].

Sea and river dikes are designed to resist the forces of large coastal storm surges in areas with urban populations and valuable economic assets. Dikes are made of various fill materials such as concrete, clay, and sand, and they are covered with a layer of resistant vegetation or armoring material such as asphalt or boulders (Figure 2). Armoring for sea dikes is more robust than river dikes because of wave impacts, and costs are consequently higher. The cost of a new sea dike in the United States is estimated by Aerts et al. [16] at $28.8 million/km, and for Vietnam at $2.3 million/ [36]. For The Netherlands, Jonkman et al. [14] estimate costs at $19.3–27.2 million/km per meter of dike raised, and for Vietnam at $0.9–1.6 million/km per meter of dike raised. These case-study results are in the same range as those determined by a large study by Prahl et al. [22], who estimate the cost of raising sea dikes in European cities at $21.8–31.2 million/km per meter of dike raised. River dikes are generally cheaper than sea dikes at $12.1–18.2 million/km for a new dike in the United States [37,38] and $5 million/km per meter of dike raised in Canada [13]. Maintenance is estimated at $0.15 million/km and $0.03 million/m for sea dikes in The Netherlands and Vietnam, respectively [14], or between 0.01% and 1% of the initial investment cost [17].

Rural earthen dikes: Rural earthen dikes are peat- or clay-filled dikes applied in rural areas, with design standards <1/100. Jonkman et al. [14] estimate the cost of a rural dike in The Netherlands at $5.1–14 million/km per meter of dike raised, while for Canada this is $2.1 million/km per meter of dike raised [13]. According to studies in developing countries, for new earthen dikes (1–3 m in height), costs range from $0.1–0.2 million/km (Mozambique) [39] to $0.9–1.5 million/km (Vietnam) [14].

Floodwalls: Other types of levees can be made from steel piles or concrete and are often designed as T-walls [40] (Figure 2). The foundations of these structures are also made from concrete or steel to provide stability and prevent seepage and piping [14]. Costs for a new 7-m T-wall are estimated at $31 million/km [40]. The cost of a deployable floodwall is $6.6 million/km [29]. Operation and maintenance costs of dikes and floodwalls vary between 0.01% and 1% [17]. Another study provides numbers detailed estimates for the operation and maintenance of two flood-protection alternatives in Texas at 0.5% of the total investment costs [41].

Other protection measures:

Breakwater: Offshore breakwaters are above-water structures parallel to the shore which reduce wave heights, provide shelter to a harbor and prevent sediment deposition in the entrance channel of a port. There are three main types of breakwaters: (1) rubble-mound breakwaters, which consist of a core of small rocks covered with large rocks or concrete elements; (2) vertical-wall breakwaters, which are filled with concrete blocks or sand; and (3) vertical-composite breakwaters, which are concrete structures founded on rubble substructures where the *caissons* (or concrete blocks) are placed on a high rubble foundation [42]. Costs range from $1.4–6 million/km in developed countries to $63 million/km for complex structures (e.g., the breakwater of the ports of Los Angeles and Long Beach [17]. For developing countries, costs vary between $0.13 and $0.5 million/km [21]

Rip-rap: Rip-rap, also known "rock armor" or "rubble," is rock or other material used to armor shorelines, streambeds, bridge abutments, pilings, and other shoreline structures against erosion (Figure 2). The unit cost for riprap used for protecting coastal zones is estimated at $292–780/m [43,44] or $80/ton [41], with maintenance costs at 2–4% [45].

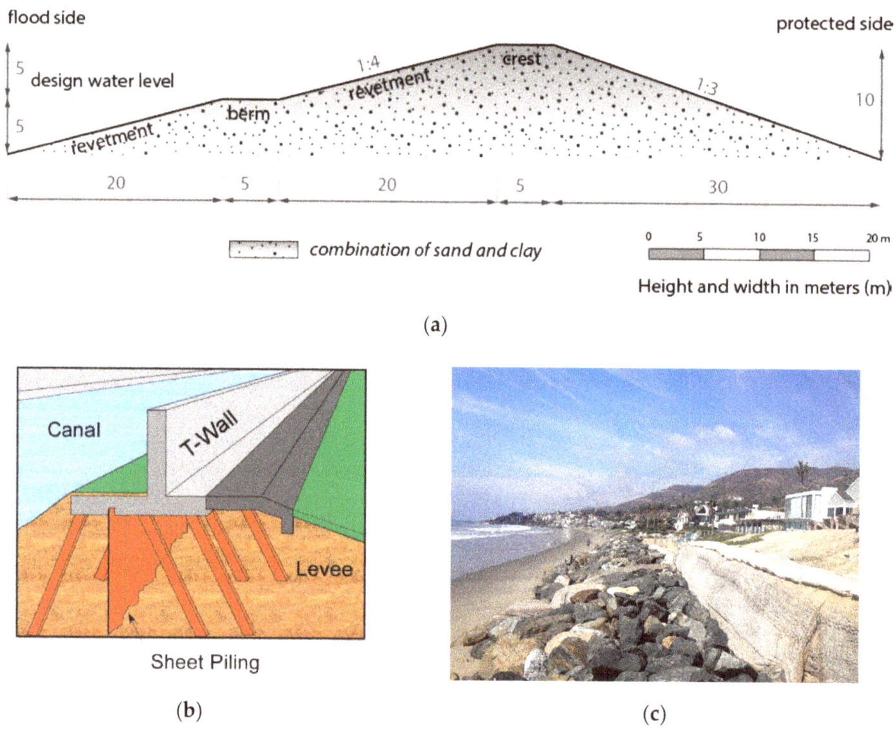

Figure 2. Types of flood protection: (**a**) a cross-section of a typical Dutch sea-dike, filled with sand and clay [14] (**b**) Two types of steel and concrete floodwalls applied [46] (**c**) Rip-rap [17].

Table 4. Types of flood-protection measures, investment costs, and maintenance costs.

Country	Flood Protection	Design Water Level (m)	Unit Cost	Year	\$2016/CPI [1]	O&M/year \$2016	Reference [2]
Global	Storm surge barrier	<7 m	\$0.27–3.6 billion/km	2012	\$0.32–4.2 billion/km	\$0.6–22 million/year	[14] [P] [16] [P]
The Netherlands	Sea dike	4–6 m	\$18.7–22.4 million/km per m dike raised	2009	\$20.8–25 million/km per m dike raised	\$0.15 million/km per year	[14] [P]
United States	Sea dike	7 m	\$25.6 million/km	2013	\$26.4 million/km	0.01–1%	[16] [P]
European cities	Sea dike	varying	\$18.6–26.7 million/km per m dike raised	2012	\$21.8–31.2 million/km per m dike raised	0.01–1%	[22] [P]
Vietnam	Sea dike	n.a.	\$2 million/km	2013	\$2.3 million/km	n.a.	[36] [P]
Vietnam	Sea dike	3–5 m	\$0.7–1.2 million/km per m dike raised	2009	\$0.9–1.7 million/km per m dike raised	\$0.03 million/km per year	[14] [P]
United States	River dike	3–6 m	\$10.5–16.2 million/km	2013	\$10.9–16.8 million/km	0.01–1%	[37] [P] [38] [P]
Laos	River dike	n.a.	n.a.	n.a	\$4.1 million/km	n.a.	[47]
Canada	River dike	n.a.	\$5 million/km per m dike raised	2012	\$5.3 million/km per m dike raised	0.01–1%	[13] [P]
The Netherlands	Rural dike	4–6 m	\$4.5–12.4 million/km per m dike raised	2009	\$5–14 million/km per m dike raised	\$0.15 million/km per year	[14] [P]
Canada	Rural dike	n.a.	\$1.8 million/km per m dike raised	2012	\$1.9 million/km per m dike raised	n.a.	[13] [P]
Philippines/ Mozambique	Small earthen dike	1 m	\$0.1–0.2 million/km	2009	\$0.1–0.2 million/km	n.a.	[36] [37] [39] [P]
Vietnam	Rural dike	n.a.	\$0.7–1.2 million/km per m dike raised	2013	\$1–1.7 million/km per m dike raised	\$0.03 million/km per year	[14] [P]
Indonesia	Rural dike	n.a.	n.a	n.a	\$3.4–3.9/m^3	n.a.	[48]
United States	Deployable floodwall	n.a.	\$5.5 million/km	2015	\$5.6 million/km	\$0.26 million/km	[29] [P]
United States	T-Wall	4 m	\$12.5 million/km	2008	\$13.8 million/km	0.01–1%	[40]; [16] [P] [17] [P] [29]
United States	T-Wall	7.2 m	\$26.5 million/km	2008	\$29.3 million/km	0.01–1%	[40]; [16] [P] [17] [P]

Table 4. *Cont.*

Country	Flood Protection	Design Water Level (m)	Unit Cost	Year	$2016/CPI [1]	O&M/year $2016	Reference [2]
Vietnam	Breakwater	(<2 m waterdepth)	$0.13–0.5 million/km	2015	$0.14–0.5 million/km	n.a.	[21] P
Developed countries	Breakwater	(<2 m waterdepth)	$1.3–4.8 million/km	2015	$1.4–6 million/km	n.a.	[21] P
United States	Breakwater (Ruble mound)	(17 m deep, 3 m high)	$60 million/km	2015	$60.8 million/km	0.01–1%	[17] P
United States	Rip-rap	2–3 m	$250–666/m	2012	$262–699/m	2–4%	[43] [44] [17] P
United States	Rip-rap	n.a.	$80/ton	2016	$80/ton	n.a.	[41]
United Kingdom	Rip-rap	n.a.	$130–330/m²	2016	$130–330/m²	n.a.	[49]
United States	Sandbag wall	+1 m	-	-	~$200.000–400.000/km	n.a.	Appendix F
United States	(Retrofit-) Bulkhead	n.a.	$10–41 million/km	2010	$12.7–51.9 million/km	0.01–1%	[16] P

[1] Values calculated using the 2016 consumer price index (CPI); [2] P = peer-reviewed; n.a. = not available.

A *bulkhead* is a retaining wall that is generally made of steel or wood which stretches ~3–10 m below the water surface and at least 1–3 m above. They are often used to protect pier walls in ports and harbors and are built to prevent soil erosion, flooding, and maintain sufficient navigation width. Recent bulkheads are made of vinyl or concrete, and wooden bulkhead pilings are usually the least expensive. Aerts et al. [16] apply a unit cost range between $12.7 and $51.9 million/km.

A *sandbag wall* is composed of individual bag that are filled with sand, often during the flood event. Though the method is considered effective, it is time- and labor-consuming. Sandbag walls have trapezoidal or triangular cross-sections, which means that, the higher wall, the more bags are needed. Estimates for a wall of +1 m in the United States are between ~$200,000 and $400,000 per km. One sandbag costs $3–6, and to protect a door opening with a sandbag wall of +1.2 m requires 72 bags at ~$210–420 (Appendix F Table A7).

Note that some flood protection related to river-bank maintenance has a link with "nature-based solutions" (see Section 3.4).

3.3. Coastal Protection by Beaches and Dunes

3.3.1. Beach Nourishment

Sub-tidal sandflats and bars, beaches, and sand dunes are natural barriers that reduce the impact of storm surges and waves along the coast [50]. Therefore, beach nourishment is widely used to combat coastal erosion [51]. Other advantages of beach nourishment are to increase and maintain coastal ecosystems, to enhance the potential for recreation, and to preserve the protective values of a coastline against storm surges. Sand-mining for beach-nourishing can be achieved by offshore dredging up to 20–30 km from the coastline, depending on the morphology of the coastal shelf. Furthermore, nourishment sand often comes from the periodic dredging of ports and harbors and is transported to nearby beaches.

The cost of material can vary greatly depending on its origin and associated transportation costs [52]. According to several cases in the United States, costs vary from ~$5 to $18/m^3 [17] (Table 5). Nourishment in The Netherlands is estimated at ~$4–8/m^3 [14] and average numbers for the EU vary from $5 to $11/m^3 [15]. Studies in Australia, South Africa, and Vietnam show cost estimates of $7.7/m^3, $20.8/m^3, and $5.8/m^3, respectively [15,53].

Table 5. Cost of coastal protection by beaches and dunes.

Country	Type	Cost	Year	$2016/CPI [1]	O&M/year $2016	Reference [2]
The Netherlands	Beach nourishment	$3–7.5/m^3$	2009	$4.3–8.4/m^3$	n.a.	[14] P
European Union	Beach nourishment	$4.2–8.4/m^3$	2009	$5.5–11/m^3$	n.a.	[15] P
United States	Beach nourishment	$4.7–17.6/m^3$	2015	$4.8–17.8/m^3$	n.a.	[17] P
South Africa	Beach nourishment	$14.3/m^3$	2009	$20.6/m^3$	n.a.	[15] P
Australia	Beach nourishment	$6.4/m^3$	2009	$7.7/m^3$	n.a.	[15] P
Vietnam	Beach nourishment	$5.6/m^3$	2015	$5.8/m^3$	n.a.	[53]
Australia/United States	Dune Restoration	n.a.	n.a.	$7.636–13.888/ha	$333–2.526/ha	Appendix C; [54] P
United States	Dune recovery from oil spill	n.a.	n.a.	$52.000–76.000/ha	$333–2.526/ha	Appendix C; [54] P
United States	Beach restoration	$10.5 million/km	2015	$10.8 million/km	$0.5 million/km	[29]
United States	Beach restoration and groins	$22.1 million/km	2015	$22.4 million/ha	$0.55 million/km	[29]
United States	Groin	$1.6 million/groin	2010	$1.8 million/groin	0.01–1%	[16] P

[1] Values calculated using the 2016 consumer price index (CPI); [2] P = peer-reviewed; n.a. = not available.

3.3.2. Dune Restoration

Dunes are usually located right at the front of the beach and are created by sand deposition due to winds, often on wider beaches of >35 m [55]. Dunes protect mainland against flooding and can provide habitat for plants, birds, and other terrestrial and beach organisms [56]. Dune restoration mostly involves re-planting native dune vegetation and the installation of sand fencing. Fences can be used on the seaward side to trap sand and help stabilize any bare sand surfaces [57]. Native vegetation may be planted to stabilize natural or artificial dunes and to promote the accumulation of sand from wind-blown sources [58]. Invasive non-native vegetation is often removed.

Costs vary from $7636 to $13,888/ha for studies in Australia and the United States (Appendix C Table A4). If dunes have been subject to erosion, dune reconstruction involves the placement of sand against the remaining dunes using bulldozers. Construction costs are for labor, new vegetation, and the sand needed to reconstruct the dune area. Some studies involve cleaning activities—for example, after a dune area has been hit by an oil spill. Such projects are more expensive through additional cleaning measures, with costs estimated at $52,000–76,000/ha (Appendix C). Maintenance costs of restored dunes are estimated at $333–2526/ha per year [54].

A *groin* is a structure which is oriented perpendicular to a shore and which reduces the flow of sediment along that shore. Retention structures (e.g., groins) can help to capture sand and sustain the lifetime of beach nourishment. Sand collects on the up-drift side of a groin until it is filled and the amount of sand on the beach stays the same [17]. Aerts et al. [16] estimate that the re-conditioning or new development of existing groins for New York City beach-nourishment projects at approximately $1.6 million per groin, including 15% contingencies. The USACE [29] provides cost estimates for nourishment including groins at $0.55 million/km.

3.4. Nature-Based Solutions for Coastal Ecosystems

Reducing flood risk by restoring or creating new coastal ecosystems is increasingly seen as an alternative to hard-engineered protection measures. (For an extensive overview of co-benefits of nature-based solutions see Morris et al. [59]) Coastal ecosystems already have a value in flood protection, and research shows that, without mangroves, 18 million people would be flooded every year [60]. However, global mangrove forests decreased by 19% over the period 1981–2005 [21], while over 60% of the world's coral reefs are declining through overfishing, coastal development, and climate change [61]. In a comparative study for the U.S. Gulf coast, nature-based adaptation options could avert up to $50 billion of the expected flood losses in 2030, with an average benefit–cost ratio >3.5 [62].

Nature-based solutions (Table 6)—including the restoration of degraded coral reefs, or coastal wetlands (e.g., seagrasses, saltmarshes and mangroves)—can reduce flood-water flow and wave height [63]. Wetlands also function like sponges, temporarily storing tidal or flood waters and slowly releasing them, thus reducing flood heights [64]. Furthermore, restoring coastal vegetation and reefs can stabilize shorelines, promote sediment deposition and biodiversity, and reduce erosion [65].

3.4.1. General Restoration Estimates

Based on a database comprised of 954 observations, Bayraktarov et al. [19] estimate median and average restoration costs of marine coastal habitats at $80,000/ha and $1,600,000/ha, respectively (U.S. dollars, 2010). However, the same study states that the real costs are probably a factor 2–4 higher when both capital and operating costs are included. The cost of restoring marine coastal ecosystems depends on three main factors: (1) the area and type of ecosystem (e.g., the type of material or vegetation used in the restoration project); (2) the economy of the country (e.g., developing/developing), which determines labor cost; and (3) the restoration technique applied.

Table 6. Cost for nature-based solutions for coastal ecosystems.

Country	Type	Costs	Year	$2016/CPI [1]	O&M/year $2016	Reference [2]
Developed countries	Wetland restoration	$67,128/ha	2010	$84,938/ha	n.a.	[19] P
United States	Wetland restoration	$228,647/ha	2015	$231,619/ha	$11.027/ha	[29]
United States	Marshland creation	$3/m^3	2007	$3.5/m^3	n.a.	[37]
Global	Salt marshes	$11,100/ha	2014	$12,005/ha	n.a.	[21] P
Developed	Salt marshes	-	-	$1191/ha	n.a.	[19] P
Developing	Salt marshes	-	-	$67,128/ha	n.a.	[19] P
Developed	Seagrass	-	-	$106,782/ha	n.a.	[19] P
Developed countries	Mangrove restoration	$38,982/ha	2010	$49,324/ha	n.a.	[19] P
Developing countries	Mangrove restoration	$1191/ha	2010	$1506/ha	$7–85/ha	[66] [67] [19] P
Global	Mangrove restoration	$1000–3000/ha	2014	$1081–3244/ha	n.a.	[21] P
Vietnam	Mangrove restoration	$25–200/m	2014	$25–200/m	$7–85/ha	[22] P [66] [67]
Developed countries	Coral reef restoration	$1,826,651/ha	2010	$2,311,296/ha	n.a.	[19] P
Global	Coral reef restoration	$1,156,200/ha	2014	$1,250,546/ha	n.a.	[21] P
Developing countries	Coral reef restoration	$89,269/ha	2010	$112,953/ha	n.a.	[19] P
Developed	Oyster Reef	$66,821/ha	2014	$72,273/ha	n.a.	[19] P
United Kingdom	Artificial reef Construction	$30,000–90,000/100 m	2015	$30,263–90,789/100 m	n.a.	[68]

[1] CPI = $2016 numbers calculated using the Consumer price index (CPI); [2] P = peer-reviewed; n.a. = not available.

3.4.2. Seagrass and Saltmarches

Marine seagrass ecosystems are mainly found in shallow bays, estuaries, and coastal waters from the mid-intertidal (shallow) region down to depths of 50–60 m. The most extensive seagrass systems grow on sand and muddy ocean-beds. Ondiviela et al. [69] report that seagrass ecosystems can reduce current velocity, dissipate wave energy, and stabilize sediment—especially in shallow waters with low wave-energy environments. Restoration costs are estimated at $106,782/ha for areas in developing countries [19].

Coastal saltmarshes occur in the intertidal zone near estuaries or lagoons and also reduce wave heights, even under extreme conditions. Creating a salt-marsh zone in front of dikes may result in a reduced dike-reinforcement task [70]. Restoration costs of coastal wetlands in general are estimated at $67,128/ha [19].

3.4.3. Mangrove Restoration

Coastal mangroves are salt-adapted trees and shrubs that grow in tropical or subtropical areas [71]. Mangrove restoration usually entails reforestation of species, and restoration success depends on local circumstances such as hydrology, length of the planting period ("time for survival"), and seedling quality. Mangrove survival also depends on the degree of salinity, which depends much on the amount of available sediment and freshwater inflow to compensate for salinity [72].

Most mangrove-restoration projects analyzed in the scientific literature have an area of 10–120,000 ha. Construction costs (material, labor) and water depth are important factors that determine mangrove restoration costs. A meta-analysis by Narayan et al. [21] shows that the average restoration costs vary from $1081 to $3244/ha. Bayraktarov et al. [19] report lower and upper limits of $1506–49,324/ha, respectively. The lower value for developing countries is confirmed by an older study by Lewis [73] and more recently by Hakim [66] for a case study in Indonesia ($858/ha). Land-purchase costs are not included in these estimates. Furthermore, in a flood-management case study in Vietnam, a comparison was made between mangrove restoration and technical solutions such as breakwaters [74]. For this study, the cost for planting mangroves at water depths of 1–1.8 m varies between $25–200/m of coastline, which is cheaper than the cost for developing breakwaters ($125–475/m coastline [21]). Maintenance costs can be high for some species, as small seedlings are vulnerable to wave impacts, which require protection measures (e.g., breakwaters) and thinning and pruning activities. For Indonesia, maintenance costs have been estimated at 10% of initial investment, or around $85/ha [66]. For Vietnam, this has been estimated at $7.1/ha [67]. Such maintenance costs need to be considered for at least four years after planting seedlings.

3.4.4. Coral, Oyster and Artificial Reef Restoration

Artificial coral reefs reduce wave energy and coastal erosion and protect shorelines against flooding (Beck et al. [60]). Their effectiveness for reducing flooding is determined by reef width (relative to the average wave length) and reef depth (relative to the average wave height) [19]. Existing natural coral reefs can be restored by planting coral on degraded areas [75]. As for transplanting new coral on degraded areas, only 65% transplanted species survive. Bayraktarov et al. [19] use an average of 54,200 coral transplants to populate one hectare. Restoration cost varies from $2,311,296/ha in developed to $112,953/ha in developing countries [19]. For the restoration of oyster reefs, used oyster and clam shells from farmers and restaurants are placed in the water, with a cost of $72,273/ha [19]. Mangroves and other coastal vegetation is often planted in conjunction with oyster restoration to provide surface area to inhabit.

An artificial reef is a man-made structure that mimics some of the characteristics of a natural reef, such as promoting marine life and biodiversity [76]. Artificial reefs can be constructed from rocks, wood, old tires, or submerged shipwrecks sunk to the sea floor. For the 2598 artificial reef projects in Florida, concrete secondary-use materials are used most (43%) followed by concrete modules (24%), steel materials (such as steel towers and military equipment) (17%), steel vessels and barges (11%), and natural rock (primarily limestone boulders) (3%) [77]. One option is to deploy fabricated modules of concrete or natural materials such as limestone boulders. For such solutions, a study in the United Kingdom estimates costs at $30,263–90,789/100 m of new reef. The cost to prepare, tow, and deploy a steel vessel is around $10,000–80,000 [68].

3.5. Channel Management and Nature-Based Solutions for Riverine Systems

Maintaining the conveyance of discharge is an important aspect of managing flood levels, as channels that have filled up via natural sedimentation processes lower discharge capacity and hence increase flood risk. Channel management refers to activities that aim at retaining flow capacity and water levels in river systems for different users (e.g., shipping, ecosystems, and agriculture). Measures to maintain discharge capacity include periodic dredging, river widening, and creating new side channels (Table 7).

3.5.1. Dredging and River Widening

Dredging is the removal of sediment from the rivers and harbors. It can be done by hydraulic (e.g., by a "suction hedger") and mechanical methods (e.g., a "bucket dredger"). Environmental regulations increasingly require the cleaning of contaminated dredged material and the safe disposal of dredged material in controlled deposit areas. Therefore, dredging costs are increasingly associated with both the excavation, treatment, and disposal of dredged material. Costs for mechanical dredging in the United Kingdom vary between $44 and $59/m^3, whereas suction dredging is cheaper at $13/m^3—both including the cost of disposal [78]. Costs for mechanical dredging and transport only in The Netherlands are estimated at $15–19/m^3 [79], which indicates that the storage of frequently contaminated material is relatively expensive. Costs for dredging in Bangladesh are estimated at ~$2/m^3 [80].

Table 7. Costs for channel management, and nature-based solutions for riverine systems.

Country	Type	Cost	Year	$2016/CPI [1]	O&M/year $2016	Reference [2]
United Kingdom	Mechanical dredging + storage	$36–48/m^3	2007	$44–59/m^3	$1.680–51.311/km	[78]
United Kingdom	Suction dredging	$10.6/m^3	2007	$13/m^3	$1.680–51.311/km	[78]
United Kingdom	Bank dredging	$48,400/km	2007	$59,805/km	n.a.	[78]
The Netherlands	Dredging and transport	$15–18.9/m^3	2013	$15.1–19.4/m^3	n.a.	[79]
Bangladesh	Dredging	$1.16/m^3	2007	$2/m^3	n.a.	[80]
United Kingdom	Channel widening	$4–16/m^3	2010	$4.5–18/m^3	n.a.	[81]
The Netherlands	River widening	$2.64 billion	2005	$3.11 billion	n.a.	[82]
United Kingdom	Eco engineered bank protection	$44,000–792,000/km	2007	$54,000–978,000/km	n.a.	[78]
Germany	Detention area (32 million m^3)	$3/m^3	2015	$3/m^3	n.a.	[83] O
Peru	Detention area (373 million m^3)	1/m^3	2005	$1.9/m^3	$5 million	[84]
Global	Inland wetlands	-	-	$45,752/ha	n.a.	[19] P
United States	Inland wetland	-	-	$10,022/ha	$785/ha	[85]

[1] Values calculated using the 2016 consumer price index (CPI); [2] P = peer-reviewed; O = online media; n.a. = not available.

In addition to the dredging of existing channels, river widening is increasingly seen as a "nature-based solution". River widening in lower stretches of a river basin decreases peak water levels, as the river is provided with "more room" to discharge its flood waters [86]. In addition, such measures restore both ecological values and biodiversity [87]. A U.K. cost estimate for widening a small rural river system by excavating river floodplains varies between $4.5 and $18/m^3 [81], mainly for excavation.

An example of a large-scale river widening program is "Room for the River" in The Netherlands [86]. This program entails 35 projects along the lower branches of the river Rhine, and the main goal is to enhance the maximum discharge capacity from approximately 15,000 to 16,000 m^3/s. Each of the individual projects address one or more measures that lower the river bed through excavation, setback of dikes, or widen the (side-) channel (s) (Figure 3). The total investment cost of this project is around $2.64 billion [82]. Some projects within the program, such as those near the city of Nijmegen, are relatively expensive, as they involve costly engineered protection measures, which were implemented in a densely populated urban environment [82].

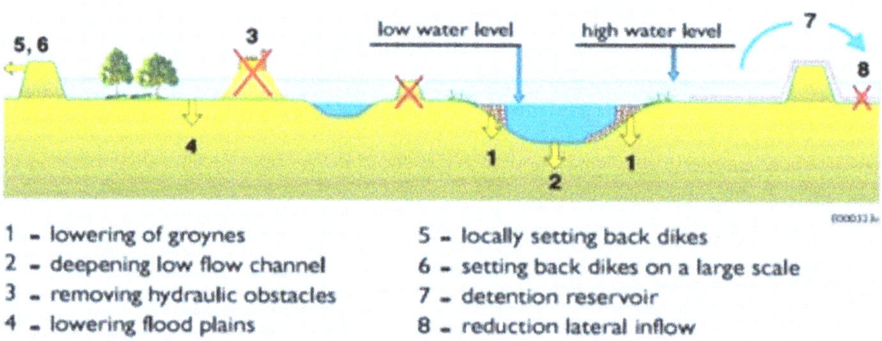

Figure 3. Measures applied in the "Room for the River" project in The Netherlands [88].

3.5.2. Operation and Maintenance

Operational activities include inspections and periodic evaluation studies, and they are estimated at $4049/km [78]. Maintenance measures include weed control (vegetation clearance), obstruction/dirt removal, de-silting, and small bank repairs. Apart from the generic factors that determine costs (e.g., labor rates), the maintenance costs depend on the "target condition" of the channels and whether channel-clearance measures are implemented manually or mechanically. For well-maintained channels ("Grade 2"; [78]), costs for manual maintenance are higher than for mechanical methods; they vary between $5730–51,311 km/year and $1680–17,096 km/year for manual and mechanical maintenance, respectively [78].

3.5.3. Nature-Based "Soft Bank" Protection and Water Buffering

In Section 3.2, a number of hard-engineered bank-protection measures (e.g., Riprap) were described. However, in more rural areas and in areas with lower protection standards, nature-based methods may often both reduce flood risk and improve environmental values such as biodiversity. Methods include implementing brush mattresses, revegetation, using biodegradable geotextiles to stabilize grade, and the application of logs or other natural materials resistant to erosive flows. Costs vary between $54,000 and $978,000/km for smaller rural river branches [49,78]. Most of these measures only have a design life of 3–15 years [89].

Other channel-related measures:

River detention areas: River detention areas are larger bath-tub-like systems (>3 million m^3) surrounded by a dike and designed to temporarily retain peak river discharges. Detention areas are located along river channels and capture floodwater above a pre-defined water-level. At this point, a control device (a pipe or a spillway) is overtopped with flood-water into the detention/retention area. Spillways are also applied to activating side channels in case of extreme water levels [90]. Along the river Rhine in Germany, multiple detention areas between 3.6–32 million m^3 are being developed to reduce flood peaks [91]. The cost of the largest "*detention polder Hordt*" (870 ha, volume: 32 million m^3) is estimated at $98 million [83]. A similar flood-detention project has been proposed in the Rio Piura in Peru, including a detention area of 2600 ha, a 20-km new dike with construction costs estimated at $429 million, and operation and maintenance estimated at $5 million/year [84]. With a dike 3 m in height, the volume would be +/−373 million m^3 with unit construction costs of $1.9/$m^3$.

Inland wetland and water buffering: Creating wetlands in upstream areas enhances the buffer-capacity of ecosystems to absorb—at least temporarily—peak rainwater before it drains into the main river channels. Wetlands also have the ability to reduce nitrogen, phosphorous, pesticides, and sediment loading in open-water systems. Restoration of inland wetlands is less expensive than restoration of coastal wetlands. The cost for the former is estimated at $45,752/ha [19]. Tyndall and Bowman [85] estimate restoration costs for inland wetlands in Iowa (United States). For this case study, costs vary depending on site planning and design, excavation activities, installation of control structures (e.g., levees), and the opportunity cost of land made unsuitable for agricultural production. Design and construction costs are $10,022/ha, and yearly maintenance costs over a time span of 40 years are $785/ha [85].

3.6. Measures Against Local Inundations

Traditionally, measures against local inundations in urban environments consist of engineered drainage systems with concrete channels, pipes, and culverts to quickly drain water under gravity towards larger water bodies such as lakes, rivers, or seas (Table 8). In locations that lie below sea level or in low-lying areas where drainage under gravity is difficult, pumping systems may assist in draining the water towards larger water bodies. The capital costs of engineered urban-drainage systems are very high, so nature-based "sustainable urban drainage solutions" (SUDS) have recently been developed to retain rainwater (see also Appendix E Table A6). Examples of SUDS are green roofs and wetlands (For an extended review of SUDS see [23,92]). Most cost estimates are made for individual measures such as pipelines or ditches. For such measures, unit costs are based on the diameter of length of the proposed storm-water drainage infrastructure [93]. The U.K. Environment Agency (EA) [94] suggest that maintenance costs for SUDS in the United Kingdom range from 0.5% to 10% of the construction cost, with the exception of an infiltration trench (20%).

Sewer pipes: Construction costs in the United Kingdom and United States for implementing sewer pipes vary between $215–453/m and $61–861/m for concrete and metal pipes, respectively. The diameter varies between ~0.15–0.45 m and 0.25–2 m for concrete and metal pipes, respectively. Maintenance (cleaning and inspection) of sewer pipes in South Africa is estimated at ~$10/m [23].

Pumping stations: Appendix D Table A5 shows cost figures for pumping systems in several countries. Factors that determine costs (length of pipes, distance to source area, type of motor, drainage area, etc.) are described by Marzouk and Ahmed [95]. Costs vary between $0.4– and $1.7 million/$m^3$/s. The numbers show that cost are not largely linked to local labor costs, as costs are sometimes higher in developing countries than in developed countries. This may indicate that these systems are highly sophisticated and require labor from external, specialized businesses.

Retention and retention ponds are designed only for flood control and are also known as dry ponds. In a city environment, a pond is intended to retain storm water for a period of time, releasing the water after the storm. An outlet pipe (or control device) is mostly placed at the bottom elevation of the detention volume to allow the pond to drain dry. Unlike dry retention ponds, wet retention ponds hold a permanent pool of water. The bottom of a wet retention pond is often excavated below the water table, thereby allowing fauna and vegetation in the water to consume nutrients and suspended pollutants to settle. Costs vary between \$15 and \$50/m^3 for cases in the United Kingdom and United States [94,96]; the higher cost is for wet retention ponds [96]. The costs of legal fees, land costs, and other unexpected or additional costs are not included in these estimates. Further reading and detailed cost estimates can be found in a report by the Environmental Protection Agency (EPA) [97].

Green roofs are mostly developed upon the flat roofs of new or existing buildings (Figure 4). They are designed to store and evaporate rainwater to reduce the run-off peak to the sewer system [98]. A green-roof consists of a vegetation layer, a substrate layer (which retains water and anchors the vegetation), and a drainage layer (to discharge water). The construction cost of a green roof is dependent on the thickness (typically 0.15 m or more) and variety of vegetation (grasses, herbs, and shrubs). Costs vary from \$32–39/m^2 in South Africa to \$114–225/m^2 in the United States.

Parks and green zones are designed to meet the recreational needs of urban populations, but also have an important role in the hydrological regulation of cities, thereby mitigating local inundations [99]. Development costs of a park in the United States are \$1521/m^2 [100]. Maintenance costs of parks in the United Kingdom vary between \$0.4 and \$2/m^2 [99].

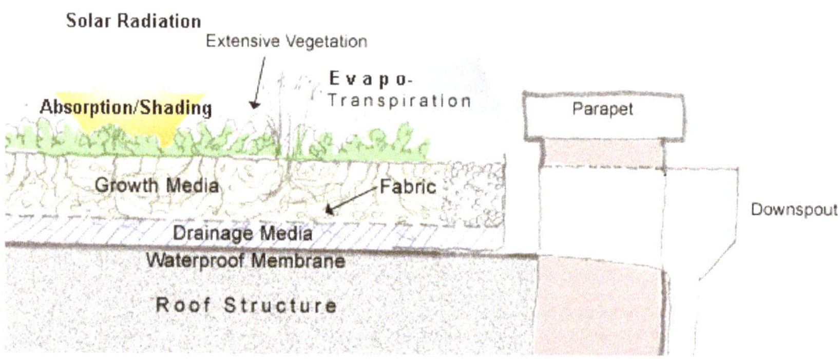

Figure 4. Example of a green roof and its main components (source: [101]).

Table 8. Cost of measures against local inundations.

Country	Type	Costs	Year	$2016/CPI [1]	O&M/year $2016	Reference [2]
United Kingdom	Sewer pipe, urban (concrete)	$183–385/m	2008	$215–453/m	0.5–10%	[94]
United States	Sewer pipe, urban (metal)	$51–723/m	2006	$61–861/m	n.a.	[93]
South Africa	Lined channel	$59/m	2013	$67/m	~$10/m	[23]
Global	Pumping station	-	-	$0.4–1.7 million/$m^3$/s	n.a.	Appendix D
United Kingdom	Retention pond	$13–22/$m^3$	2015	$15–26/$m^3$	0.5–10%	[94]
United States	Retention area	$17.5–35/$m^3$	1999	$24.8–50.6/$m^3$	<1% (dry); 3–6% (wet)	[96] [93]
United Kingdom	Wetland	$22–26/$m^3$	2008	$26–30/$m^3$	0.5–10%	[94]
United Kingdom	Green roof	$70–79/$m^2$	2008	$82–93/$m^2$	0.5–10%	[94]
South Africa	Green roof	$27–33/$m^2$	2013	$32–39/$m^2$	n.a.	[23]
United States	Green roofs	$107–211/$m^2$	2010	$114–225/$m^2$	$2.3–3.5/$m^2$	[102] [103] P
Singapore	Green roof	$54–120/$m^2$	2003	$74–164/$m^2$	n.a.	[104] P
Phillipines	Green roof	$59–64/$m^2$	2001	$67–73/$m^2$	n.a.	[105]
United States/United Kingdom	Park construction	$1110/$m^2$	2008	$1227/$m^2$	$0.4–2/$m^2$	[100] [99] P
United Kingdom	Infiltration trench	$65–87/m	2015	$77–102/m	20%	[94]
United Kingdom	Concrete storage tank	$395–455/$m^3$	2015	$465–535/$m^3$	0.5–10%	[94]

[1] Values calculated using the 2016 consumer price index (CPI); [2] P = peer-reviewed; n.a. = not available.

4. Discussion and Conclusions

This study has collected empirical data on the cost of flood adaptation measures, using case studies and project-based information from various sources in the literature. The focus is on construction and maintenance costs for six categories of flood adaptation measures. The amount and quality of the data varies considerably, though recent research on specific flood-management issues has advanced the empirical basis of cost data. Data quality and uncertainty have been addressed by using cost intervals with upper and lower limits and an indication of the data quality of non-peer reviewed reports.

One of the issues with finding reliable costs is what cost categories are included in the aggregate cost estimate [13]. The studies used in this report are often unclear as to what cost components have been addressed in the final estimate, or have different components. This shows that cost estimates should be handled with care and with certain error margins; however, in most cases the estimates are probably on the conservative side, as some cost components have not been valued [14].

The issue of uncertainty in cost estimates also plays a role in the stage of the project at which the cost estimate was made: The error margins of cost estimates in the design and planning phases are obviously much higher than those of cost figures derived after project implementation. Wright and Pierce [33], for example, estimate 20% contingencies in a design for the dry flood-proofing of pumping stations. Addressing unexpected costs in the design phase holds especially for complex engineering projects such as the development of storm surge barriers, where the contingencies in the design phase are about 50% [106]. These projects suffer from rapid cost increases during the design phase, as requirements may change (e.g., a request for a higher protection standard), unforeseen complications in geographical conditions may arise (e.g., geological stability of the underground), or weather conditions which cause delays may occur.

Uncertainties also pertain to estimating the cost of operation and maintenance [13]. Such costs depend, for example, on the frequency of inspections and maintenance requirements that change over time due to the aging of the structure [26]. In the extensive review by Bayraktarov et al. [19], only a few of the underlying studies differentiate between cost components (e.g., between capital and operating costs) or provide information on other cost factors, such as planning, land acquisition, financing, monitoring, and repair/replacement. Some of the studies in this review estimate the annual operation and maintenance costs as a percentage of the construction costs but do not provide the data that underpins these estimates. Though yearly maintenance costs are low at first sight compared to construction costs, they add up quickly in an economic analysis, as they are valued over the lifetime of the measure.

Additional temporal aspects play a role in the interpretation and use of cost estimates [107]. While in existing studies, cost estimates are addressed as one-time investment numbers, such investments are often phased out over time. Examples are cost-benefit analyses for the planning of levee-reinforcement programs [106,108]. Furthermore, the lifetime of the proposed measures plays an important role in economic analysis. The expected lifetime of the larger investments (levees, storm surge barriers, sewer systems) are usually >50 years, whereas the lifetime is usually shorter (20–30 years) for cheaper measures such as flood-proofing buildings [26]. However, the required lifetime of an investment differs per country and even per project: The assumed lifetime for measures in a flood-management study in New York City is 50 years [29], whereas it is 100 years for measures in a comparable study in The Netherlands [109].

In addition to empirical data, studies exist that use modeling techniques to fit empirical cost data against explanatory variables. For example, a study by Mauer et al. [110] applies a model that calculates the length and size distribution of the sewer pipes in an urban area on the basis of rainfall intensity, housing densities, and area. Future research could expand such approaches. This also addresses the concerns of some researchers [13] that applying unit cost estimates assumes a fixed linear relationship between, for example, dike cost and some variable such as "meter height raised". Such costs may increase non-linearly with increasing dike heights, and non-linear models are needed to describe such relations—especially in the face of future climate change. Research by Lenk et al. [13], however, shows, on the basis of empirical data in The Netherlands and Canada, that a unit cost expression is adequate to express the cost of flood protection per height raised or per unit length.

Research on the economic evaluation of the cost and benefits for nature-based solutions (for example, in urban drainage) compared to hard-engineered drainage measures is at an early stage. This is due to a few factors: (1) Nature-based solutions, for example, in urban drainage, require irrigation and possible replanting until the vegetation is fully established, and these costs are still difficult to accurately determine [23]; (2) developers and city planners may be concerned that natural drainage options may decrease the area suitable for economic production; and (3) it is difficult to find data on the hydrologic benefits of the measures as reflected, for example, in a design standard. This is more straightforward in flood protection projects, where the design standard refers to a return period of maximum water level that should be incorporated in the design of an embankment. Though nature-based solutions are basically designed to do the same (i.e., lowering water levels, absorb wave energy, store water, etc.), more research and modeling is needed on the "hydrologic and hydraulic return" of a dollar investment in nature-based solutions.

Future research may address these issues, and expand the research with estimating both the cost and benefits of flood adaptation measures, and assess the benefit cost ratios (BCRs) of such measures. With such numbers, a comparative study can be conducted.

Funding: This work was conducted under VICI Grant nr. 453-13-006, from the Dutch Science Foundation.

Acknowledgments: I thank Sanne van Amsterdam for the detailed preparatory work and Lars de Ruig and Max Tesselaar for reading earlier versions of this manuscript.

Conflicts of Interest: The author declares no conflict of interest.

Appendix A

Table A1. Flood-proofing buildings.

Country	Building Type	Measure	Unit	year	Source
Bangladesh/ Vietnam	Wooden frame	Elevation: Stilts Bamboo/Reinforced concrete	$1250/2500	2015	[32]/http://de.phaidon.com/agenda/architecture/articles/2013/april/17/vietnams-flood-proof-bamboo-houses/
United Kingdom	Residential	Dry flood-proofing	$1950–5759	2008	[26]
United Kingdom	Shop	Dry flood-proofing	$1989–8632	2008	[26]
United Kingdom	Office	Dry flood-proofing	$2990–9399	2008	[26]
United Kingdom	Residential	Wet flood-proofing	$8073–18369	2008	[26]
United Kingdom	Shop	Wet flood-proofing	$11,063–17,706	2008	[26]
United Kingdom	Office	Wet flood-proofing	$14,937–24,895	2008	[26]

Appendix B

Table A2. Flood-proofing-specific measures.

Country	Type of Flood-Proofing Measure	Costs Expressed Per	Costs	Reference
United States	Sprayed-on cement	Linear foot of wall covered	U.S.$16.80 (2009)	[25]
United States	Waterproof membrane	Linear foot of wall covered	U.S.$5.70 (2009)	[25]
United States	Asphalt	Linear foot of wall covered	U.S.$12.00 (2009)	[25]
United States	Drainage line around house	Linear foot	U.S.$31 (2009)	[25]
United States	Plumbing check valve	Each	U.S.$1060 (2009)	[25]
United States	Sump and sump pump (with back-up battery)	Lump sum	U.S.$1710 (2009)	[25]
United States	Metal flood shield	Linear foot of shield surface	U.S.$375 (2009)	[25]
United States	Wooden flood shield	Linear foot of shield surface	U.S.$117 (2009)	[25]

Table A3. Individual flood-proofing measures (Env. Agency [111]).

Country	Type of Flood Proofing Measure	Cost	Comment
United Kingdom	Periphery wall (based on a 40-m length)	£3500–4500	May require ancillary pumps (maintenance costs required) Depends on size of curtilage
United Kingdom	Periphery wall residential gate (1.2 m)	£2500–4500	
United Kingdom	Raise threshold	£1200–1500	
United Kingdom	Storm porch (per door)	£5800–8800	Includes additional cost of locking mechanism
United Kingdom	Flood resistant door (per door)	£875–2500	
United Kingdom	Periscope airbricks (assumes 12 per property)	£2500–3000	Includes installation costs
United Kingdom	Flood resistant door	£750–2500+	
United Kingdom	Automatic door guards (domestic 2 m opening)	£8000	Costs exclusive of ground work and construction
United Kingdom	Free-standing barriers (for detached house)	£5000–12,000	Ancillary pumps may also be required
United Kingdom	Flood skirt (per house)	£10,000–35,000	Costs include construction, fitting, and training
United Kingdom	Sump and pump	£50–2500	Costs depend on pump capacity and sump size
United Kingdom	Anti-flood valves	£50–500	Costs excluding labor to fit and construct an inspection chamber

Appendix C

Table A4. Cost nature-based solution, coast.

Country	Measures	Year	Unit	Source
Australia	Dune restoration	2014	$7636/ha	http://www.environment.nsw.gov.au/resources/coasts/130083Merimbula.pdf
United States	Dune restoration	2014	$10,000/ha	https://www.nature.org/media/florida/natural-defenses-in-southeast-florida.pdf
United States	Dune enhancement		$13,888/ha	54 acres = 22 ha; $300,000 Enhancing vegetation to improve flood protection https://www.nature.org/media/florida/natural-defenses-in-southeast-florida.pdf maintenance: $5045/ha per year
United States	Dune restoration	2012	$52,089/ha	55 acres = 22.3 ha/$1,145,976; including clean up from oil spill https://www.estuaries.org/pdf/2012conference/room19/session3/Reynolds_RAE_2012_pres.pdf
United States	Dune restoration	2013	$76,404/ha	Dune restoration with oil spill 20 acres = 8 ha total restoration surface https://eu.pnj.com/story/news/2018/03/04/project-restore-six-miles-perdido-key-dunes-starts-monday/391857002/ http://www.gulfspillrestoration.noaa.gov/sites/default/files/wp-content/uploads/Escambia_FS.pdf)
United States	Dune restoration	2014	$96,875/ha	20 acres = 8 ha; $775,000; this is inclusive of 83,000 cubic yards of new sand https://www.nature.org/media/florida/natural-defenses-in-southeast-florida.pdf
United States	Coastal habitat enhancement	2014	$607,142/ha	https://www.nature.org/media/florida/natural-defenses-in-southeast-florida.pdf
United States	Coastal habitat enhancement	2014	$925,925/ha	https://www.nature.org/media/florida/natural-defenses-in-southeast-florida.pdf

Appendix D

Table A5. Cost, pumping stations.

Pump Station	Capacity	Capacity	Project	Cost	Cost	Year	Cost 2016	Note
	Cubfeet/s	m^3/s		($ million)	($million/$m^3$/s)		($million/$m^3$/s)	
Henderson Bayou P. station	1000	28.3	East Ascension	15.8	0.6	2011	0.7	1
Bayou Trepagnier p. station	800	22.7	Pontchartrain Levee District	11.5	0.5	2004	0.8	1
Dwyer Road Pump station	875	24.8	New Orleans S&WB	13.6	0.5	2010	0.7	1
Ijmuiden		260	Noordzeekanaal	68	0.3	2003	0.4	2, 3
Katwijk		40	Boezemkanaal	46.8	1.2	2014	1.3	4
Egypt-1		0.55	Urban drainage	0.58	1.06	2011	1.3	5
Egypt-2		1.1	Urban drainage	0.75	0.7	2011	0.9	5
Uzbekistan		91.5	Bhukara 1	139	1.5	2013	1.7	6

1. (The Gulf Coast Community Protection and Recovery District), Storm Surge Surpression Study. 2. https://www.technischweekblad.nl/nieuws/supergemaal-in-ijmuiden/item5913. 3. https://www.trouw.nl/home/capaciteit-van-gemalen-vergroot~a820c6a5/. 4. https://www.kivi.nl/uploads/media/5652f45f281f5/Oppervlaktewaterbeheer_Hoogheemraadschap_Leiden_18_september_2014.pdf. 5. Marzuk and Ahmed (2011). 6. ADB https://www.adb.org/sites/default/files/project-document/81008/44458-012-tacr-01e.pdf.

Appendix E

Table A6. Operation and maintenance costs for sustainable urban drainage solutions (SUDS) (Env, Agency [94]).

SUDS Option	Maintenance Cost	Source
Green roofs	£2500 per year for first 2 years for covered roof with sedum mat, £600 per year after. £1250 per year for first 2 years for covered roof with biodiverse roof, £150 per year after	Bamfield (2005)
Simple rainwater harvesting (water butts)	Negligible	
Advanced rainwater harvesting	£250 per year per property for external maintenance contract	RainCycle 2005
Permeable paving	£0.5–1/m^3 of storage volume	HR Wallingford, 2004
Filter drain/perforated pipes	£0.2–1/m^2 of filter surface area	HR Wallingford, 2004
Swales	£0.1/m^2 of swale surface area, £350 per year	HR Wallingford, 2004
Infiltration basin	£0.1–0.3/m^2 of detention basin area £0.25–1/m^3 of detention volume	HR Wallingford, 2004
Soakaways	£0.1/m^2 of treated area	HR Wallingford, 2004
Infiltration trench	£0.2–1/m^2 of filter surface area	HR Wallingford, 2004
Filter strip	£0.1/m^2 of filter surface area	HR Wallingford, 2004
Constructed wetland	£0.1/m^2 of wetland surface area. Annual maintenance of £200–250/year for first five years (declining to £80–100/year after three years).	HR Wallingford, 2004
Retention (wet) pond	£0.5–1.5/m^2 of retention pond surface area, £0.1–£2/m^3 of pond volume	HR Wallingford, 2004
Detention basin	£0.1–0.3/m^2 of detention basin area, £0.25–1/m^3 of detention volume, £250–1000 per basin	HR Wallingford, 2004
		HR Wallingford, 2004

Bamfield (2005). Whole Life Costs & Living Roofs. The Springboard Centre, Bridgewater. A Report By The Solution Organisation for Sarnafil. Available from http://livingroofs.org/; HR Wallingford (2004). Whole Life Costing for Sustainable Drainage. Report SR 627; Raincycle (2005). Rainwater Harvesting Hydraulic Simulation and Whole Life Costing Tool v2.0. User Manual. SUDS Solutions.

Appendix F

Table A7. Unit cost for sand bags and the sand bag flood barrier.

Country	Measures	Unit	$2016	Source
United States	Sandbag		$3–6/bag	http://www.aquadam.net/Flood_Control/fldcntrl.html
United States	Sandbag wall +1 m high		~300,000/km	http://www.aquadam.net/Flood_Control/fldcntrl.html
United States	Sandbag wall +1.2 m high		$760,000/km	http://geocellsystems.com/brochures/pdfs/True_Cost_of_Sandbags.pdf
United States	Sandbag wall +1.2 m high	#bags	#72,000/km	https://www.sandbaggy.com/blogs/articles/sandbags-calculator

References

1. Munich, R.E. Natural Catastrophe Review: Series of Hurricanes Makes 2017 Year of Highest Insured Losses Ever. Available online: https://www.munichre.com/en/media-relations/publications/press-releases/2018/2018-01-04-press-release/index.html (accessed on 2 July 2018).
2. Jongman, B.; Ward, P.J.; Aerts, J.C. Global exposure to river and coastal flooding: Long term trends and changes. *Glob. Environ. Chang.* **2012**, *22*, 823–835. [CrossRef]
3. Hall, J.W.; Grey, D.; Garrick, D.; Fung, F.; Brown, C.; Dadson, S.J.; Sadoff, C.W. Water Security: Coping with the curse of freshwater variability. *Science* **2014**, *346*, 429–430. [CrossRef] [PubMed]
4. Dawson, R.J.; Dickson, M.E.; Nicholls, R.J.; Hall, J.W.; Walkden, M.J.; Stansby, P.K.; Mokrech, M.; Richards, J.; Zhou, J.; Milligan, J.; et al. Integrated analysis of risks of coastal flooding and cliff erosion under scenarios of long term change. *Clim. Chang.* **2009**, *95*, 249–288. [CrossRef]
5. Nicholls, R.J.; Cazenave, A. Sea-Level Rise and Its Impact on Coastal Zones. *Science* **2010**, *328*, 1517–1520. [CrossRef] [PubMed]
6. Hall, J.W.; Watts, G.; Keil, M.; De Vial, L.; Street, R.; Conlan, K.; O'Connell, P.E.; Beven, K.J.; Kilsby, C.G. Towards risk-based water resources planning in England and Wales under a changing climate. *Water Environ. J.* **2012**, *26*, 118–129. [CrossRef]
7. Kreibich, H.; Van Den Bergh, J.C.; Bouwer, L.M.; Bubeck, P.; Ciavola, P.; Green, C.; Hallegatte, S.; Logar, I.; Meyer, V.; Schwarze, R.; et al. Costing Natural Hazards. *Nat. Clim. Chang.* **2014**, *4*, 303–306. [CrossRef]
8. Thieken, A.; Cammerer, H.; Dobler, C.; Lammel, J.; Schöberl, F. Estimating changes in flood risks and benefits of non-structural adaptation strategies—A case study from Tyrol, Austria. *Mitig. Adapt. Strateg. Glob. Chang.* **2016**, *21*, 343–356. [CrossRef] [PubMed]
9. Ward, P.; Aerts, J.C.J.H.; Botzen, W.J.; Bates, P.; Kwadijk, J.; Hallegatte, S.; Scussolini, P.; Winsemius, H. Future costs and benefits of river flood protection in urban areas: A global framework. *Nat. Clim. Chang.* **2017**, *6*, 381–385. [CrossRef]
10. Merz, B.; Aerts, J.C.; Arnbjerg-Nielsen, K.; Baldi, M.; Becker, A.; Bichet, A.; Blöschl, G.; Bouwer, L.M.; Brauer, A.; Cioffi, F.; et al. Floods and climate: Emerging perspectives for flood risk assessment and management. *Nat. Hazards Earth Syst. Sci.* **2014**, *14*, 1921–1942. [CrossRef]
11. Aerts, J.C.; Botzen, W.W.; Emanuel, K.; Lin, N.; de Moel, H.; Michel-Kerjan, E.O. Evaluating flood resilience strategies for coastal megacities. *Science* **2014**, *344*, 473–475. [CrossRef] [PubMed]
12. Kind, J.M. Economically efficient flood protection standards for The Netherlands. *J. Flood Risk Manag.* **2014**, *7*, 103–117. [CrossRef]
13. Lenk, S.; Rybski, D.; Heidrich, O.; Dawson, R.J.; Kropp, J.P. Costs of sea dikes—Regressions and uncertainty estimates. *Nat. Hazards Earth Syst. Sci.* **2016**, *17*, 765–779. [CrossRef]
14. Jonkman, S.N.; Hillen, M.M.; Nicholls, R.J.; Kanning, W.; van Ledden, M. Costs of adapting coastal defences to sea-level rise—New estimates and their implications. *J. Coast. Res.* **2013**, *29*, 1212–1226. [CrossRef]
15. Linham, M.M.; Green, C.H.; Nicholls, R.J. *Costs of Adaptation to the Effects of Climate in the World's Largest Port Cities*; AVOID: London, UK, 2010; p. 225.
16. Aerts, J.C.; Botzen, W.W.; de Moel, H.; Bowman, M. Cost estimates for flood resilience and protection strategies in New York City. *Ann. N. Y. Acad. Sci.* **2013**, *1294*, 1–104. [CrossRef] [PubMed]
17. Aerts, J.C.; Barnard, P.L.; Botzen, W.J.; Grifman, P.; Finzi Hart, J.; De Moel, H.; Newton Mann, A.; de Ruig, L.; Sadrpour, N. Sea Level Rise, Flood Risk, and Adaptation Options in Los Angeles. *Ann. N. Y. Acad. Sci.* **2018**. [CrossRef] [PubMed]
18. Lasage, R.; Veldkamp, T.I.E.; de Moel, H.; Van, T.C.; Phi, H.L.; Vellinga, P.; Aerts, J.C.J.H. Assessment of the effectiveness of flood adaption strategies for HCMC. *Nat. Hazards Earth Syst. Sci.* **2014**, *14*, 1441–1457. [CrossRef]
19. Bayraktarov, E.; Saunders, M.; Abdullah, S.; Mills, M.; Beher, J.; Possingham, H.P.; Mumby, P.J.; Lovelock, C.E. The cost and feasibility of marine coastal restoration. *Ecol. Appl.* **2016**, *26*, 1055–1074. [CrossRef] [PubMed]
20. Lamond, J.; Rose, C.; Bhattacharya-Mis, N.; Joseph, R. Evidence Review for Property Flood Reliance Phase 2 Report. 2017. Available online: https://www.floodre.co.uk/wp-content/uploads/UWE-report_Evidence-review-for-PFR_Phase-2-report.pdf (accessed on 10 November 2018).

21. Narayan, S.; Beck, M.; Reguero, B.; Losada, I.; Van Wesenbeeck, B.; Pontee, N.; Sanchirico, J.; Ingram, J.; Lange, G.; Burks-Copes, K. The Effectiveness, Costs and Coastal Protection Benefits of Natural and Nature-Based Defences. *PLoS ONE* **2016**, *11*, e0154735. [CrossRef] [PubMed]
22. Prahl, B.F.; Rybski, D.; Kropp, J.P.; Burghoff, O.; Held, H. Applying stochastic small-scale damage functions to German winter storms. *Geophys. Res. Lett.* **2012**, *39*, L06806. [CrossRef]
23. Armitage, N.; Vice, M.; Fisher-Jeffes, L.; Winter, K.; Spiegel, A.; Dunstan, J. *Alternative Technology for Stormwater Management. The South African Guidelines for Sustainable Drainage Systems*; No. TT 558/13; Water Research Commission: Gezina, South Africa, 2013.
24. Jongman, B.; Winsemius, H.; Aerts, J.C.J.H.; Ward, P.J. Declining vulnerability to river floods and the global benefits of adaptation. *Proc. Nat. Acad. Sci. USA* **2015**, E2271–E2280. [CrossRef] [PubMed]
25. FEMA. *Homeowner's Guide to Retrofitting*, 2nd ed.; US Department of Homeland Security, Federal Insurance and Mitigation Administration (FEMA): Washington, DC, USA. Available online: https://www.fema.gov/homeowners-guide-retrofitting (accessed on 17 August 2018).
26. Keating, K.; May, P.; Pettit, A.; Pickering, R. Cost Estimation for Household Flood Resistance and Resilience Measures—Summary of Evidence. Environment Agency, 2015. Available online: http://evidence.environment-agency.gov.uk/FCERM/Libraries/FCERM_Project_Documents/SC080039_cost_house_resist_and_resilience.sflb.ashx (accessed on 4 April 2018).
27. Kreibich, H.; Christenberger, S.; Schwarze, R. Economic motivation of households to undertake private precautionary measures against floods. *Nat. Hazards Earth Syst. Sci.* **2011**, *11*, 309–321. [CrossRef]
28. Kousky, C. Financing Flood Losses: A Discussion of the National Flood Insurance Program. 2017. Available online: http://www.rff.org/files/document/file/RFF-DP-17-03.pdf (accessed on 10 November 2018).
29. USACE. North Atlantic Coast Comprehensive Study: Resilient Adaptation to Increasing Risk, Final Report. January 2015, p. 144. Available online: http://www.nad.usace.army.mil/Portals/40/docs/NACCS/NACCS_main_report.pdf (accessed on 10 November 2018).
30. Jones, C.P.; Coulbourne, W.L.; Marshall, J.; Rogers, S.M. Evaluation of the National Flood Insurance Program's Building Standards. American Institutes for Research and the NFIP Evaluation Working Group, 2006. Available online: http://www.fema.gov/library/viewRecord.do?id=2592 (accessed on 10 November 2018).
31. Home Advisor. How Much Does It Cost to Raise a House? 2016. Available online: https://www.homeadvisor.com/cost/foundations/raise-a-foundation/#cost (accessed on 14 September 2018).
32. Biswas, S.; Hasan, M.A.; Islam, M.S. Stilt Housing Technology for Flood Disaster Reduction in the Rural Areas of Bangladesh. *Int. J. Res. Civ. Eng.* **2015**, *3*, 1–6.
33. Wright and Pierce. Waste Water Pump Station Flooding Vulnerability Evaluation. 2016. Available online: http://www.waterfordct.org/sites/waterfordct/files/final_waterfordpsfloodreport_0.pdf (accessed on 1 May 2018).
34. Noble, N.; Roy, D.C. *Flood Resistant Housing. Low-Cost Disaster-Resistant Housing in Bangladesh*; Practical Action Technical Brief: Warwickshire, UK, 2010.
35. Swiss RE. Flood Risk in Brazil: Prevention, Adaptation and Insurance. Available online: https://www.researchgate.net/profile/David_Bresch/publication/253327670_Flood_risk_Brazil_full_study/links/00b4951f7f5db0793d000000/Flood-risk-Brazil-full-study.pdf (accessed on 10 November 2018).
36. Thanh Danh, V.; Viet Khai, H. Using a Risk Cost-Benefit Analysis for a Sea Dike to Adapt to the Sea Level in the Vietnamese Mekong River Delta. *Climate* **2014**, *2*, 78–102. [CrossRef]
37. Dijkman, J. *A Dutch Perspective on Coastal Louisiana Flood Risk Reduction and Landscape Stabilization*; United States Army—European Research Office of the US Army: London, UK, 2007; p. 226.
38. Smith, D.L.; Miner, S.P.; Theiling, C.H.; Behm, R.; Nestler, J.M. Levee Setbacks: An Innovative, Cost-Effective, and Sustainable Solution for Improved Flood Risk Management. 2017. Available online: https://pdfs.semanticscholar.org/ed2f/6d8f9db6fb7af40ce1ad17999bdb3fdd2eea.pdf (accessed on 10 November 2018).
39. van Ogtrop, F.F.; Hoekstra, A.Y.; van der Meulen, F. Flood management in the lower Incomati river basin, Mozambique: Two Alternatives. *J. Am. Water Resour. Assoc.* **2005**, *41*, 607–619. [CrossRef]
40. Bos, A.J. Optimal Safety Level for the New Orleans East Polder, A Preliminary Risk Analysis. Master's Thesis, University of Amsterdam, Amsterdam, The Netherlands, 2008.
41. GCCPRD. Storm Surge Surpression Study, Phase 3. 2016. Available online: http://www.gccprd.com/pdfs/GCCPRD%20Phase%203%20Report%20-%20Recommended%20Actions.pdf (accessed on 10 November 2018).

42. Tutuarima, W.H.; d'Angremond, W.H. Cost Comparison of Breakwater Types. *Coast. Eng.* **1998**, 1934–1944. [CrossRef]
43. PPIC. California Coastal Management with a Changing Climate. Available online: http://www.ppic.org/content/pubs/report/R_1108GMR.pdf (accessed on 1 May 2018).
44. ERG Economic and Social Impacts of a Changing Coastline in California. Available online: ftp://reef.csc.noaa.gov/pub/socioeconomic/NSMS/California/Economic%20and%20Social%20Impacts%20CA_3_30_12_final.docx (accessed on 1 May 2018).
45. Heberger, M.; Cooley, H.; Herrera, P.; Gleick, P.; Moore, E. The Impacts of Sea-Level Rise on the California Coast. Report nr. CEC-500-2009-024-D. California Climate Change Center of the Pacific Institute, 2009. Available online: http://www.pacinst.org/reports/sea_level_rise/report.pdf (accessed on 10 November 2018).
46. Nelson, S.A. Hurricane Katrina—What Happened? Available online: http://www.tulane.edu/~sanelson/Katrina/katrina_images.htm (accessed on 14 September 2018).
47. JCS. US$65-Million Riverside Project Set to Begin Construction: Pakse. Available online: http://jclao.com/us65-million-riverside-project-set-begin-construction-pakse/ (accessed on 10 November 2018).
48. Ardhiansyah-Arifin, M.F. Cost analysis for heavy equipment in earthfill work—An optimization of heavy equipment fleet (Case study: Jabung ring dike project). *AIP Conf. Proc.* **2017**, *1818*, 020005. [CrossRef]
49. SEPA. *Bank Protection, Rivers and Lochs. Engineering in the Water Environment, Good Practice Guide WAT-SG-23*; Scottish Environment Protection Agency: Stirling, Scotland, 2008. Available online: https://www.sepa.org.uk/media/150971/wat_sg_23.pdf (accessed on 10 November 2018).
50. Hanley, M.E.; Hoggart, S.P.G.; Simmonds, D.J. Shifting sands? Coastal protection by sand banks, beaches, and dunes. *Coast. Eng.* **2014**, *87*, 136–146. [CrossRef]
51. Dean, R.G.; Houston, J.R. Determining shoreline response to sea level rise. *Coast. Eng.* **2016**, *114*, 1–8. [CrossRef]
52. Zhu, X.; Linham, M.M.; Nicholls, R.J. *Technologies for Climate Change Adapatation—Coastal Erosion and Flooding*; TNA Guidebook Series; Danmarks Tekniske Universitet, Risø Nationallaboratoriet for Bæredygtig Energi: Roskilde, Denmark, 2010.
53. Boos, S.; Dahlstrom, A. Coastal Evolution at Nha Trang Bay, Vietnam. Water Resources Engineering TVVR-15/5006. Lund University, 2015. Available online: http://lup.lub.lu.se/luur/download?func=downloadFile&recordOId=7448002&fileOId=7448014 (accessed on 10 November 2018).
54. Verburg, R.W.; Hennen, W.H.G.J.; Puister, L.F.; Michels, R.; van Duijvendijk, K. *Estimating Costs of Nature Management in the European Union; Exploration Modelling for PBL's Nature Outlook*; Wageningen University, Natuur & Milieu: Wageningen, The Netherlands, 2017.
55. Garcia-Lozano, C.; Pinto, J. Current status and future restoration of coastal dune systems on the Catalan shoreline (Spain, NW Mediterranean Sea). *J. Coast. Conserv.* **2018**, *22*, 519–532. [CrossRef]
56. Martinez, M.L.; Gallego-Fernández, J.B.; Hesp, P. *Restoration of Coastal Dunes*; Springer: Berlin, Germany; New York, NY, USA, 2013.
57. Nordstrom, K.F.; Arens, S.M. The role of human actions in evolution and management of foredunes in The Netherlands and New Jersey, USA. *J. Coast. Conserv.* **1998**, *4*, 169–180. [CrossRef]
58. Rozé, F.; Lemauviel, S. Sand dune restoration in North Brittany, France: A 10-year monitoring program. *Restor. Ecol.* **2004**, *12*, 29–35. [CrossRef]
59. Morris, R.; Konlecher, T.; Ghisalberti, M.; Swearer, S. From grey to green: Efficacy of eco-engineering solutions for nature-based coastal defence. *Glob. Chang. Biol.* **2017**, *24*, 1827–1842. [CrossRef] [PubMed]
60. Beck, M.W.; Losada, I.J.; Menéndez, P.; Reguero, B.G.; Díaz-Simal, P.; Fernández, F. The global flood protection savings provided by coral reefs. *Nat. Commun.* **2018**, *9*, 2186. [CrossRef] [PubMed]
61. Burke, L.; Reytar, K.; Spalding, M.; Perry, A. *Reefs at Risk Revisited*; World Resources Institute: Washington, DC, USA, 2011; 130p, Available online: http://www.wri.org/publication/reefs-risk-revisited (accessed on 10 November 2018).
62. Reguero, B.; Beck, M.; Bresch, D.; Calil, J.; Meliane, I. Comparing the cost effectiveness of nature-based and coastal adaptation: A case study from the Gulf Coast of the United States. *PLoS ONE* **2018**. [CrossRef] [PubMed]
63. United States Geological Survey (USGS). Wetland Restoration and Creation. Available online: https://water.usgs.gov/nwsum/WSP2425/restoration.html (accessed on 4 April 2018).

64. United States Environmental Protection Agency (EPA). Wetlands Protection and Restoration. Available online: https://www.epa.gov/wetlands (accessed on 4 April 2018).
65. Spalding, M.D.; Ruffo, S.; Lacambra, C.; Meliane, I.; Hale, L.Z.; Shepard, C.C.; Beck, M.W. The role of ecosystems in coastal protection: Adapting to climate change and coastal hazards. *Ocean Coast. Manag.* **2014**, *90*, 50–57. [CrossRef]
66. Hakim, L.L. Cost and Benefit Analysis for Coastal Management. A Case Study of Improving Aquaculture and Mangrove Restoration Management in Tambakbulusan Village Demak Indonesia. 2016. Available online: http://edepot.wur.nl/421310 (accessed on 10 November 2018).
67. Marchand, M. Mangrove Restoration in Vietnam: Key Considerations and a Practical Guide. TU Delft Report T2666. 2008, p. 36. Available online: https://repository.tudelft.nl/islandora/object/uuid:98b5ba43-1452-4631-81dc-ad043ef3992c?collection=research (accessed on 10 November 2018).
68. Climate-ADAPT (European Climate Adaptation Network). Groynes, Breakwaters and Artificial Reefs. Available online: http://climate-adapt.eea.europa.eu/metadata/adaptation-options/groynes-breakwaters-and-artificial-reefs (accessed on 14 September 2018).
69. Ondiviela, B.; Losada, I.J.; Lara, J.L.; Maza, M.; Galván, C.; Bouma, T.; van Belzen, J. The role of seagrasses in coastal protection in a changing climate. *Coast. Eng.* **2014**, *87*, 158–168. [CrossRef]
70. Van Loon-Steensma, J.M. Salt marshes to adapt the flood defences along the Dutch Wadden Sea coast. *Mitigat. Adapt. Strateg. Glob. Chang.* **2015**, *20*, 929–948. [CrossRef] [PubMed]
71. Beck, M.W.; Narayan, S.; Trespalacios, D.; Pfliegner, K.; Losada, I.J.; Menéndez, P.; Espejo, A.; Torres, S.; Díaz-Simal, P.; Fernandez, F.; et al. *The Global Value of Mangroves for Risk Reduction*; Summary Report; The Nature Conservancy: Berlin, Geramny, 2018.
72. United Nations Environment Program (UNEP). Mangrove Conservation and Restoration. Available online: http://web.unep.org/coastal-eba/content/mangrove-conservation-and-restoration (accessed on 4 April 2018).
73. Lewis, R.R. Mangrove Restoration—Costs and Benefits of Successful Ecological Restoration. In Proceedings of the Mangrove Valuation Workshop, Universiti Sains Malaysia, Penang, Malaysia, 4–8 April 2001; Beijer International Institute of Ecological Economics: Stockholm, Sweden, 2001. Available online: http://www.fao.org/forestry/10560-0fe87b898806287615fceb95a76f613cf.pdf (accessed on 10 November 2018).
74. Losada, I.J.; Menéndez, P.; Espejo, A.; Torres, S.; Díaz-Simal, P.; Abad, S.; Beck, M.W.; Narayan, S.; Trespalacios, D.; Pfliegner, K.; et al. *The Global Value of Mangroves for Risk Reduction*; Technical Report; The Nature Conservancy: Berlin, Germany, 2018; p. 176.
75. Reef Resilience Network (RRN). Restoration. Retrieved. Available online: http://www.reefresilience.org/restoration/ (accessed on 4 April 2018).
76. National Oceanic and Atmospheric Administration (NOAA). What Is an Artificial Reef? Available online: https://oceanservice.noaa.gov/facts/artificial-reef.html (accessed on 4 April 2018).
77. EPA. Guidelines and Management Practices for Artificial Reef Siting, Use, Construction, and Anchoring in Southeast Florida. Southeast Florida Coral Reef Initiative. TP-176. 2011, p. 162. Available online: http://nsgl.gso.uri.edu/flsgp/flsgph11003.pdf (accessed on 10 November 2018).
78. EA, Environment Agency. Cost Estimation for Channel Management—Summary of Evidence. Report—SC080039/R3. 2015, p. 67. Available online: http://evidence.environment-agency.gov.uk/FCERM/Libraries/FCERM_Project_Documents/SC080039_cost_channel_mgmt.sflb.ashx (accessed on 10 November 2018).
79. Timmermans, M. Beheerplan Baggerwerken Waterschap Hollandse Delta. 2014, p. 32. Available online: https://www.wshd.nl/binaries/content/assets/wshd---website/common/agendapunt_12_beheerplan_baggerwerken_2014-2018.pdf (accessed on 10 November 2018). (In Dutch)
80. World Bank. *People's Republic of Bangladesh Revival of Inland Water Transport: Options and Strategies Bangladesh*; Development Series Paper No. 20; World Bank: Washington, DC, USA, 2007. Available online: http://siteresources.worldbank.org/INTBANGLADESH/Resources/BDS20.pdf (accessed on 10 November 2018).
81. City of Bath. Bath and North East Somerset Flood Risk Management Strategy Report. Appendix J. Available online: http://www.bathnes.gov.uk/sites/default/files/sitedocuments/Planning-and-Building-Control/Planning-Policy/Evidence-Base/Flood-Risk/FRMSAppendixJ.pdf (accessed on 17 August 2018).

82. CPB. Kosteneffectiviteit van maatregelen en pakketten Kosten-batenanalyse voor Ruimte voor de Rivier. J. Ebregt (In Dutch). C.J.J. Eijgenraam en H.J.J. Stolwijk. CPB Document 832005. Available online: https://www.cpb.nl/sites/default/files/publicaties/download/kosteneffectiviteit-van-maatregelen-en-pakketten-kosten-batenanalyse-voor-ruimte-voor-de.pdf (accessed on 10 November 2018).
83. Die Rheinpfalz, Mehr Geld vom Bund für Hochwasserschutz. Available online: https://www.rheinpfalz.de/lokal/artikel/mehr-geld-vom-bund-fuer-hochwasserschutz/ (accessed on 4 April 2018). (In German)
84. Mechler, R. *Cost-Benefit Analysis of Natural Disaster Risk Management in Developing Countries*; Disaster Risk Management in Development Cooperation; GTZ: Bonn, Germany, 2005; p. 126.
85. Tyndall, J.; Bowman, T. *Iowa Nutrient Reduction Strategy Best Management Practice Cost Overview Series: Constructed Wetlands*; Department of Ecology & Natural Resource Management, Iowa State University: Ames, IA, USA, 2016.
86. Klijn, F.; Asselman, N.; Wagenaar, D. Room for Rivers: Risk Reduction by Enhancing the Flood Conveyance Capacity of The Netherlands' Large Rivers. *Geosciences* **2018**, *8*, 224. [CrossRef]
87. Palmer, M.A.; Hondula, K.L.; Koch, B.J. Ecological Restoration of Streams and Rivers: Shifting Strategies and Shifting Goals. *Annu. Rev. Ecol. Evol. Syst.* **2014**, *45*, 247–269. [CrossRef]
88. STOWA, Stichting Toegepast Onderzoek Waterbeheer. Deltaproof. Room for the River. Available online: http://deltaproof.stowa.nl/Publicaties/deltafact/Room_for_the_river?subject=9 (accessed on 14 September 2018).
89. Roca, M. Green Approaches in River Engineering, HR Wallingford. Available online: http://www.hrwallingford.com/ (accessed on 14 September 2018).
90. Schielen, R.M.; Jesse, P.; Botwidt, L.J. On the use of flexible spillways to control the discharge ratio of the Rhine in The Netherlands: Hydraulic and morphological. *Neth. J. Geosci.* **2007**, *86*, 77–88.
91. Wierig, H.P.; Kretschmer, W. Reserve Area for Flood Retention Hördt. Alfa Project Presentation. Available online: http://alfa-project.eu/_userdata/files/2%20SGD%20S%C3%BCd%20-%20Wierig%20-%20ALFA%20conference%20%5BCompatibiliteitsmodus%5D.pdf (accessed on 14 September 2018).
92. Woods Ballard, W.; Wilson, S.; Udale-Clarke, H.; Illman, S.; Scott, T.; Ashley, R.; Kellagher, R. The SUDS Manual. CIRIA C697, London. 2015. Available online: www.scotsnet.org.uk/documents/NRDG/CIRIA-report-C753-the-SuDS-manual-v6.pdf (accessed on 10 November 2018).
93. Narayanan, A.; Pitt, R. Costs of Urban Stormwater Control Practices. University of Alabama, 2006. Available online: http://rpitt.eng.ua.edu/class/International%20urban%20water%20systems/Arvind%20Costs%20of%20Urban%20Stormwater%20Control%20Feb%2005%202006%20clean%20copy.htm (accessed on 10 November 2018).
94. Environment Agency. Cost Estimation for SUDS—Summary of Evidence. Report–SC080039/R9; 2015. Available online: http://evidence.environment-agency.gov.uk/FCERM/Libraries/FCERM_Project_Documents/SC080039_cost_SUDS.sflb.ashx (accessed on 10 November 2018).
95. Marzouk, M.M.; Ahmed, R.M. A case-based reasoning approach for estimating the costs of pump station projects. *J. Adv. Res.* **2011**, *2*, 289–295. [CrossRef]
96. Environmental Protection Agency. *EPA Preliminary Data Summary of Urban Storm Water Best Management Practices*; EPA-821-R-99-012; Environmental Protection Agency: Washington, DC, USA, 1999; p. 214.
97. Environmental Protection Agency (EPA). Storm-Water Wet Pond and Wetland Management Handbook. EPA 833-B-09-11; 2009; p. 80. Available online: https://www3.epa.gov/npdes/pubs/pondmgmtguide.pdf (accessed on 10 November 2018).
98. Mentens, J.; Raes, D.; Hermy, M. Green roofs as a tool for solving the rainwater runoff problem in the urbanized 21st century? *Landsc. Urban Plan.* **2006**, *77*, 217–226. [CrossRef]
99. Tempesta, T. Benefits and costs of urban parks: A review. *Aestimum* **2015**, *67*, 127–143.
100. DuMoulin, A.; Welle, B.; Harnik, P.; Pierson, C.; DeCoster, K.; McGillivray, B.; Evers, T. *Downtown Parks: Funding Methods, Management Structures, and Costs*; City of Minneapolis: Minneapolis, MN, USA, 2008; p. 123.
101. Dvorak, B.; Volder, A. A Green Roof Pilot Study at the Texas A&M Architecture Center Langford Building a Final Report, January 29 2010. Available online: https://faculty-legacy.arch.tamu.edu/bdvorak/research.htm (accessed on 10 November 2018).
102. GSA. The Benefits and Challenges of Green Roofs on Public and Commercial Buildings. United States General Services Administration, 2011; p. 129. Available online: https://www.gsa.gov/cdnstatic/The_Benefits_and_Challenges_of_Green_Roofs_on_Public_and_Commercial_Buildings.pdf (accessed on 10 November 2018).

103. Niu, H.; Clark, C.; Zhou, J.; Adriaens, P. Scaling of economic benefits from green roof implementation in Washington, DC. *Environ. Sci. Technol.* **2010**, *44*, 4302–4308. [CrossRef] [PubMed]
104. Wong, N.H.; Tay, S.F.; Wong, R.; Leng Ong, C.; Sia, A. Life cycle cost analysis of rooftop gardens in Singapore. *Build. Environ.* **2003**, *38*, 499–509. [CrossRef]
105. BI. The Difficult Beginnings of Green Roofs, Walls in PH. Available online: https://business.inquirer.net/2649/the-difficult-beginnings-of-green-roofs-walls-in-ph#ixzz5V0xOfZSj (accessed on 10 November 2018).
106. Kind, J.; Gauderis, J.; van Velzen, E.; Silva, W. Kengetallen Kosten-batenanalyse Waterveiligheid 21e eeuw. Report 2008.044. Dutch Ministry of Water and Transport, 2008. Available online: https://repository.tudelft.nl/islandora/object/uuid%3A8834660b-e573-4c0b-a2c2-ab771223094f (accessed on 10 November 2018). (In Dutch)
107. Maharjan, M.; Pathirana, A.; Gersonius, B.; Vairavamoorthy, K. Staged cost optimization of urban storm drainage systems based on hydraulic performance in a changing environment. *Hydrol. Earth Syst. Sci.* **2009**, *13*, 481–489. [CrossRef]
108. Eijgenraam, C.; Brekelmans, R.; den Hertog, D.; Roos, K. Optimal Strategies for Flood Prevention. *Manag. Sci.* **2017**, *63*, 1644–1656. [CrossRef]
109. Aerts, J.C.J.H.; Sprong, T.; Bannink, B. *Aandacht voor Veiligheid*; VU University Press: Amsterdam, The Netherlands, 2008; p. 213. Available online: https://research.vu.nl/en/publications/aandacht-voor-veiligheid-leven-met-water-klimaat-voor-ruimte-dg-w (accessed on 10 November 2018). (In Dutch)
110. Maurer, M.; Scheidegger, A.; Herlyn, A. Quantifying costs and lengths of urban drainage systems with a simple static sewer infrastructure model. *Urban Water J.* **2013**, *10*, 234–254. [CrossRef]
111. (EA) Environment Agency. Available online: http://evidence.environment-agency.gov.uk/FCERM/Libraries/FCERM_Project_Documents/SC080039_cost_house_resist_and_resilience.sflb.ashx (accessed on 10 October 2018).

water

MDPI

Review

Permeable Pavements Life Cycle Assessment: A Literature Review

Lucas Niehuns Antunes *, Enedir Ghisi and Liseane Padilha Thives

Department of Civil Engineering, Federal University of Santa Catarina, Laboratory of Energy Efficiency in Buildings, Florianópolis-SC 88040-900, Brazil; enedir@labeee.ufsc.br (E.G.); liseane.thives@ufsc.br (L.P.T.)
* Correspondence: lucas_niehuns@hotmail.com (L.N.A.); Tel.: +55-48-3721-2115; Fax: +55-48-3721-5191

Received: 28 September 2018; Accepted: 1 November 2018; Published: 3 November 2018

Abstract: The number of studies involving life cycle assessment has increased significantly in recent years. The life cycle assessment has been applied to assess the environmental performance of water infrastructures, including the environmental impacts associated with construction, maintenance and disposal, mainly evaluating the amount of greenhouse gas emissions, as well as the consumption of energy and natural resources. The objective of this paper is to present an overview of permeable pavements and show studies of life cycle assessment that compare the environmental performance of permeable pavements with traditional drainage systems. Although the studies found in the literature present an estimate of the sustainability of permeable pavements, the great heterogeneity in the evaluation methods and results is still notable. Therefore, it is necessary to homogenize the phases of goal and scope, inventory analysis, impact assessment and interpretation. It is also necessary to define the phases and processes of the evaluation, as well as the minimum amount of data to be considered in the modelling of life cycle assessment, in order to avoid heterogeneity in the functional units and other components. Thus, more consistent results will lead to a real evaluation of the environmental impacts caused by permeable pavements. Life cycle assessment studies are essential to guide planning and decision-making, leading to systems that consider increasing water resources and reducing natural disasters and environmental impacts.

Keywords: permeable pavements; life cycle assessment; stormwater management; sustainability

1. Introduction

The increase in the frequency of flooding in urban areas related to the increase of impermeable surfaces highlights the inadequacy of traditional urban drainage systems. According to Min et al. [1], it is expected that the frequency of high intensity and short duration rainfall events will increase in the coming decades as a consequence of climate change. Wasko and Sharma [2] identified a strong correlation between peak precipitation intensity and high temperatures, and concluded that global warming may lead to an increase of floods of short duration. In addition, Luo et al. [3] report that flash floods have occurred more frequently in Asian cities, with recent increases in urbanization and extreme rainfall, causing significant damage to infrastructure, communities and the environment. This increase in the number of floods shows that it is necessary to use sustainable urban drainage systems capable of restoring the natural hydrological cycle in urban areas and allowing an increase in evapotranspiration and infiltration capacity. Permeable pavements are examples of systems that fulfill this function [4].

According to Scholz and Grabowiecki [5], the management of stormwater in urban areas was observed in a more ecological way due to the emergence of sustainable drainage systems that collect, store, treat and redistribute or recycle water. Compared to the traditional drainage system, stormwater retention and infiltration is a sustainable and cost-effective process, which is suitable for urban areas. In addition, these systems have benefits such as the reduction of runoff, groundwater recharge, saving water through recycling, and preventing pollution.

Permeable pavements are considered sustainable drainage systems because they are pavements that support the demands of mechanical efforts and at the same time allow the percolation and temporary accumulation of water, reducing surface runoff without causing damage to their structures [6]. Several studies have shown the advantages of using this type of pavement. In comparison to conventional pavements, permeable pavements provide runoff reductions of up to 42% [5]. According to Pagotto et al. [7], the quality of stormwater is improved by the use of permeable pavements for most pollutants. Heavy metals are reduced by up to 74%, solids are retained at a rate of 87% and hydrocarbons are intercepted at an even higher rate (90%).

Brattebo and Booth [8] examined the long-term efficacy of four permeable pavement systems in the United States. The study showed a significantly better performance for permeable pavements, both for water quality, which had lower toxic levels, and for stormwater infiltration. In the four systems, practically all of the precipitation was infiltrated. The levels of copper and zinc obtained in the water samples collected from the conventional asphalt concrete runoff were alarming: toxic concentrations were reached in 97% of the samples. However, in 31 out of 36 water samples infiltrated in permeable pavements, the concentrations were below the detectable toxic level.

According to Maiolo et al. [9], there is a need to have a methodology capable of providing an accurate estimate of the sustainability of drainage systems. In fact, this assessment may not only be tied to environmental benefits related to lifespan, but assessments are necessary in the steps that precede and follow the lifespan. A valid criterion for the verification of the sustainability of a product or system is the life cycle assessment (LCA). LCA presents an opportunity to evaluate and compare projects and choose the most appropriate drainage systems, quantifying a variety of environmental impacts and benefits. LCA has been effectively applied to assess the environmental performance of the water infrastructure, including the environmental impacts associated with the construction, maintenance and disposal of various green infrastructure technologies, such as permeable pavements [10]. This assessment is based primarily on the amount of greenhouse gas emissions, as well as the consumption of energy and natural resources. Some parameters significantly affect the evaluation, such as local climatic patterns, regulatory requirements, quality of infiltrated stormwater, lifespan and treatment efficiency of the systems [11].

As stated by the Electric Power Research Institute [12], at a national scale, the transport and treatment of water and wastewater accounts for nearly 4% of the US electricity demand. Such dependency of water infrastructure on electric utility infrastructure leads to serious environmental impacts. In this way, decentralized water management brings benefits not only as a means of reducing stresses on the water treatment infrastructure but also as a strategy to reduce the demand that water companies impose on the regional energy system, and on reducing the carbon footprint [13]. As a point of reference, the City of New York [14] estimates that systems of water treatment, supply, and sewage along with the methane escaping into the atmosphere (generated by the sewage treatment process) add up to 17% of New York's greenhouse gas emissions.

De Sousa et al. [15] evaluated the environmental performance of green infrastructures (permeable pavements and bioretention basins) by comparing them to water storage and treatment scenarios using traditional drainage systems (grey infrastructure). The results showed that green infrastructures emitted 75% to 95% less greenhouse gases, mainly due to the lower use of electricity during the life cycle. Wang et al. [11] showed by means of a case study in China that 73.48% of energy consumption, 46.70% of greenhouse gas emissions, 98.33% of lead emissions and 99.70% of zinc emissions could be avoided by using permeable pavement instead of conventional pavement.

While understanding the life cycle implications of sustainable drainage systems is only in its early stages, LCA studies are important in guiding planning and decision-making when considering multiple objectives such as increased water resources and reduction of natural disasters and environmental impacts [16]. Thus, the objective of this paper is to present an overview of permeable pavements and show studies of LCA that compare the environmental performance of permeable pavements with traditional drainage systems, in order to provide scientific instructions for the choice of more sustainable drainage systems and thus improve the sustainable management of stormwater in urban areas.

2. Permeable Pavements

2.1. Definition

Permeable pavements are pavements that simultaneously support the demands of mechanical stresses and rolling conditions, whose structure allows the percolation and temporary accumulation of water, reducing surface runoff without causing damage to their structure [17]. In this type of pavement, the structure is composed of a combination of layers, which are: permeable sub-base, permeable base, permeable bedding layer (when applicable) and permeable surface, dimensioned to withstand traffic loading, distribute stresses on the subgrade and allow the percolation of water. The base and sub-base of the pavement consist of open granulometry materials with aggregates that do not contain fines, or with a small amount of fines, resulting in a relatively large void ratio after compaction [18].

Permeable pavements can be modelled with various types of permeable surfaces, such as porous asphalt, pervious concrete, and permeable interlocking concrete [19]. They can be used as an alternative to conventional impervious hard surfaces, such as roads, car parks, footpaths and pedestrian areas [20].

As for the infiltration system, permeable pavements can be designed in three different ways: with total infiltration of the stormwater, partial infiltration or without infiltration, as shown in Figure 1.

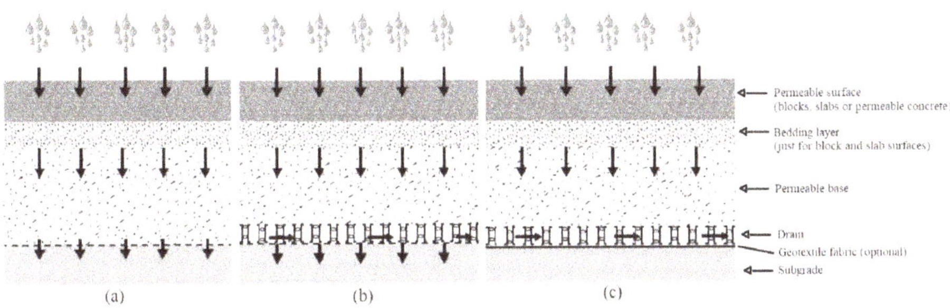

Figure 1. Examples of permeable pavements systems: (**a**) with total infiltration; (**b**) with partial infiltration; (**c**) without infiltration. **Source:** Based on ABNT [17].

2.2. Permeability, Infiltration and Quality of Infiltrated Stormwater

Several studies have demonstrated the benefits of using permeable pavements, such as reducing runoff, groundwater recharge, saving water through recycling and preventing pollution by improving the quality of the infiltrated stormwater [21–26]. The Ramsey-Washington Metro Watershed District [27] conducted a study that aimed to implement a permeable pavement system in a 650 m^2 parking lot in Oregon. The investment was aimed at infiltrating and storing precipitation, reducing runoff from stormwater, maximizing permeability of the area and improving water quality, retaining heavy metals and toxins. The cost for the implantation of permeable pavement was US$102/m^2 and was designed to have 100% infiltration in precipitations of up to 51 mm. Thus, any precipitation up to this figure would not generate runoff. On the other hand, the implementation cost of a conventional pavement system would vary from US$35/m^2 to US$46/m^2, and for this type of pavement the runoff would be 15,000 litres for a 25 mm precipitation.

Legret and Colandini [28] compared the pollution contained in the drainage of stormwater collected from a permeable pavement to the pollution contained in the drainage from a traditional pavement located in the city of Rezé, France. The retentions of suspended solids, lead, cadmium and zinc were, respectively, 59, 84, 77 and 73% higher in the permeable pavement.

Pratt et al. [29] studied the ability of a permeable pavement reservoir structure to retain and treat petroleum-derived pollutants through in situ microbial bio-degradation. The authors constructed a full-scale model permeable pavement in a laboratory. The pavement comprised pre-formed concrete blocks bedded on clean gravel, with vertical drainage provided through gravel-filled inlets between the blocks. A geotextile membrane separated the block bed from the underlying sub-base, comprising 600 mm depth of washed 20–50 mm granite. The entire structure rested on an additional geotextile underlay, supported by a stainless-steel mesh, allowing effluent to flow into a collection funnel located at the base of the tank. The model was subjected to prolonged low-level hydrocarbon contamination, representative of typical loadings to urban surfaces such as highways and car parks. Water quality was monitored by means of oil and grease concentration, chemical oxygen demand (COD), and pH. The retention efficiency of oil in the permeable pavement was 97.6%. The construction materials had a buffering effect, maintaining an effective pH of about 7.0, which is beneficial to microbial growth. With the benefits shown by the results, the study demonstrated that the structure can be used as an effective in situ aerobic bio-reactor. Also, the development of permeable pavements as pollution treatment devices offers a potential solution to the problem of uncontrolled discharge of contaminant loads associated with stormwater.

Pagotto et al. [7] compared the hydraulic behaviour and the quality of the stormwater drained by a section of a highway in the city of Nantes, France, first using a conventional pavement and finally after the replacement of the conventional pavement with a permeable pavement. Regarding the hydraulic behaviour, the permeable pavement system obtained excellent results. Response times (time elapsed between the beginning of the rain and the beginning of the runoff) were, on average, twice as long on this type of pavement. The delay caused the maximum flow rates to be reduced (6.2 litres/s in the conventional pavement and 5.5 litres/s in the permeable pavement) and the discharge time was higher (average discharge duration was 1.15 times greater for permeable pavements).

There was a great difference between the two types of pavements in the quality of the stormwater drained. The percentage of hydrocarbons decreased by 92% and the total suspended solids decreased by 81%. Regarding metals, the reduction ranged from 35% (copper) to 78% (lead). For all metals, the particulate forms are retained at a high rate (greater than 70%). However, metals in the dissolved form are retained with greater difficulty. These results explain the considerable level of retention of zinc, cadmium and lead (mainly present in particle form) by weight in percentage terms and the lower retention of copper (mainly present in dissolved form). The study also showed that in each rainfall event, on average, 0.28 kg of sediment was retained in the permeable pavement, against more than 4.1 kg in the conventional pavement [7].

James [30] has shown that traffic on highways is a major source of pollutants and that these are charged to rivers and streams when precipitation occurs. A survey by the Forth River Purification Board indicates that more than 14% of unsatisfactory river water is due to stormwater runoff in urban areas. The quality of the water drained by permeable and conventional pavements was compared and the results obtained are shown in Table 1. It is possible to perceive that the permeable pavement has great participation in the process of treatment of stormwater, being able to be a great facilitator in the development of sustainable drainage systems.

Table 1. Reduction of pollutants when using permeable pavements compared to conventional pavements.

Parameter	Reduction of Pollutants (%)
Suspended solids	80–99
Phosphorus	65–71
Nitrogen	75–85
Total organic carbon	82
Lead	50–98
Zinc	62–99
Chrome	87–88
Cadmium	0–34
Copper	42
Heavy metals	90–99
Biochemical oxygen demand	80–83
Chemical oxygen demand	88
Hydrocarbons	95
Oil	97–98

Source: Based on James [30].

Gilbert and Clausen [31] evaluated the amount of stormwater drained in two types of sidewalks: one with typical asphalt surface and other covered with paver. Paver driveways were constructed with stone blocks (115 by 230 mm) interlocking concrete permeable pavement. Pavers were hand installed over 5 cm compacted and screeded coarse sand on top of 15 cm processed gravel. Drainage voids comprised 12% of the surface area and were filled with 3–6 mm peastone. The reduction in the runoff from asphalt to the paver was 72%. The mean infiltration was zero for the asphalt and 11.2 cm/h for the paver. However, the rate of infiltration of the paver pavement decreased with time due to pore obstruction by fine particles. The water drained by the paver sidewalk contained significantly less pollutants compared to the asphalt pavement. Considering the benefits in reducing the runoff and the high infiltration rates, the use of paver in the construction of sidewalks over the traditional asphalt material is more advantageous.

Hou et al. [32] evaluated the infiltration rate of three different types of permeable pavement systems compared to a conventional pavement system. For rainfall rates less than 59 mm/h, the runoff coefficient was zero for the permeable pavement, while the conventional pavement coefficient was 0.85. In addition to the better infiltration rate, it was also verified that the runoff start time after the rain event was higher for the permeable pavement (73 min later). Consequently, the discharge time of stormwater was also higher, which reduces the risk of flooding caused by heavy precipitation.

Eck et al. [33] evaluated the use of Permeable Friction Course (PFC) in the states of Texas and North Carolina in the USA. PFC is a layer of porous asphalt laid in thicknesses of 25 to 50 mm overlaying conventional impermeable pavement. PFC is a type of permeable pavement made of coarse and fine aggregates, asphalt binders, and stabilizing additives, but it does not encourage infiltration and reduces flow volume, such as the full depth permeable pavement. Instead, PFC layers remove rainfall from the road surface and allow it to flow through the porous layer to the roadside. With the use of PFC, the total suspended solids had a reduction of up to 96% when compared to conventional pavement, and good results were found for other parameters such as phosphorus (reduction of up to 78%), copper (69%), lead (above 90%) and zinc (90%). The performance of the Permeable Friction Course can be compared to that of a sand filter because the particulate substances are well filtered while the dissolved substances have little or no retention. Regarding the runoff, 29% to 47% of the total precipitation was retained.

2.3. *Application of Stormwater Collected from Permeable Pavements for Non-Potable Uses in Buildings*

As seen in the previous section, permeable pavements have the ability to retain pollutants and improve the quality of stormwater. Some studies have evaluated the possibility of using this water for non-potable uses in buildings, such as toilet flushing, garden watering, car washing, among others.

Pratt [34] performed a case study at a UK-based hostel whose building had 400 m^2 of roof area and 325 m^2 of parking area. Stormwater precipitated on both surfaces would be stored in the parking sub-base. The parking surface contained permeable blocks that allowed infiltration of stormwater into the sub-base. The water stored in the sub-base was connected to a tank in the hostel and used for toilet flushing. The water storage capacity on the pavement was approximately equal to 34 m^3.

Antunes et al. [35] evaluated the possibility of using stormwater from permeable pavements in non-potable uses in residential, commercial and public buildings in the city of Florianópolis, Brazil. In the study, two models of porous asphalt concrete modified with rubber and Styrene-Butadiene-Styrene (SBS) polymer were assessed. The mean percentage of infiltration found for the models was 85%. In this way, the potential for potable water savings ranged from 1 to 18% in the residential sector, 2 to 57% in the public sector, and 6 to 69% in the commercial sector, depending on the tank size.

Hammes et al. [36] evaluated the performance of two permeable pavements in terms of quantity and quality of infiltrated stormwater, aiming at its use in activities that allow the use of non-potable water. The pavements structures are shown in Figure 2 (models A and B). The permeable pavements tested had a mean of 70 and 80% infiltration, respectively. The lower infiltration value for the model A was mainly due to the presence of the filter course. A positive influence of the pavements was observed in some parameters of water quality. However, the need for an additional treatment of the water to adapt it to the expected quality for use was verified. In addition, it was proposed to use the permeable pavement in a parking lot of the Federal University of Santa Catarina (Brazil) for stormwater infiltration, storage and subsequent use in toilets and urinals flushing. The potential for potable water saving would be at least 53%.

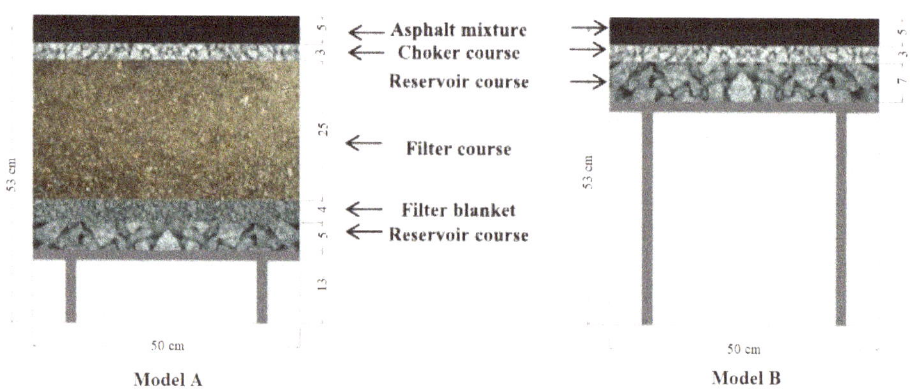

Figure 2. Permeable pavement models tested. **Source:** Hammes et al. [32].

Thives et al. [37] conducted a study to determine the infiltration capacity and the quality of stormwater infiltrated by permeable pavements with drainage asphalt concrete surface. The concentrations of phosphorus, iron, aluminium, zinc, nitrite, chromium, copper and pH increased after the infiltration in the pavements studied, while the ammonia concentration decreased. However, only phosphorus and aluminium concentrations exceeded the limits required for non-potable uses. It was also found that at least 84% of stormwater could be infiltrated and would be available for non-potable uses.

Thives et al. [38] carried out a study to estimate the potential for potable water savings in multifamily buildings using stormwater collected from paved streets in an area of the city centre of Florianópolis, southern Brazil. For a paved area equal to 9058 m^2 and a stormwater tank capacity of 1000 m^3, the potential for potable water savings ranged from 17 to 33% according to the water demand for non-potable purposes.

Although pollutant removal rates vary according to climatic conditions and permeability parameters, the studies mentioned in this review demonstrate the efficiency of permeable pavements in reducing stormwater runoff, as well as improving water quality infiltrated through the pavement. However, the literature still lacks publications related to the real sustainability of permeable pavements, which should relate the benefits brought by these systems to the environmental impacts produced during all phases, from material extraction to the end of the pavement lifespan.

3. Life Cycle Assessment

Awareness regarding the forecasting and prevention of environmental impacts related to construction is increasing. In this way, interest in developing methods to better understand and deal with these impacts has been increasing. One of the techniques in development for this purpose is life cycle assessment (LCA). LCA can identify opportunities for improving the environmental performance of services at various points in their life cycles, as well as selecting relevant environmental performance indicators, assisting decision-makers in governmental or nongovernmental organizations, for example defining priorities and strategic planning [39].

LCA focuses on potential environmental impacts, such as the use of resources and the consequences of releases to the environment throughout the life cycle of a product or service, from the acquisition of raw materials, production, use, post-use treatment, recycling until final disposal. LCA studies are composed of four phases: goal and scope; life cycle inventory; impact assessment; and interpretation [40].

3.1. Pavements Life Cycle Assessment

This section presents a brief literature review about traditional pavements life cycle assessment, showing some of the various studies and giving the reader an overview about the subject. AzariJafari et al. [41] highlight the large increase in the number of studies on the life cycle assessment of conventional pavements. Current literature demonstrates a wide range of environmental load implications associated with pavements [42–44]. Chiu et al. [45] demonstrated that actions aimed at sustainable development in pavement construction projects can lead to the reduction of greenhouse gas emissions and their life cycle cost. However, there are still immature concepts, which require more research in the coming years, in different stages of the evaluation of the pavement life cycle. One of the fields still little explored is that of permeable pavements. Few studies regarding the life cycle of these pavements and the environmental benefits that can be achieved through the retention of water and consequent reduction of the problems related to floods and water recharge are found in the literature.

LCA is an appropriate tool that can help designers deal with the environmental aspects of their pavements to achieve the goal of building more sustainable pavements. In fact, LCA helps to quantify, analyse and compare the environmental impacts of different types of pavement, from material extraction to the end of its lifespan [19].

AzariJafari et al. [41] compared publications involving LCA of several types of pavements. The results show a significant heterogeneity of functional units and other components. LCA standards, such as ISO 14040 and 14044, do not have technical details on, for example, phases and processes that should be included in the assessment, the lifespan to be analysed, or what the minimum amount of data is that should be considered in modelling LCA. In addition to inconsistencies between publications, significant differences in calculated life cycle environmental impact outcomes make comparisons of results simply impossible.

Approximately US$150 billion and 320 million tonnes of building materials are invested annually in the construction, rehabilitation and maintenance of pavements in the United States. However, very little is known about the environmental damages caused by the construction of these pavements [46]. Some studies have shown that the type of pavement can influence vehicle fuel consumption [47,48]. Taylor and Patten [48] have shown that Portland cement-based concrete pavements can decrease the amount of fuel consumed when compared to pavements constructed with hot-mix asphalt concrete (HMA).

Huang et al. [49] developed a life cycle assessment tool for the construction and maintenance of asphalt pavements. The structure of LCA was composed of process parameters (energy consumed in transport, material production and pavement construction), pavement parameters (size, materials used, lifespan), unit, project inventory and characterization results. The results are divided into different categories, such as depletion of minerals and fossil fuels, depletion of the ozone layer, global warming, acidification, photo-oxidant formation, human toxicity, eco-toxicity, eutrophication, among others. The study proposed a method for grouping and weighting categories, according to the "Eco-points" developed by the Building Research Establishment (UK) for the construction sector, as shown in Figure 3.

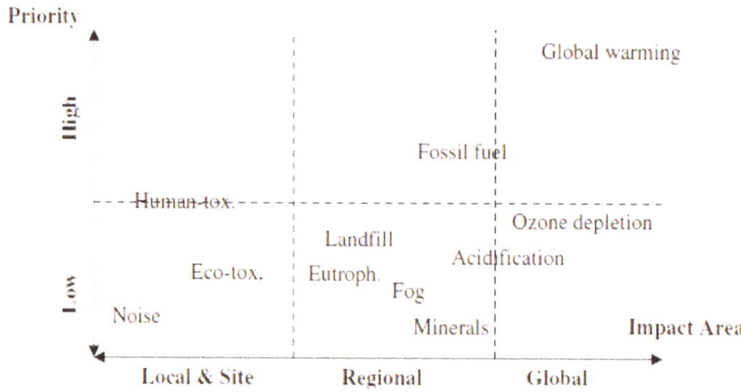

Figure 3. Grouping and weighting of LCA environmental impact categories. **Source:** Huang et al. [49].

Huang et al. [49] used the proposed LCA methodology to conduct a case study investigating the environmental impacts of the asphalt pavement life cycle on a highway, in which the natural aggregates were partially replaced with glass residues and incineration ash. The results were compared to conventional pavement of the same size and function, but using only virgin aggregates. Asphalt mixing, bitumen and aggregates production consumed, respectively, approximately 62%, 23% and 6% of the total energy and consequently produced more emissions than the other processes. The use of recycled materials reduced the consumption of asphalt binder by about 7%. Another significant benefit of recycling was the saving of 5766 tonnes of aggregates and the recycling of 579 and 989 tonnes of glass waste and incineration ash respectively. Aggregate transport accounted for more than 61% of all diesel use, due to the long transport distance (193 km). Trains with higher fuel efficiency (0.17 MJ/t.km) than trucks (0.46–0.94 MJ/t.km) were used to transport aggregates. Glass and ash were obtained from local sources and the use of diesel to transport asphalt was only 17%, as the highway was located very close to the asphalt plant (6.4 km). The results of this study show the great dependence of the location of the road and the materials used in the pavement structure, which significantly interfere with the environmental impacts of the life cycle.

Santero and Horvath [50] evaluated the global warming potential of conventional pavements in the United States, analysing several components such as: extraction and production of materials, transportation, equipment used, carbon absorption, heat islands, surface roughness of the pavement, rolling resistance, albedo, among others. Figure 4 shows the emission of carbon dioxide (in Mg CO_2e) per kilometre of road over 50 years obtained by Santero and Horvath [50]. Grey bars show variations of global warming potential, while black bars show the extreme values of each component. The results demonstrate the wide range of possible impacts to the components of the pavement life cycle. This impact ranges from insignificantly small to 60,000 Mg CO_2e per kilometre of road over 50 years.

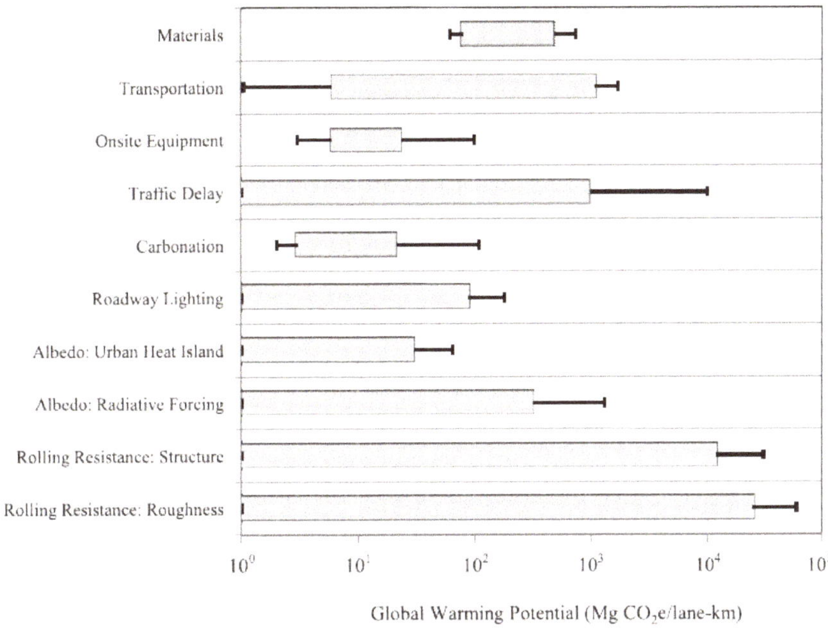

Figure 4. Impact of the global warming potential for components of the pavement life cycle. **Source:** Santero and Horvath [50].

3.2. Permeable Pavements Life Cycle Assessment

In recent years, the use of permeable concrete as paving material in low volume road applications has gained importance due to its positive environmental aspects. Due to the increased use of permeable concrete in the pavement industry, there is large scope for future research to better understand the material, which will make it a promising material for sustainable future roads [51]. Wang et al. [11] developed a model of LCA that can be applied to permeable pavements of both asphalt and concrete in order to evaluate the environmental impacts caused by these types of pavement. The impacts investigated in the study were related to urban floods, stormwater recycling and water purification. The authors compared the use of a permeable asphalt pavement with a conventional asphalt pavement on a typical four-lane secondary highway. The results showed that in 10 km of the modelled highway, 49 TJ of energy consumption, 6700 tonnes of CO_2e emissions, 0.1 tonne of lead emission and 1.0 tonne of zinc emission could be avoided if permeable pavements were used in place of conventional pavement. The study showed that the most significant reduction in energy consumption, greenhouse gas emissions, lead emissions and zinc emissions occurs during the use phase of the pavement. In addition, in an area of 200,000 m^2 (10 km $\times$ 20 m), the volume of stormwater recycled to the subgrade annually using the permeable pavement is 154,000 m^3.

Spatari et al. [13] examined the reduction of energy consumption and the reduction of greenhouse gas emissions through selected Low Impact Development (LID) strategies using the LCA in an urban watershed model. The LID strategies consisted of a retrofit in the conventional sidewalks (with impervious surface), these being replaced with permeable pavements. An annual energy reduction of 7.3 GJ and a 0.4 tonne reduction in greenhouse gas emissions were estimated for the strategy implemented in a neighbourhood of New York City. Examining the materials for the LID strategy, the rubber mats and concrete sidewalk components contribute most to the embodied energy (31% and 28%, respectively) and greenhouse gas (GHG) emissions (34% and 27%, respectively), while transportation energy accounts for approximately 10% of the construction materials' life cycle energy and 17% of life cycle GHG emissions. The annual savings are small compared to the energy intensity and greenhouse gases of LID materials, resulting in slow environmental return (paybacks ranged from 70 to 180 years). This preliminary analysis suggests that if implemented along an urban watershed, LID strategies can have significant energy cost savings for water pollution control facilities, and may advance in reducing their carbon footprint.

A study by the Brazilian Council for Sustainable Construction [52] carried out the evaluation of the modular life cycle of concrete blocks for interlocking pavements, which can be used as surface of permeable pavements. The study estimated indicators such as material use, water and energy consumption, CO_2 emission and waste generation in the production process. The data were collected in 33 block factories, located in different regions of Brazil. The results showed the great variability in the consumption, depending mainly on the type of production adopted by the factories and also on the dimensions of the blocks. Energy consumption ranged from 50 to 810 MJ/m^2. The CO_2 emission varied from 10 to 70 $kgCO_2/m^2$. Water consumption, in turn, varied from 0.01 to 0.91 litres/piece. The waste generated by the factories is diverse, such as wood, plastic, paper, oil, steel and cementitious material. The percentage of recycling practiced by the factories ranges from 67% to 100%.

Li et al. [53] evaluated the life cycle of different sustainable drainage systems: permeable pavements, green roofs and wetlands. Indicators at all stages of the life cycle (construction, operation, maintenance and final disposal) were evaluated. The results showed that the abiotic depletion potential, the acidification potential and the global warming potential of the three drainage systems obtained the greatest impacts in each category: resource depletion, ecosystems and human health, respectively. The impact on human health is related to the concrete used in construction, directly impacting the exhaustion of resources. Resource depletion has also contributed significantly to ecosystem damage, while high abiotic depletion is mainly due to the transport of materials. The study also showed that permeable pavements contributed significantly to flood reduction, with a runoff control rate of 67.5%. However, permeable pavements obtained the highest abiotic depletion potential, mainly due to the greater use of building materials in their structure.

Maiolo et al. [9] developed a methodology based on the sustainability index to evaluate the life cycle of permeable pavements and green roofs implemented in Italy. Figure 5 shows the structure of the permeable pavement used in the study. The application of the LCA highlighted that there are substantial contributions to the layers made up of natural material (sand, gravel), which have an impact due to transportation from the place of origin to the place of execution of the system. In addition, the life cycle of polymeric materials is the same for both drainage systems because of non-renewable sources of energy supply and transport types whose energy class is not particularly competitive. A confirmation of this fact is that the contribution of carbon dioxide has a higher percentage than the emissions of other gases (methane and dinitrogen monoxide), as shown in Figure 6. In conclusion, the authors state that the comparison between the sustainability indices shows that the green infrastructures are technologies that adequately reflect the objective of reducing the environmental impact produced by drainage systems.

A study conducted by the Center for Neighborhood Technology (CNT) and American Rivers [54] showed that air temperature can be reduced by permeable pavements, which absorb less heat than conventional pavements. By reducing the heat island effect in urban areas, such cooling can reduce diseases and fatalities related to excessive heat during extreme events of high temperatures and heat waves.

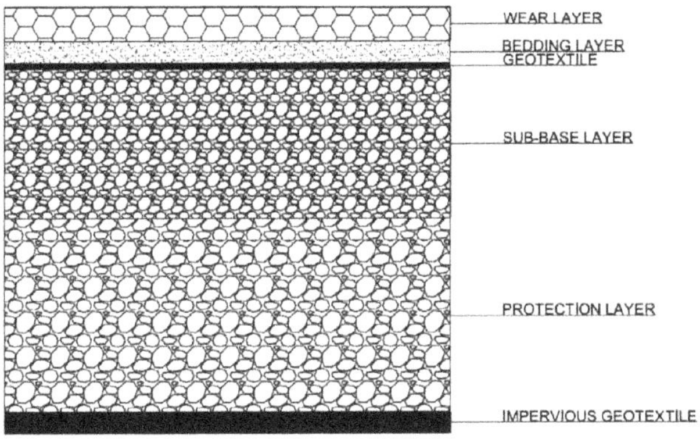

Figure 5. Structure of the permeable pavement used in the study. **Source:** Maiolo et al. [9].

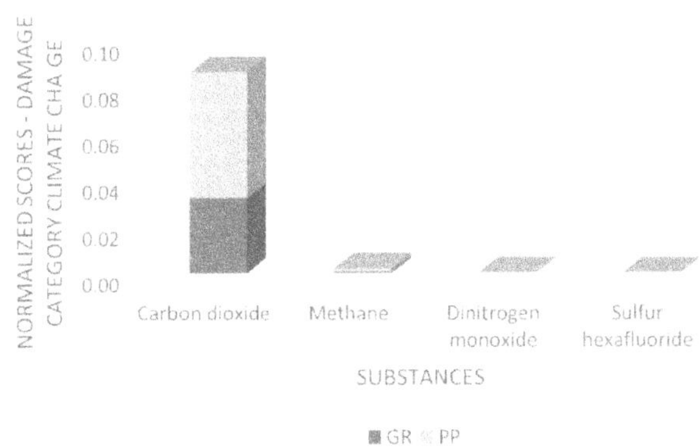

Figure 6. Gases emitted during the life cycle of permeable pavement (PP) and green roof (GR). **Source:** Maiolo et al. [9].

De Sousa et al. [15] compared the life cycle assessment of three different drainage systems in the United States. System 1 consisted of green infrastructures including 27.12 ha of permeable pavements, 1.18 ha of bioretention basins, 2.80 ha of infiltration plants, 1.06 ha of rain gardens and 8.54 ha of cisterns in the subgrade. This combination was collectively sized to capture the first 2.5 cm runoff generated from approximately one-third of the total drainage area. The infrastructure of system 1 occupied about 5% of the total area. Systems 2 and 3 were grey infrastructures. System 2 only retains the runoff in a storage tank and launching it into the Bronx River, while system 3 also performed the treatment prior to launching into the river. The installation of system 1 emitted 20,000 t CO_2e, compared to 31,500 t CO_2e of system 2 and 100,000 t CO_2e of system 3. Of the total emissions associated with the construction of green infrastructures, the major contributions came from transport (8500 t CO_2e), followed by the production of cement and concrete (8400 t CO_2e).

The study also presented a cumulative emission estimate for the phase of operation and maintenance of the systems in a period of 50 years. The net emissions of the green strategy were 19,000 t CO_2e, while grey strategies emitted 85,000 t CO_2e (detention) and 400,000 t CO_2e (detention and treatment). These results were significantly influenced by the emissions associated with the operation and maintenance activities required for systems 2 and 3, and by the sequestration of carbon provided by vegetation in system 1. Thus, it is noted that green infrastructures have a superior environmental performance when compared to grey infrastructure systems.

Yuan et al. [55] compared the environmental and economic impacts of manufacturing permeable paving blocks (with at least 10% porosity) compared to conventional paving blocks in China.

The functional unit used in the study was 1 m^2. All inputs of raw materials, energy consumption, transport, waste and effluent discharge were calculated using the functional unit as a baseline for the two types of block production processes. Only the phase of production of the blocks and the phase of acquisition of the raw materials were considered. The economic cost to produce blocks of conventional pavement and permeable blocks was 24.26 RMB (in October 2018, 1 Chinese Yuan (RMB) is equal to 0.15 United States Dollar (US$)) and 29.68 RMB per m^2, respectively. The results showed that cement was the material that caused the greatest environmental impact on the permeable blocks. This impact could be optimized by reducing consumption. The result of the calculation showed that if cement consumption were reduced by 5%, the overall environmental impact would be reduced by about 2.21%, and the cost of production would be reduced by 1.02 RMB. The coefficient of permeability of the blocks was 1.8×10^{-2} m/s. Thus, during a 3-year service period, the blocks would have a stormwater infiltration capacity of 2.01 m^3 per 1 m^2 of area.

3.3. Life Cycle Cost Analysis

There is little published data on the life cycle cost analysis (LCCA) of permeable pavements that include actual costs and performance. Most studies are limited to comparative initial cost analyses for permeable pavements compared to conventional pavements, which indicates that the cost of permeable pavements is greater than the cost of conventional pavements; however, some studies indicate that the initial total costs are similar or lower because permeable pavements do not require stormwater drainage systems [19].

According to Mei et al. [56], rulers face the increasingly difficult task of planning water management systems in urban areas, especially in relation to uncertainties of climate and socioeconomic changes, which requires decision-makers to plan the water management infrastructure from economic and adaptation points of view. For a specific area, considering draining a region, several green infrastructure options are possible within the scope of planning. However, the systems have different impacts and hydrological costs, making assessments necessary to integrate the sustainability and cost-benefit of these systems.

Wang et al. [19] conducted a life cycle cost analysis to understand the cost implications of building and maintaining permeable pavements. The input data for the models were obtained from laboratory research and computer performance modelling. A detailed life cycle assessment could not be performed due to insufficient available data on the construction, long-term performance, maintenance and salvage value of permeable pavements and alternative Best Management Practices (BMPs) currently used for stormwater management. Two scenarios were considered in the study: a shoulder retrofit of a high-speed highway, and a low-speed highway or parking lot/maintenance yard. Both scenarios compared conventional pavements with conventional treatment BMP versus the use of permeable pavements. The results indicate that permeable pavements are potentially more cost effective than currently available BMP technologies. These results were used to prepare preliminary paving projects for pilot studies of permeable pavement in California and to identify under what conditions they are appropriate for use. Although a more comprehensive life cycle assessment should be undertaken after the completion of the pilot studies.

Kluck et al. [57] point out that, in Holland, pavements with permeable surface are used in order to reduce noise produced by the traffic. However, permeable pavements have a shorter lifespan than traditional pavements, causing frequent maintenance and supposedly increasing costs and thus causing economic and environmental damage. The study conducted by the authors aimed to replace the traditional binder used in the permeable pavement by synthetic binders in order to increase the lifespan of the system. Considering the net present value of the investment, it was concluded that the permeable pavement produced with the synthetic binders costs the same as the conventional pavement, but with a life cycle up to ten times greater, which brings environmental and economic benefits for the drainage system of the Dutch urban areas.

The economic benefits of permeable pavements can be appreciated when life cycle cost analysis is performed. However, due to the lack of large-scale testing, long-term performance data, and construction and maintenance cost data, life cycle cost analysis has been difficult to perform, requiring several assumptions. Wang et al. [19] compared permeable pavement systems with conventional stormwater management systems used at the road shoulders. Permeable pavements reduced life cycle costs by up to 30%. In another study, conducted by Terhell et al. [58], based on data obtained from several agencies, it was found that permeable pavements can save up to US$64,649, considering installation costs, and US$3,788,856 considering stormwater treatment benefits over 25 years for 1/2 acre area compared to conventional pavement. The reduction in the cost of construction is attributed mainly to the fact that permeable pavements do not require side drains, overlays and so on.

To compare flood control efficacy and cost-benefit of green infrastructures, Mei et al. [56] evaluated the implementation of permeable pavements, green roofs, wetlands and bioretention basins in China. The increasing order of effectiveness of flood control was: green roof, permeable pavement, wetland and bioretention basin. This sequence is related both to the characteristics of the study area and to the properties of the specific practices of the green infrastructures. Implementation of the combination of the four practices would result in a peak flow reduction of 80.62%. The study also contemplated the life cycle cost of the systems, considering the phases of design, planning, construction, operation and benefits brought by the strategies. The increasing order of life cycle cost was wetland (US$31.72/$m^2$), permeable pavement (US$98.48/$m^2$), bioretention basin (US$186.90/$m^2$), and green roof (US$317.10/m^2). As a conclusion, it was found that the combination of permeable pavements with bioretention basins and wetlands is recommended as the best strategy for flood control and cost-benefit for the study site.

Chui et al. [59] verified that the life cycle cost of drainage systems depends on the place where they are implemented, and in the case studied, the life cycle cost of the systems were lower in the city of Hong Kong (China) when compared to Seattle (USA). The effective costs for the reduction of runoff were 0.02 L/10^3 US$, 0.15 L/$10^3$ US$, and 0.93 L/$10^3$ US$, for green roof systems, bioretention basin and permeable pavement in the city of Hong Kong, while in the city of Seattle, the figures were 0.03 L/10^3 US$, 0.29 L/$10^3$ US$ and 1.58 L/10^3 US$, respectively. It is noted that the results found by Chui et al. [59] show an opposite cost-benefit order when compared to the study published by Mei et al. [56]. Chui et al. [59] concluded that the relation between the reduction of the stormwater runoff and the cost of the permeable pavement forms an "S" curve; that is, the permeable pavement ideal design tends to have a smaller area and a thinner pavement surface. However, for more intense rainfall events, it is cheaper to expand the surface than to increase depth. The permeable pavement obtained the best cost-benefit for the reduction of the runoff between the three structures studied. Therefore, this type of pavement is recommended for places where stormwater management is the main objective.

3.4. Final Remarks

The review presented in this paper shows that there are several studies whose results prove the sustainability brought by green infrastructures, including permeable pavements, as well as their cost-benefit. However, it can be seen that LCA studies still present a significant heterogeneity of functional units, evaluation limits, phases, processes, parameters and minimum data evaluated, among other components. Thus, the results are often inconsistent, especially when compared to each other, and do not lead to an accurate assessment of the environmental impacts caused by these systems during their life cycles.

4. Conclusions

Due to the increase of impermeable areas and the consequent increase of floods in urban areas, the inadequacy of traditional urban drainage systems is increasingly notable. The trend is that the flood events and other problems related to the recharge and pollution of water resources will grow in the coming years due to global warming and man-made changes. In this way, the importance of using new sustainable drainage systems increases in order to enhance the permeability of surfaces and restore the natural hydrological cycle. These systems include permeable pavements, which were the focus of this paper.

The literature reviewed shows that permeable pavements are capable of filtering and storing stormwater. When compared to the traditional drainage system, they are sustainable and cost efficient, being fully adequate for urban areas, bringing benefits such as reducing stormwater runoff, as well as improving the quality of water infiltrated through the pavement. The LCA studies reviewed were able to provide an estimate of the sustainability of permeable pavements. However, there is still a need for a methodology capable of providing more precise results regarding the environmental impacts caused by these pavements. Thus, the evaluation should not be linked only to environmental benefits related to their lifespan, but assessments are necessary in the steps that precede and follow the lifespan.

Various parameters, such as local weather patterns, regulatory requirements, infiltrated stormwater quality, lifespan and treatment efficiency of systems, should be taken into account. The phases of goal and scope, inventory analysis, impact assessment and interpretation should be more homogeneous, defining phases and processes of the evaluation and the minimum amount of data to be considered in the modelling of LCA. Thus, heterogeneity in the functional units and other components should be avoided, bringing more consistent results and leading to a real evaluation of the environmental impacts caused by permeable pavements.

Although life cycle studies on permeable pavements still present several immature concepts, being only in their early stages, LCA is essential to guide planning and decision-making, leading to systems that consider the increase of water resources and the reduction of natural disasters and environmental impacts.

Author Contributions: L.N.A. wrote the first draft of the article; L.P.T. and E.G., contributed ideas concerning the structure and content of the article, and wrote the final version.

Funding: This research did not receive any specific grant from funding agencies in the public, commercial, or not-for-profit sectors.

Conflicts of Interest: The authors declare no conflict of interest.

References

1. Min, S.K.; Zhang, X.; Zwiers, F.W.; Hegerl, G.C. Human contribution to more-intense precipitation extremes. *Nature* **2011**, *470*, 378–381. [CrossRef] [PubMed]
2. Wasko, C.; Sharma, A. Steeper temporal distribution of rain intensity at higher temperatures within Australian storms. *Nat. Geosci.* **2015**, *8*, 527–529. [CrossRef]
3. Luo, P.; Mu, D.; Xue, H.; Ngo-Duc, T.; Dang-Dinh, K.; Takara, K.; Nover, D.; Schladow, G. Flood inundation assessment for the Hanoi Central Area, Vietnam under historical and extreme rainfall conditions. *Sci. Rep.* **2018**, *8*, 12623. [CrossRef] [PubMed]
4. Brunetti, G.; Simunek, J.; Piro, P. A comprehensive numerical analysis of the hydraulic behavior of a permeable pavement. *J. Hydrol.* **2016**, *540*, 1146–1161. [CrossRef]
5. Scholz, M.; Grabowiecki, P. Review of permeable pavement systems. *Build Environ.* **2007**, *42*, 3830–3836. [CrossRef]
6. Ball, J.E.; Rankin, K. The hydrological performance of a permeable pavement. *Urban Water J.* **2010**, *7*, 79–90. [CrossRef]
7. Pagotto, C.; Legret, M.; Le Cloirec, P. Comparison of the hydraulic behaviour and the quality of highway runoff water according to the type of pavement. *Water Res.* **2000**, *34*, 4446–4454. [CrossRef]
8. Brattebo, B.O.; Booth, D.B. Long-term stormwater quantity and quality performance of permeable pavement systems. *Water Res.* **2003**, *37*, 4369–4376. [CrossRef]
9. Maiolo, M.; Carini, M.; Capano, G.; Piro, P. Synthetic sustainability index (SSI) based on life cycle assessment approach of low impact development in the Mediterranean area. *Cogent Eng.* **2017**, *4*, 1410272. [CrossRef]

10. Glick, S.; Shuler, S.; Guggemos, A.A. Life cycle analysis for sustainable development: A case study of parking lot pavements. *Int. J. Construct. Educ. Res.* **2013**, *9*, 226–236. [CrossRef]
11. Wang, Y.; Li, H.; Ghadimi, B.; Abdelhady, A.; Harvey, J. Initial evaluation methodology and case studies for life cycle impact of permeability of permeable pavements. *Int. J. Transp. Sci. Technol.* **2018**, *7*, 169–178. [CrossRef]
12. Electric Power Research Institute. *Water & Sustainability: U.S. electricity consumption for water supply & treatment e the next half century, 4, Topical Report*; Electric Power Research Institute: Palo Alto, CA, USA, 2002.
13. Spatari, S.; Yu, Z.; Montalto, F.A. Life cycle implications of urban green infrastructure. *Environ. Pollut.* **2011**, *159*, 2174–2179. [CrossRef] [PubMed]
14. City of New York. *PlaNYC Sustainable Stormwater Management Plan 2008*; City of New York: New York, NY, USA, 2008.
15. De Sousa, M.R.C.; Montalto, F.; Spatari, S. Using life cycle assessment to evaluate green and grey combined sewer overflow control strategies. *J. Ind. Ecol.* **2012**, *16*, 901–913. [CrossRef]
16. Ghimire, S.R.; Johnston, J.M.; Ingwersen, W.W.; Sojka, S. Life cycle assessment of a commercial rainwater harvesting system compared with a municipal water supply system. *J. Clean. Prod.* **2017**, *151*, 74–86. [CrossRef] [PubMed]
17. Mullaney, J.; Lucke, T. Practical review of pervious pavement designs. *Clean. Soil Air Water* **2014**, *42*, 111–124. [CrossRef]
18. Associação Brasileira de Normas Técnicas (ABNT). *NBR 16416/2015: Pavimentos Permeáveis de Concreto—Requisitos e Procedimentos (NBR 16416/2015: Pervious Concrete Pavement—Requirements and Procedures)*; Associação Brasileira de Normas Técnicas (Brazilian Association of Technical Standards) (ABNT): São Paulo, Brazil, 2015.
19. Wang, T.; Harvey, J.T.; Jones, D. *A Framework for Life-Cycle Cost Analyses and Environmental Life-Cycle Assessments for Fully Permeable Pavements*; Technical Memorandum, Institute of Transportation Studies, Caltrans Document No.: CTSW-TM-09-249.03 UCPRC Document No.: UCPRC-TM-2010-05; Institute of Transportation Studies: Davis, CA, USA, 2010.
20. Beecham, S.; Lucke, T.; Myers, B. Designing porous and permeable pavements for stormwater harvesting and reuse. In Proceedings of the First European IAHR Congress, International Association for Hydro-Environment Engineering and Research, Edinburgh, UK, 4–6 May 2010.
21. Jia, H.; Lu, Y.; Yu, S.L.; Shen, Y. Planning of LID–BMPs for urban runoff control: The case of Beijing Olympic Village. *Sep. Purif. Technol.* **2012**, *84*, 112–119. [CrossRef]
22. Sansalone, J.; Kuang, X.; Ying, G.; Ranieri, V. Filtration and clogging of permeable pavement loaded by urban drainage. *Water Res.* **2012**, *46*, 6763–6774. [CrossRef] [PubMed]
23. Park, D.G.; Sandoval, N.; Lin, W.; Kim, H.; Cho, Y.H. A case study: Evaluation of water storage capacity in permeable block pavement. *KSCE J. Civ. Eng.* **2014**, *18*, 514–520. [CrossRef]
24. Knappenberger, T.; Jayakaran, A.D.; Stark, J.D.; Hinman, C.H. Monitoring porous asphalt stormwater infiltration and outflow. *J. Irrig Drain Eng.* **2017**, *143*, 04017027. [CrossRef]
25. Pratt, C.J.; Mantle, J.D.G.; Schofield, P.A. UK research into the performance of permeable pavement, reservoir structures in controlling stormwater discharge quantity and quality. *Water Sci. Technol.* **1995**, *32*, 63–69. [CrossRef]
26. Pratt, C.J.; Mantle, J.D.G.; Schofield, P.A. Urban stormwater reduction and quality improvement through stormwater reduction and quality improvement through the use of permeable pavements. *Water Sci. Technol.* **1989**, *21*, 769–778. [CrossRef]
27. Ramsey-Washington Metro Watershed District. Porous Asphalt Parking Lot. Little Canada, 2006. Available online: http://rwmetrowatershed.govoffice.com/vertical/Sites/%7BAB493DE7-F6CB-4A58-AFE0-56D80D38CD24%7D/uploads/%7B495C857D-FED3-409E-B58A-E6A16D827DEA%7D.PDF (accessed on 12 July 2018).
28. Legret, M.; Colandini, V. Effects of a porous pavement with reservoir structure on runoff water: Water quality and fate of heavy metals. *Water Sci. Technol.* **1999**, *39*, 111–117. [CrossRef]
29. Pratt, C.J.; Newman, A.P.; Bond, P.C. Mineral oil bio-degradation within a permeable pavement: Long term observations. *Water Sci. Technol.* **1999**, *39*, 103–109. [CrossRef]
30. James, E. A Literature Review on the Effect of Porous Asphalt Roads on Water Pollution. In *Sustainable Road Surfaces for Traffic Noise Control*; Silvia Project Report; European Commission: Ottawa, ON, Canada, 2013.

31. Gilbert, J.K.; Clausen, J.C. Stormwater runoff quality and quantity from asphalt, paver, and crushed stone driveways in Connecticut. *Water Res.* **2006**, *40*, 826–832. [CrossRef] [PubMed]
32. Hou, L.; Feng, S.; Ding, Y.; Zhang, S.; Huo, Z. Experimental study on rainfall-runoff relation for porous pavement. *Hydrol. Res.* **2008**, *39*, 181–190. [CrossRef]
33. Eck, B.J.; Winston, R.J.; Hunt, W.F.; Barrett, M.E. Water Quality of Drainage from Permeable Friction Course. *J. Environ. Eng.* **2012**, *138*, 174–181. [CrossRef]
34. Pratt, C.J. Use of permeable, reservoir pavement constructions for stormwater treatment and storage for re-use. *Water Sci. Technol.* **1999**, *39*, 145–151. [CrossRef]
35. Antunes, L.N.; Thives, L.P.; Ghisi, E. Potential for potable water savings in buildings by using stormwater harvested from porous pavements. *Water* **2016**, *8*, 110. [CrossRef]
36. Hammes, G.; Thives, L.P.; Ghisi, E. Application of stormwater collected from porous asphalt pavements for non-potable uses in buildings. *J. Environ. Manag.* **2018**, *222*, 338–347. [CrossRef] [PubMed]
37. Thives, L.P.; Ghisi, E.; Brecht, D.G.; Pires, D.M. Filtering capability of porous asphalt pavements. *Water* **2018**, *10*, 206. [CrossRef]
38. Thives, L.P.; Ghisi, E.; Silva, N.M. Potable water savings in multifamily buildings using stormwater runoff from impermeable paved streets. *Eur. J. Sustain. Dev.* **2018**, *7*, 120–130.
39. Guinée, J.B.; Heijungs, R.; Huppes, G. Life cycle assessment: Past, present, and future. *Environ. Sci. Technol.* **2011**, *45*, 90–96. [CrossRef] [PubMed]
40. International Organization for Standardization (ISO). *ISO 14040/2006: Environmental Management—Life-Cycle Assessment—Principles and Framework*; International Organization for Standardization (ISO): Geneva, Switzerland, 2006.
41. AzariJafari, H.; Yahia, A.; Amor, M.B. Life cycle assessment of pavements: Reviewing research challenges and opportunities. *J. Clean. Prod.* **2016**, *112*, 2187–2197. [CrossRef]
42. Praticò, F.G.; Ammendola, R.; Moro, A. Factors affecting the environmental impact of pavement wear. *Transp. Res. Part D-Transp. Environ.* **2010**, *15*, 127–133. [CrossRef]
43. Santero, N.J.; Masanet, E.; Horvath, A. Life-cycle assessment of pavements. Part I: Critical review. *Resour. Conserv. Recycl.* **2010**, *55*, 801–809. [CrossRef]
44. Giustozzi, F.; Crispino, M.; Flintsch, G. Multi-attribute life cycle assessment of preventive maintenance treatments on road pavements for achieving environmental sustainability. *Int. J. LCA* **2012**, *17*, 409–419. [CrossRef]
45. Chiu, C.T.; Hsu, T.H.; Yang, W.F. Life cycle assessment on using recycled materials for rehabilitating asphalt pavements. *Resour. Conserv. Recycl.* **2008**, *52*, 545–556. [CrossRef]
46. Holtz, K.; Eighmy, T.T. Scanning European advances in the use of recycled materials in highway construction. *Public Roads* **2000**, *64*, 34–40.
47. Santos, J.; Ferreira, A.; Flintsch, G. A life cycle assessment model for pavement management: Road pavement construction and management in Portugal. *Int. J. Pavement Eng.* **2014**, *16*, 315–336. [CrossRef]
48. Taylor, G.; Patten, J. *Effects of Pavement Structure on Vehicle Fuel Consumption*; Centre for Surface Transportation Technology (CSTT): New Delhi, India; National Research Council of Canada (NRC): Ottawa, ON, Canada, 2002.
49. Huang, Y.; Bird, R.; Heidrich, O. Development of a life cycle assessment tool for construction and maintenance of asphalt pavements. *J. Clean. Prod.* **2009**, *17*, 283–296. [CrossRef]
50. Santero, N.J.; Horvath, A. Global warming potential of pavements. *Environ. Res. Lett.* **2009**, *4*, 034011. [CrossRef]
51. Chandrappa, A.K.; Biligiri, K.P. Pervious concrete as a sustainable pavement material—Research findings and future prospects: A state-of-the-art review. *Constr. Build. Mater.* **2016**, *111*, 262–274. [CrossRef]
52. Conselho Brasileiro de Construção Sustentável (CBCS). *Sustentabilidade na Indústria de Blocos e Pavimento de Concreto: Avaliação do ciclo de vida Modular (Sustainability in the Concrete Block and Pavement Industry: Modular Life Cycle Assessment)*; Conselho Brasileiro de Construção Sustentável (Brazilian Council for Sustainable Construction) (CBCS): São Paulo, Brazil, 2014.
53. Li, Y.; Huang, Y.; Quanliang, Y.; Zhang, W.; Meng, F.; Zhang, S. Multi-objective optimization integrated with life cycle assessment for rainwater harvesting systems. *J, Hydrol.* **2018**, *558*, 659–666. [CrossRef]
54. Center for Neighborhood Technology (CNT); American Rivers. *The Value of Green Infrastructure, a Guide to Recognizing Its Economic, Environmental and Social Benefits*; CNT: Chicago, IL, USA, 2010.

55. Yuan, X.; Tang, Y.; Li, Y.; Wang, Q.; Zuo, J. Environmental and economic impacts assessment of concrete pavement brick and permeable brick production process—A case study in China. *J. Clean. Prod.* **2018**, *171*, 198–208. [CrossRef]
56. Mei, C.; Liu, J.; Wang, H.; Yang, Z.; Ding, X.; Shao, W. Integrated assessments of green infrastructure for flood mitigation to support robust decision-making for sponge city construction in an urbanized watershed. *Sci. Total Environ.* **2018**, *639*, 1394–1407. [CrossRef] [PubMed]
57. Kluck, E.; Ven, M.V.; Baggen, J.; Wee, B.V.; Hofman, R. Environmental life cycle cost for durable porous surface layers with synthetic binders. *Int. J. Pavement Eng.* **2010**, *3*, 142–148.
58. Terhell, S.L.; Cai, K.; Chiu, D.; Murphy, J. *Cost and Benefit Analysis of Permeable Pavements in Water Sustainability*; University of California Agriculture and Natural Resources: Davis, CA, USA, 2015.
59. Chui, T.F.M.; Liu, X.; Zhan, W. Assessing cost-effectiveness of specific LID practice designs in response to large storm events. *J. Hydrol.* **2016**, *533*, 353–364. [CrossRef]

water

MDPI

Review

River Channel Relocation: Problems and Prospects

Alissa Flatley [1,*], Ian D Rutherfurd [1] and Ross Hardie [2]

[1] School of Geography, University of Melbourne, 221 Bouverie Street, Carlton, VIC 3053, Australia; idruth@unimelb.edu.au

[2] Alluvium Consulting, Level 1, 105–115 Dover Street, Cremorne, VIC 3013, Australia; ross.hardie@alluvium.com.au

* Correspondence: alissa.flatley@unimelb.edu.au; Tel.: +61-408-708-940

Received: 28 August 2018; Accepted: 26 September 2018; Published: 29 September 2018

Abstract: River relocation is the diversion of a river into an entirely new channel for part of their length (often called river diversions). Relocations have been common through history and have been carried out for a wide range of purposes, but most commonly to construct infrastructure and for mining. However, they have not been considered as a specific category of anthropogenic channel change. Relocated channels present a consistent set of physical and ecological challenges, often related to accelerated erosion and deposition. We present a new classification of river relocation, and present a series of case studies that highlight some of the key issues with river relocation construction and performance. Primary changes to the channel dimensions and materials, alongside changes to flow velocity or channel capacity, can lead to a consistent set of problems, and lead to further secondary and tertiary issues, such as heightened erosion or deposition, hanging tributaries, vegetation loss, water quality issues, and associated ecological impacts. Occasionally, relocated channels can suffer engineering failure, such as overtopping or complete channel collapse during floods. Older river relocation channels were constructed to minimise cost and carry large floods, and were straight and trapezoidal. In some countries, modern relocated channels represent an exciting new challenge in that they are now designed to replicate natural rivers, the success of which depends on understanding the characteristics, heterogeneity, and mechanisms at work within the natural channel. We discuss shortcomings in current practice for river relocation and highlight areas for future research for successful rehabilitation of relocated rivers.

Keywords: river relocation; river channel; engineering; geomorphology; rehabilitation

1. Introduction

The majority of the world's rivers are now modified by humans [1–3], and many of those modifications affect the form and character of the river channels themselves. These changes have been well documented through research on: channelisation [4–6], dams [7–10], flow impoundment [11], and levees [12]. However, there is a type of river channel modification that has not been well documented, and this is the diversion of rivers into entirely new channels for part of their length. The lack of research surrounding this type of river channel change is compounded by the lack of formal description or classification of this type of channel modification.

The term 'river diversion' is commonly used to describe various engineered changes to channels and is routinely used to describe diversions of water out of a channel, such as for irrigation or for inter-basin transfers. For example, the Chinese recently completed the South North Water Transfer, the world's largest water diversion [13]. However, this paper is not concerned with this type of water diversion, where a proportion of water is essentially decanted out of a waterway (thus, we do not consider aqueduct systems, the many canal bypass channels that cross much of Europe [14,15], or the irrigation networks that are so common across the world's lowlands). Instead, here we are concerned with the physical relocation of a river channel to a new position. For this reason, we refer to these as 'river relocations'. This channel change is distinct from diversion of the water, or channelization of the river in position. Thus, our interests relate to engineering and geomorphology more than hydrology. In addition, the relocation of a river has been described in many ways, including: watercourse diversion [16], river realignment [17], channelization [18], water diversion [19], river deviation [20], and river flow control works [21], which are frequently used interchangeably. In our definition of river relocation, river flow is redirected into a new, purpose-built channel, and returned either to the original channel downstream, a new channel, such as a neighbouring watercourse, or a river mouth in the downstream position. In this definition of river relocation, the water within the channel is typically neither used in any consumptive sense, nor stored with the intention of being used or treated [22,23].

Fundamentally, a relocated channel replaces a natural section of a river with a short section of artificial (man-made) channel. The artificial channel is usually different from the natural channel in several ways: it is often shorter and steeper, has different bed and bank material, has no floodplain, and cuts across tributaries. These differences then lead to secondary effects including erosion, flooding, and barriers to fish passage. Thus, relocated channels are not just engineering problems, as they affect every aspect of river geomorphology and ecology.

Many river reaches across the world have been relocated (see Figure 1 for a small selection), but there is little research into the impacts of their relocation, their construction, or subsequent performance. To some extent this is because rivers are often relocated in places where they receive little scrutiny, such as for mining. This paper (a) classifies different types of river relocations, (b) presents case studies to illustrate key engineering, construction, and performance issues that arise from river relocation, (c) reviews the key consequences and challenges of relocating a natural stream, and (d) suggests guidelines for their design and subsequent rehabilitation.

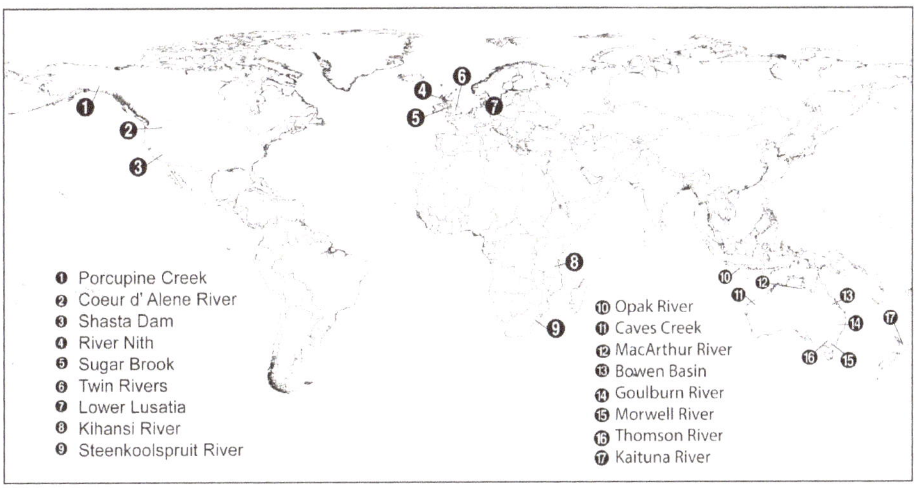

Figure 1. Global map of river relocation case studies considered in this paper.

2. Purposes of River Relocation

Ancient civilisations, such as the Egyptians and Mesopotamians, modified watercourses for consumptive purposes from the Neolithic period [24]. However, the earliest true river relocation that we have found is the ninth century diversion of the Opak River for the construction of the Loro Jonggrang temple within the Prambanan Temple compound in Indonesia [25]. There were almost certainly earlier channel relocations than this. Modern river relocation is carried out for a wide variety of purposes, and following are some examples.

1. Temporary river relocation for construction, such as temporary river relocation channels for dams [26], in which the river channel is temporarily diverted (Figure 2a), and the original channel is dried out to facilitate the construction of the dam or other structure across the river. Rivers can be temporarily diverted to clean up contaminants (e.g., relocation of the Coeur d'Alene River in Idaho for the clean-up of contaminated tailings (Figure 2b)).
2. Permanent relocation channels to make way for infrastructure. Examples are highway construction (such as the diversion of the Wraysbury River for construction of the M25 in the UK [27]) and airport runway expansion [28] or golf courses [29].
3. A particular class of infrastructure relocation is around open-cut mining operations (Figure 2c), in which rivers are relocated to gain access to mineral reserves or materials stored in paleochannels, and to minimise flood risk to adjacent infrastructure [30].
4. Rivers are relocated in association with artisanal mining practices (small-scale open-cast mining) [31,32] either to gain access to valuable deposits within the river bed, or to obtain a supply of water.
5. River relocation is carried out for flood control (e.g., Kaituna River, NZ [33]) to alter the location of the river channel to minimise damage from flooding.
6. Rivers are relocated for land reclamation, such as marsh [34] and wetland restoration [35], by reintroducing freshwater and sediments to enhance vertical accretion to degraded habitats.

Figure 2. Examples of river relocation channels. (**a**) River relocation for the construction of the Shasta Dam on the Sacramento River (1943) (Photo: California State University); (**b**) the temporary relocation of the Coeur d'Alene River, Idaho, to allow the clean-up of contaminated tailings; (**c**) permanent river diversion of the Goulburn River for coal mining in NSW, Australia (Photo: Cathy Toby).

3. River Relocation Classification

Relocated channels can either be temporary or permanent [36], with a varying effort to replicate the original river's natural condition. The new channel can be cut across a floodplain, blasted through bedrock, or in some cases, constructed as an embankment. Broadly, relocated river channels can be lined or unlined (Figure 3, Table 1).

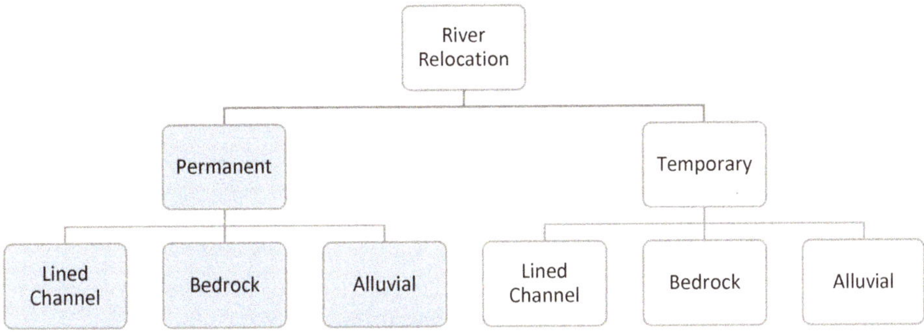

Figure 3. Classification of river relocation.

Table 1. Examples of river relocation channels.

River Relocation	Purpose	Classification	Year Constructed
Porcupine Creek, Alaska, USA	Alluvial mining	Temporary, Lined Channel	1907–1918 [37]
Coeur d' Alene, Idaho, USA	River restoration	Temporary, Lined Channel	2005 to present [38]
Shasta Dam, California	Dam construction	Temporary, Bedrock	1938–1940 [39]
River Nith, Scotland	Coal mining	Permanent, Alluvial	2000–2004 [40]
Sugar Brook, England	Airport construction	Permanent Alluvial	1998–1999 [41]
Twin Rivers, England	Airport construction	Permanent, Lined	2002–2004 [42]
Wraysbury River, England	Motorway construction	Permanent, Lined	1986 [43]
Lower Lusatia, Germany	Coal mining	Permanent, Lined	1975–1976 [44]
Kihansi River, Tanzania	Dam construction	Permanent, Lined	1999 [45]
Steenkoolspruit River, South Africa	Coal mining	Permanent, Lined	1992 [20]
Opak River, Indonesia	Temple construction	Permanent, Lined	Ninth century [25]
Caves Creek, WA, Australia	Iron ore (open-pit) mining	Permanent, Alluvial	2014 [46]
MacArthur River, Australia	Lead and zinc mining	Permanent, Alluvial	2008 [47]
Bowen Basin, Australia	Coal mining	Permanent, Alluvial/Bedrock	1970–2012 [48]
Goulburn River, Australia	Coal mining	Permanent, Bedrock	1981 [49]
Morwell River, Australia	Coal mining	Permanent, Lined	Multiple modifications made from 1977–2012 [50]
Thomson River, Australia	Alluvial mining	Permanent, Bedrock	1911–1912 [51]
Kaituna River, New Zealand	Flood prevention	Permanent, Alluvial	Modifications made in 1926, 1956, and 1995 [52]

Unlined channels use the underlying natural materials to create the new channel (such as relocated channels located within a floodplain or through bedrock). These materials can vary among bedrock, alluvial sediments, or a combination of these materials, occasionally reinforced by hard engineering in places along the channel. Lined channels are constructed using artificial materials, such as timber, synthetic geotextiles, covered pipes (concrete inverted siphons), or hard engineering, such as concrete or rip-rap along the channel. In some cases, a new channel is engineered on an embankment that sits higher than the surrounding landscape (for example, through a mining pit) and can also be accompanied by a series of drop structures to maintain the energy and velocity of the river flow within the relocated channel.

This paper concentrates on the issues surrounding permanent river relocation channels (highlighted in blue in Figure 3), as this type of relocation usually presents the most management challenges. Note that artificially cutting off river meanders is a form of relocation, but these short relocations are only considered here where they are cut through bedrock. Note also that the definition of full flow river relocation can be complicated, depending on how much of the flood flows are diverted by the constructed channel. Some just divert up to the bankfull flow and allow the flood flows to continue to pass down the old channel/floodplain section.

We now describe examples of lined, bedrock and alluvial river relocations (with examples of each) from the above classification, and the problems associated with each type. From these examples, and other sources, we summarise the key management challenges associated with these diversion channels. We begin with fully artificial relocated channels (i.e., lined channels) and move to river relocation channels cut in natural rock or sediments.

4. Case Studies of River Relocation

4.1. Lined Channel Relocation

Lined channel relocation has been carried out from as early as the 1800s as a consequence of mining. Early watercourse modification was typically for the consumptive use of water through race construction and sluicing [53]. However, many river channels were also relocated to gain access to alluvial materials underneath the channel, such as alluvial gold deposits. The majority of early river relocation efforts were local, small-scale, and predominantly unrecorded [54]; they were comparable to modern artisanal and small-scale open-cast mining river relocations. Larger river relocation channels were constructed using large timber flumes (Figure 4), whereas smaller river relocations were dug as ditches into the surrounding landscape. Historic river relocation flumes lacked geomorphic characteristics of the natural channel; they were fully artificial and were prone to failure during large floods.

(**a**) (**b**)

Figure 4. River relocation flume (circa 1907–1918) in the Porcupine Mining District, Alaska. (**a**) The flume was 2.4 km long, up to 2.4 m deep, and 7–12 m wide; (**b**) early flumes were prone to failure during floods (Source: Sheldon Museum).

4.2. Case Study: Morwell River Relocation, Victoria, AUS

A particularly challenging type of relocation is where the diverted channel is carried in an elevated flume or channel. A good example of this type of relocation is the Morwell River relocation (MRR) in eastern Victoria, Australia, constructed to access coal reserves at the Yallourn coalmine, Australia's largest open-cut coal mine [55]. The MRR is a 3.5 km channel carried in an elevated embankment that relocated the river through the middle of the open-cut mine pit to connect with the Latrobe River downstream [56] (Figure 5). The embankment was constructed using engineering fill from 13 million cubic meters of overburden that was stripped from the mine itself [56]. The Morwell river has been previously relocated for coal mining at the Hazlewood coal mine, and its present course is the result of multiple relocation attempts.

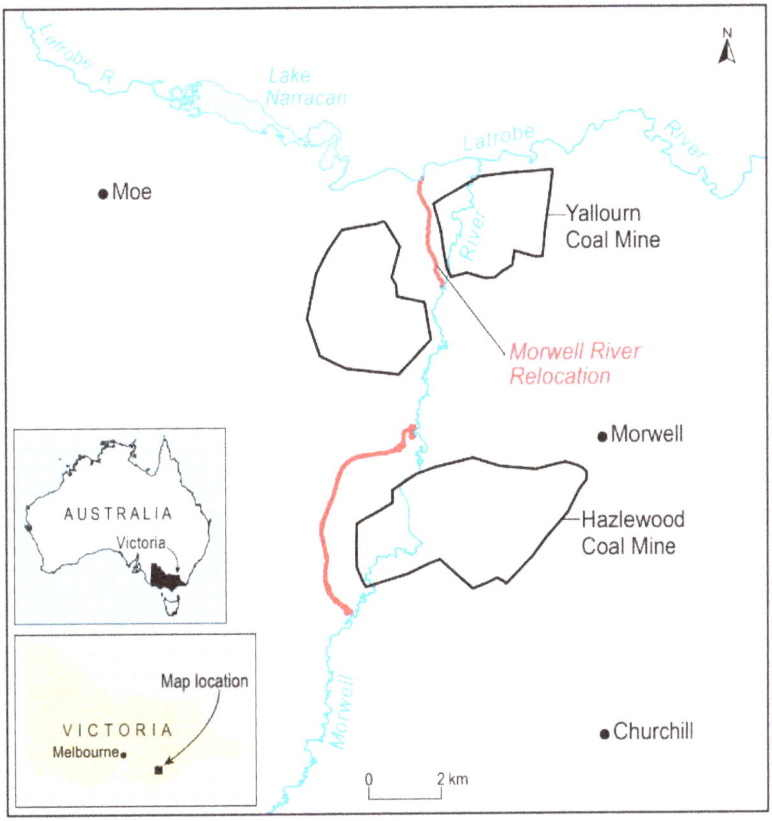

Figure 5. Map of the Morwell River Relocation. The Yallourn coal mine is located to the north of Hazlewood coal mine, where additional river relocation has previously taken place.

The design of the river relocation channel included an artificially meandering channel and a floodplain, with a width that varied between 40 m and 60 m, compared with the 1000 m width of the original natural floodplain [57]. The embankment collapsed during a large flood on 6 June 2012, diverting the Morwell river into the adjacent Yallourn open-cut mine pit (Figure 6). The downstream Latrobe River reversed direction and flowed up the diversion and into the mine pit. The flooding stopped coal production in the state's major coal mine and threatened power supplies. The mine had considerable trouble disposing of the millions of litres of polluted water that entered the mine [58,59]. The flooding reduced production from the state's major coal mine to 25% capacity for 4 weeks [60], and the total cost to repair the MRR was between 109 and 150 million AUD [60,61]. Despite its meandering morphology, the diverted channel developed no natural channel characteristics before its eventual failure.

Figure 6. The 2012 Morwell River collapse. Note the meandering relocated channel and associated embankment collapse (Source: Environment Victoria).

4.3. Bedrock Relocation Tunnels

The simplest river relocation channels are found when a new channel has been blasted through bedrock, commonly as a tunnel through horseshoe bends within sections of river. Thirteen such channel relocation tunnels were constructed for historical gold mining purposes in Victoria, Australia (Figure 7) [62]. The purpose was to dry out the meander bend to allow easy access for alluvial mining. These relocation channels were typically short and utilised the natural features of the watercourse to minimise the cost or distance of the relocated channel. The introduction of dynamite in 1867 [63] allowed for more substantial channels to be constructed. These tunnels disrupt sediment supply through the reach (sediment tends to build-up upstream of the relocation). They also act as barriers to fish passage. In this case, the nationally threatened Australian grayling cannot traverse the high flow velocities in the steep bedrock channel [64].

(a) (b)

Figure 7. (**a**) Thomson River relocation, also known as Horseshoe Bend [62] (Source: Victorian Heritage Database); (**b**) diagram of residual reach and relocation tunnel (Source: West Gippsland CMA, 2010).

One of the most common types of river relocation are open channels cut through bedrock. This type of relocation is most common in open-cut mining operations. This type of mining usually takes place higher in the catchment, where floodplains are narrower, and any channel has to be cut into the bedrock valley walls. The purpose is, firstly, to divert tributaries around the mine to avoid flooding, and secondly, to divert the river away from areas that can be mined. The Goulburn River diversion in New South Wales (Australia) is an example of a bedrock river relocation constructed in 1981 to relocate 4 km of the Goulburn river around a coal mine (Figure 2c). The relocated channel is cut 10–20 m deep into a deeply weathered saprolite [65]. The central reaches of the channel relocation were constructed to have a box-like canyon form with benches constructed on the channel banks [66].

The relocated bedrock channel is a simple rectangular channel with vertical walls. Compared to the natural reaches upstream, the relocated channel is steeper and hydraulically smooth, with high stream power. As a result, the channel experiences high erosion rates with dispersive subsoils exposed throughout the channel [67]. The new channel also has simple morphology, with a flat floor, and an unnaturally dense covering of reeds [66]. Also, the bedrock channel cuts across tributaries, producing 'hanging' tributaries at each junction. These hanging tributaries become waterfalls during storms and can form gullies. They also completely disrupt up-and-downstream migration of fish, and any form of riverine connectivity between the river and the tributaries. Current rehabilitation strategies are being implemented to improve the stability and design of this relocated channel [68].

Bedrock Diversions for Coal Mining in the Bowen Basin, Queensland, Australia

Since the 1970s, over 60 full bedrock river relocation channels have been constructed in the Bowen Basin, Queensland (Figure 8), a major coal mining region [69]. Dynamic meandering channels were replaced with relocated channels that were straight and of trapezoidal form to reduce construction costs and maximise the discharge capacity of the river channel [69]. Designs from the 1980s onwards incorporated drop structures to compensate for reduced channel lengths and the accompanying increase in bed slope [69,70].

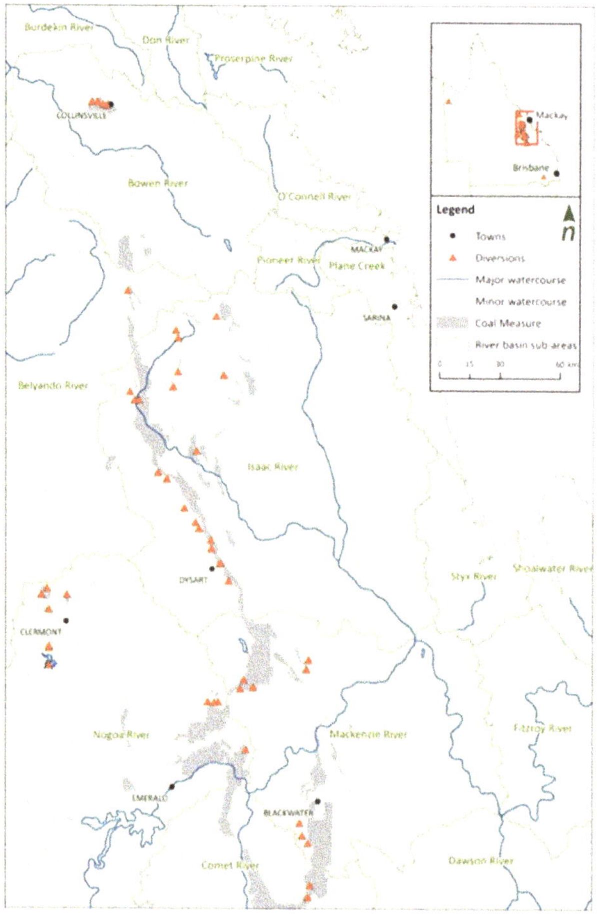

Figure 8. The Bowen Basin (Queensland) with locations of river relocation channels [69].

The Australian Coal Industry Research Program (ACARP) reviewed the performance of river relocations [69,70]. Some relocated channels experienced high erosion rates due to inadequate design widths, increased bed slopes from shortened channel lengths, and increased velocities exacerbated by an absence of vegetation, but conversely, a smaller number experienced high sedimentation [69]. Some were at risk of eroding into adjacent open-cut pits or associated mining infrastructure. The poor performance of the relocated channels (Figure 9) led to a temporary moratorium on the approval of river relocation construction by the Queensland government [71] which lasted for 5 years.

Overall, five key factors were identified that consistently limited the performance of the Bowen Basin relocated channels. These factors were sediment supply and transport, vegetation condition, the occurrence of major flood events in the early years of diversion establishment, overland flow drainage, and the transition between the relocated channel and the natural watercourse [72]. Improved design standards have dramatically improved the performance of Bowen Basin relocations, and these are discussed in Section 6.2 below.

(c)

Figure 9. Changes in the design of relocated channels over time in the Bowen Basin, Queensland: (**a**) 1970s river relocation with a trapezoidal design and exposed banks; (**b**) 1990s river relocation with limited vegetation establishment, increased channel dimensions, and rill erosion on batter banks; (**c**) 2000s river relocation with a more natural bank batter and more natural morphology in the bed [69,70].

4.4. *Alluvial River Relocation Channels*

Alluvial river relocations are carried out using natural channel materials, such as in situ alluvial sediments, and they cut across a floodplain rather than into bedrock. In some instances, they can be sculpted to maintain similar channel dimensions and bed grade to the natural channel. Other alluvial river relocation channels also incorporate the floodplain into the channel design.

4.4.1. Case Study: Twin Rivers Relocation, Heathrow Terminal 5, UK

Rivers are frequently relocated for construction purposes, either as a temporary measure while building bridges or dams, or permanent diversions for development. There are many examples of airport river relocations, including the River Mole diversion for Gatwick Airport, UK [73], Sugar Brook relocation for Manchester Airport, UK [28], the Twin Rivers relocation for the expansion of Heathrow Terminal 5 [74], and the planned relocation of the Ulwe and Gadhi rivers for the construction of Navi Mumbai airport, India [75].

Many airports are located in areas with limited available land for expansion, resulting in increased pressure to utilise river corridors for continued airport growth, with the economic return of expansion outweighing the cost of river relocation. There are specific management issues for the relocation of rivers for airport construction. As part of the construction process, valleys are backfilled to bring ground levels up to required elevations for runway construction [23]. In addition, contaminants, such as jet fuel and de-icer, can flow into the river system, presenting a significant source of pollution. Birds present an additional challenge, as open bodies of water such as rivers attract avian communities but represent a hazard to aircraft safety.

The expansion of Terminal 5 at Heathrow airport in the UK required two rivers to be relocated. The Duke of Northumberland's River and the Longford River (known collectively as the Twin Rivers) flowed through the middle of the Terminal 5 project site. Both rivers have a long history, and were originally man-made, constructed to supply royal estates located on the banks of the River Thames [74,76].

To facilitate expansion, the Twin Rivers were relocated around the western perimeter of the airport. The relocated channels were designed to ensure they had the capacity to convey peak flows of 3 m^3/s and 1.5 m^3/s for the Duke of Northumberland and the Longford, respectively [74]. River relocation developments need to comply with local and national policies, where present. In this instance, the Twin Rivers relocation needed to comply with the EU Floods Directive (Directive 2007/60/EC) and the Flood and Water Management Act (2010), which relay the overarching message that all development must consider and mitigate flood risk, ensuring that this risk is not increased because of river relocation construction [77].

The new channel design saw an increase of open channel, with 95% of the relocation channel occurring in an open channel, compared to 50% in the previous diversion design [78]. This 'daylighting' of the channel aimed to enhance the environment, with the inclusion of habitat features designed to provide a minimum of environmental equivalence compared to pre-diversion standards [74]. Habitat features included modifications for fish passage, the addition of in-channel wood, 8000 m^2 of pre-planted vegetation, and the provision of alternating berms with rock-filled gabions and logs [74]. A bird exclusion net was also added throughout the entire length of the Twin Rivers channel (Figure 10). A 0.3 mm diameter, lightweight polypropylene netting with a mesh size of 75 mm was selected, as it excludes all hazardous birds but allows exit and entry for a range of invertebrates, such as the emperor dragonfly [74].

(a)

Figure 10. *Cont.*

(b)

Figure 10. Twin Rivers relocation at Heathrow. (**a**) Aerial view of the Twin Rivers Diversion, UK; (**b**) bird exclusion netting across channel (Source: HAHL Airports Limited).

4.4.2. Case Study: Kaituna River Relocation, New Zealand

The Kaituna River is an example of a laterally active river located on the Bay of Plenty, New Zealand. This case study illustrates the cumulative impacts that can arise from river relocation. The Kaituna River is a 50 km long [79] modified river that has been relocated on several occasions. The original course of the river passed through the Papahikahawai Channel into the Ōngātoro/Maketū Estuary. In 1907, during a flood, the Kaituna River broke from its original course due to an avulsion through a sandspit at Te Tumu [80], stopping flow to an adjacent coastal estuary.

Between 1926 and 1928 [81], two parallel chutes were cut to relocate water back into the estuary, becoming known as Ford's Cut (Figure 11a). Ford's Cut enabled the river to return to the estuary, however, during the same period, natural channel migration caused the river's flow to migrate eastward, returning to the estuary via the Papahikahawai Channel [81].

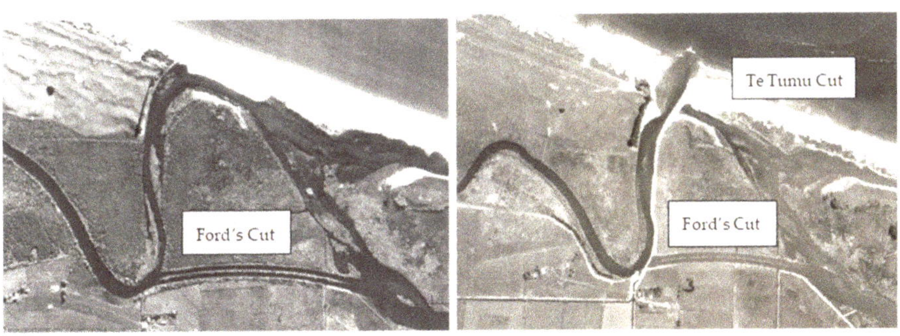

(a) (b)

Figure 11. (**a**) Ford Twin Cuts in 1948. This photo was taken before the Te Tumu cut was opened to the sea in 1957; (**b**) Ford Twin Cuts in 1957, this photo is 2 years after the Te Tumu cut was opened in 1959 (Source: The Ford Collection).

Additional serious flooding occurred again in 1949 and 1951 [80], which resulted in a management response to construct a new mouth for the Kaituna River. The new mouth of the river was commissioned in 1956 and was named the Te Tumu Cut (Figure 11b). The objective of this river relocation was to reduce the frequency and severity of flooding on the Te Puke lowlands, former wetlands that are now surrounding agricultural land [79,82]. The Te Tumu relocation channel could reinforce the natural 'second mouth' that occurs during flood events [82].

The previously engineered Ford's Cut and the Papahikahawai Channel were blocked with a causeway at their upstream ends to maintain full river flow throughout the Te Tumu cut and to potentially reclaim the Maketū Estuary [81]. However, a secondary response to the blocking of Ford's Cut and the Papahikahawai Channel was the reduction of flow through the old river estuary. The flow reduction caused tertiary issues, such as an increase in salinity which destroyed wetland and reduced the estuary's ability to flush out sand and mud [82]. The Te Tumu river relocation contributed to sediment infilling and general ecological decline of the estuary [79] and has been described as a venture carried out for the benefit of the farming community at the expense of another community: estuary users [80].

Since the Te Tumu relocation, there have been attempts to restore flow to the estuary and growing support for another channel relocation to combat increased sedimentation and the closing of the estuary mouth [83]. In 1995, the construction of four culverts was undertaken at Ford's Cut [81] to resupply water back to the Maketū estuary via flapgates following years of concern about the closure of all previous river paths. Domijan [84] estimated that this flow restoration resulted in an additional volume of 100,000 m^3 of water entering the estuary per tidal cycle. The addition of water into the estuary was hoped to reduce sediment infilling and restore some of the declining habitat and restore fish stocks or "kaimoana" [85]. The addition of water did assist in reducing the salinity in the upper estuary but there has been no measurable reduction in sedimentation rates [28], with continued poor overall hydrodynamic and ecological improvement [83]. There are plans to construct an additional relocation channel through to the Maketū Estuary to create new wetlands and maximise both community and ecological benefits [85].

5. Implications and Challenges of River Relocation

Lined, bedrock, and alluvial channel relocations have been introduced, with case studies providing examples of each of these types. Each case study highlights some of the challenges surrounding the design, construction, and performance of river relocation channels. In the past, relocated channels were considered successful if they passed all flood flows, did not erode excessively, and did not degrade the river reaches up and downstream. Since the 1990s, higher standards have been demanded of the relocated channels, and they are now expected to maintain biological and aesthetic values in both the channel and adjacent reaches up-and-downstream. In general, relocation channel construction can cause a series of primary changes (defined here as physical changes around the diverted channel), which can then lead to the generation of secondary issues (defined here as physical and biological connectivity issues caused by the primary changes), and then tertiary issues (defined as linked, but perhaps surprising, consequences on biology and human communities caused by secondary issues) (Table 2). The issues arising from channel relocation can be broadly characterised as either fundamental engineering problems, or issues relating to the ability of the relocated channel to behave in a comparable way to a natural channel. As such, the performance of river relocation channels can be considered through the lens of successful engineering, but also in relation to the natural characteristics of the channel that they replaced.

Table 2. Key issues reported with river relocation.

Key Issue	Type	Example
	Primary	
Change in channel dimensions based on new channel design.	Physical	Morwell River, AU [57]
Changes in flow velocity.	Physical	Bowen Basin, QLD [69,86]

Table 2. *Cont.*

Key Issue	Type	Example
Reduced capacity to carry flows.	Physical	Bowen Basin, QLD [70]
Increased erosion: both bed and bank erosion (prompting headcut migration up the upstream channel, and sedimentation in the downstream channel).	Physical	Bowen Basin, QLD [69]
Unstable banks; rill erosion, piping on banks.	Physical	Bowen Basin, QLD [69]
Diversion of accumulated flow into a new tributary.	Physical	Rainy River Mine Diversion, Canada [87]
Relocation channel collapse.	Physical	Morwell River, AU [59]
	Secondary	
Reduced sediment supply to the new channel and downstream reach.	Physical	Bowen Basin, QLD [46]
Increased deposition, sedimentation in the relocated channel.	Physical	Te Tumu river diversion, Kaituna, NZ [76,79]
Increased backwater effect upstream of the artificial channel requiring armoured grade control.	Physical	Caves Creek Relocation, WA [46]
Erosion in hanging tributary junctions.	Physical	Goulburn River Relocation, NSW [65]
Lowering of water tables.	Physical/Chemical	Mining river relocation (Lower Lusatian Mining Area) [88]
Loss of vegetation in channel and on banks.	Biological/Physical	Bowen Basin, QLD [69,70].
Lining of channel as a barrier to hyporheic exchange.	Physical/Biological/Chemical	River Nith, Scotland—Blocking of river flow and permeable ground [40]
	Tertiary	
Disruption to biological connections (including fish passage).	Biological	Increased velocity in the diversion, inclusion of culverts, weirs, and hanging tributaries [36]
Water Quality Changes/Contamination.	Chemical/Biological	Chemical pollution from runway detergent and de-icer; River Mole, Gatwick Airport, UK [89]
Noise and dust pollution.	Physical/Chemical/Biological	During the construction of relocation channel [36]
Loss of biodiversity (flora and fauna).	Biological	Decline in avifauna assemblages in the Kihansi river relocation [90] and bird habitat loss—Twin Rivers relocation [74]
Disruption to river continuity and navigation.	Physical	Increase of artificial engineering structures [36]
Infrastructure damage due to a leaking lined channel.	Physical	Steenkoolspruit River relocation, Witbank Coalfield, South Africa [20]

5.1. Fundamental Engineering Performance

Some of the fundamental engineering issues surrounding river relocation involve changes to the dimensions or gradient of the channel, materials, built-in engineered structures such as culverts, and the lining and length of the channel. From an engineering perspective, when a channel is relocated, the fundamental concerns are the ability of the channel to convey flood flow, and the overall structural stability of the channel. Failure of the relocated channel from an engineering perspective consequently means the structural collapse of channel elements, such as culvert failure, embankment breaching, or the overtopping of the structure during flood flow events.

River Relocation Channel Dimensions

Channel conveyance, alongside the sizing of hard engineering materials and culverts, is ultimately determined by the discharge of the river. Most modern relocation channels continue to be trapezoidal in design, developed from size and stability criteria derived from European or North American rivers. These designs focus on the relocation channel being robust and capable of conveying a certain flow efficiently.

Relocated channels are often designed to convey the 100-year average recurrence interval (ARI) flood without overtopping [36]. For river relocation channels designed for mining, a more conservative estimate of rainfall and discharge is typically used to avoid water entering the mining pit. River relocation channels constructed in and around mine sites are designed to withstand a flood with a 100-year average return interval, or even an event once every 1000 or 10,000 years [91]. Conservative design flood standards can lead to artificial channels that are constructed with enlarged flood protection bunds, and channel dimensions that exceed the size of the original channel. Engineering failure within river relocation channels often occurs when the artificial channel is poorly sized, or with materials that do not withstand large floods.

All river relocation channels present an artificial discontinuity between natural sections of a river. This artificial channel seldom has the identical physical characteristics of the adjoining upstream and downstream reaches [64]. River relocation channels tend to be straighter and shorter than the original channel, with a higher bed slope and different channel dimensions (width and depth). River relocation is often expensive, particularly when cutting through bedrock or reinforcing the channel with artificial structures. Because of this, engineers will often attempt to minimise the length and cross-sectional size of the relocated channel, resulting in a new channel that is often substantially shorter and smaller than the original.

Even if the channel dimensions and boundary materials are the same (which might be the case with an alluvial river relocation), the channel will usually be straighter, steeper, or feature a reduced floodplain width [57], prompting heightened erosion within the channel. These issues can be further intensified through the feedback loops of secondary and tertiary problems [76,79]; in other words, a change in channel dimensions can cause increased erosion and unstable banks. These unstable banks can fail, prompting vegetation loss, lower channel roughness, and further channel erosion.

Increased erosion within the channel can lead to amplified incision of adjoining tributaries alongside erosive tributary junctions (where the artificial channel re-joins the natural channel). This can cause sustained secondary issues, such as knickpoint migration from hanging tributaries [65], and increased sediment supply to the main channel. These changes produce tertiary issues, such as disruption of fish passage [36]; loss of habitat [74], species diversity, or assemblages [90]; and reduced water quality [85]. Secondary and tertiary issues can impact adjoining reaches, propagating the impacts of channel relocation both upstream and downstream. In the past, diversion channels were expected to remain as simple engineered channels that carried major floods. Vegetation would typically be removed from the channels to maintain conveyance. More recently, channels have been designed to gradually develop more natural morphology and vegetation, and to have more natural rates of erosion. We now turn to this issue of designing more natural channels.

5.2. Replicating Natural Channels

Government agencies and regulators now demand higher standards of river relocations. This is evident in several of the case studies presented above. Not only must the diversion not harm the river environment up-and-downstream, but the river relocation itself must eventually behave and function like a natural river channel. There is typically a conflict between establishing these natural values in diversions, and the functionality or engineering stability of the river relocation channel [91]. For example, smooth uniform channel beds do not encourage species diversity within the channel [92] yet provide the most efficient flow conveyance.

6. Improving River Relocation Designs

The poor performance of river relocation channels has prompted greater awareness of new channel designs to fulfil both engineering and channel replication requirements. Here, we present examples of recent best management practice approaches to relocation design.

6.1. Case Study: Sugar Brook Relocation, Manchester Airport, UK

The Sugar Brook Valley relocation is an example of an alluvial river relocation that considered the geomorphology and characteristics of a natural river. The Sugar Brook relocation is located next to Manchester Airport, UK. The construction of Manchester airport required the relocation of the Bollin River to facilitate the widening of the first runway [93]. The Bollin River relocation was 780 m long and passed underneath a 25 m embankment. The Sugar Brook relocation is one of two smaller rivers that were relocated to construct a second runway at the airport.

The river relocation was designed to ensure that the majority of the channel is open with a comparable gradient to the original watercourse. A consistent and similar channel gradient is favoured to avoid increased erosion and heightened flow rates within the channel. An appropriate gradient is also essential for maintaining sediment continuity within the channel and maintaining the appropriate stream energy.

The original design of the relocated channel was problematic. Initially, the Sugar Brook relocation required a significant excavation depth to construct the required channel bed level with the resulting excavation (Figure 12a) producing a narrow deep canyon. This design was considered to be geomorphically unreliable due to clay soils and likely undercutting of the toe of slopes, which could accelerate the collapse of high banks [28].

(a)

(b)

Figure 12. *Cont.*

(c)

Figure 12. Sugar Brook relocation at Manchester Airport. (**a**) Initial river diversion design; (**b**) newly constructed valley (in September 1999); (**c**) river relocation 2 years after construction (November 2001) [28].

To improve the stability and long-term recovery of the relocated river channel, a new design was used which considered the larger surrounding landscape in which the river is situated. A new river valley and floodplain was sculpted into an acceptable form (Figure 12b), and then a small meandering river channel was constructed within the new valley floor [28]. The Sugar Brook relocation acts as a larger valley-wide river diversion which looked more natural and stable, facilitating overall positive rehabilitation of the channel (Figure 12c).

6.2. *Improved Design Using Geomorphic Criteria: Example of the Bowen Basin Mining Relocations*

Mining river relocation channels have received increased scrutiny owing to high-profile cases of failure and poor performance. Contemporary mining is now heavily regulated, but despite rigorous engineering practices, the performance of river relocation channels is a concern to mine regulators. In general, there is a risk of failure during mine operations, and secondly there is the long-term stability and subsequent rehabilitation of the river channel to consider. Mining river relocation channels face increasing scrutiny to fulfil long-term environmental objectives. In particular, mining river relocation presents a noteworthy case study, as there is an emerging conflict between establishing natural values within relocated channels, and the functionality or engineering stability of relocation. The risks associated with mining and river relocation have prompted a series of case studies examining the improved design of river relocation channels.

ACARP Geomorphic Criteria

Mining impacts on both the quality and quantity of water are a highly contentious aspect of most mining projects [22,94]. The performance of river relocation channels was studied by a series of ACARP initiatives within the Bowen Basin, Queensland (Figure 8). The result of these investigations was the establishment of specific hydraulic and geomorphic design criteria for these regional watercourses.

Hardie and Lucas [86] assessed 35 natural reaches of streams that had not been altered within the region and identified significant relationships between the hydraulic parameters in three variable stream types (incised, limited capacity, and bedrock controlled). These distinct stream parameters could then be used in the design of new relocated channels, and the rehabilitation of existing channels that were poorly performing or degraded [86]. These hydraulic parameters act as guidelines and establish the ideal range of conditions within each stream type within the region (Table 3).

Table 3. Characteristic values for stream sample reaches [86].

Stream Type	Stream Power (W/m²)		Velocity (m³/s)		Shear Stress (N/m²)	
	2-Year ARI	50-Year ARI	2-Year ARI	50-Year ARI	2-Year ARI	50-Year ARI
Incised	20–60	50–150	1.0–1.5	1.5–2.5	<40	<100
Limited capacity	<60	<100	0.5–1.1	0.9–1.5	<40	<50
Bedrock Controlled	50–100	100–350	1.3–1.8	2.0–3.0	<55	<120

The hydraulic parameters in the guideline were refined in an additional study that evaluated the performance of 60 relocated channels, where 17 had been constructed following the guidelines. The 17 artificial channels constructed using the guidelines were found to be in better overall condition than the rest of the relocated channels [48]. An outcome of these ACARP projects was the production of a series of updated stream parameter guidelines (Table 4) that provide a design approach for relocated alluvial and bedrock channels. Additional elements were also considered, including the level of sediment supply to the relocation channel, and channel and planform variability [70,72]. This integrated design increased the likelihood of successful vegetation establishment.

Table 4. Revised criteria for river relocation designs [70,72].

Stream Type	Sediment Transport Status	Stream Power (W/m²)	
		2-Year ARI	50-Year ARI
Alluvial	Supply Limited (Low sediment supply)	15–35	50–100
	Transport limited (High sediment supply)	35–60	80–150
Bedrock controlled channels	n/a	50–100	100–350

In 2014, the Government of Queensland consolidated the earlier principles of design for river relocated channels, based on the ACARP recommendations, to produce a series of design objectives (Table 5). These objectives indicate that relocated channels should be self-sustaining, include geomorphic and vegetation features similar to the regional watercourses, positively contribute to river health values, and impose no long-term liability on the state, the proponent, or the community [16].

Table 5. Government of Queensland Key Principles of design for river relocation channels [16].

1. Permanent watercourse diversion incorporates natural features (including geomorphic and vegetation) present in landscapes and in local watercourses
2. The permanent watercourse diversion maintains the existing hydrologic characteristics of surface water and groundwater systems
3. The hydraulic characteristics of the permanent watercourse diversion are comparable with other local watercourses and suitable for the region in which the watercourse diversion is located.
4. The permanent watercourse diversion maintains sediment transport and water quality regimes that allow the watercourse diversion to be self-sustaining, while minimizing any impacts on upstream and downstream reaches
5. The permanent watercourse diversion and associated structures maintain equilibrium and functionality and are appropriate for all substrate conditions they encounter.

7. Long-Term River Relocation Rehabilitation

The previous sections introduced the conflict of establishing natural values within river relocation channels whilst also ensuring engineering functionality and stability. River relocation channels can be criticised for their lack of long-term stability and lack of ecological and environmental attributes in comparison to the original channel. More recently, the importance of identifying river behaviour [95] and geomorphic processes [96] has been highlighted as a necessity for long-term stability of constructed channels.

The overall long-term objective of river relocation channels varies depending on the river's location, and previous modifications. There are many river relocation channels (such as the Twin Rivers at Heathrow, UK) that have had a long history of modification and are channelised or are constructed on restricted floodplains, so that it is challenging for them to possess all the attributes of a natural river system. Many of the constructed channels have substantially altered boundary conditions, and it may not be appropriate or feasible to rehabilitate the river to its pre-disturbance condition. Instead, it might be more relevant to strive to maximise the beneficial features of the river in its new setting if irreversible of systematic change has occurred [97]. This section focuses on river relocation for mining and the objective of long-term river rehabilitation.

Many environmental impact assessments (or equivalent thereof) now require evidence of long-term river relocation objectives. Rivers relocated for mining purposes are subject to long-term rehabilitation objectives, including a rehabilitation plan for the site to include channel stability and positive environmental outcomes. In Australia, mining river relocation licenses can only be relinquished (that is, returned to the responsibility of the government) once they have proven that the relocation has met the outcome-based conditions stipulated in the mining license [98]. However, difficulty arises, as river relocation channels have both a temporary and permanent role throughout the mine life-cycle. They represent a key element of engineered infrastructure to ensure both the functionality of the mine during its operation and its subsequent rehabilitation after mining has ceased. Stable river relocation designs are important throughout all stages of the mine-cycle and as such pose enormous challenges for water resource managers [82], not least of which is the danger that, at some time over centuries or millennia, the relocated channel could permanently divert into the mining pit.

Post-mining, most river relocation channels are left in their new position, a few are redirected back to their original course [99,100], or in some instances, the river channel is engineered into a pit lake as riverine through-flow to maintain or improve pit lake water quality [101,102]. Rehabilitation programmes are typically designed to ensure safety and minimise potential negative impacts of the closed mine [103]. Robust and stable engineered designs are crucial for flood conveyance during mine operation, with the ecological and geomorphic components of the river course developing more importance for the implementation of rehabilitation programmes.

Consideration of relocated river channel rehabilitation often begins in the design phase. Permanent river relocation channels present a new challenge in that they are designed for the long term, with channels now constructed with an attempt to replicate the natural channel they replace. This is typically carried out using a design criteria approach (e.g., ACARP geomorphic and hydraulic criteria) where available or a reference reach approach. The design criteria approach will use specific hydraulic and hydrologic targets to create a design standard designed to create the required hydraulic conditions within the channel to enable vegetation recovery and the establishment of geomorphic forms. The reference reach approach will use natural channels to establish closure criteria [104]. Blanchette et al. [104] suggest that reference sites should lie within a river's normal variability and are both sustained and tracked over time. Both approaches advance the historic form of river relocation channels, which have tended to be trapezoidal and lacking geomorphic complexity.

8. Future Research

Many shortcomings in the current practice for river relocations have been highlighted in high-profile failures, such as the Morwell River collapse in Victoria, Australia, or in poor attainment of rehabilitation objectives. White et al. [70,72] highlight the need to revise most current river relocation designs to reduce subsequent impacts to adjacent waterways.

The regional characteristics of natural rivers should be considered during the design of new relocated channels [102]. This is particularly true in the rivers demonstrating behaviour that do not fit the planform or criteria found in European and North American rivers. Greater distinction between perennial and ephemeral watercourses is needed to fully understand the mechanisms that control major channel adjustment, such as flash flooding [81,91]. This is particularly relevant where mining operations are located in arid areas with unusual geomorphology and hydrology. The recovery of vegetation in diverted channels should also be a specific area of research.

Globally, the majority of literature surrounding river relocation is derived from grey literature, or environmental impact assessments, with minimal long-term assessment or evaluation of these projects. Our ability to construct an artificial natural channel is a measure of our understanding of natural channels, which can be limited by poor understanding of various river planforms, such as the anabranching channel, or relocated channels constructed in settings that do not fit European or North American perennial rivers. As such, river relocation channels can be considered as large-scale geomorphic experiments. Within Australia, the ACARP guidelines provide criteria for river relocation designs with explicit consideration of stream type and geomorphology, proving hydraulic reference values for future relocation channel designs. However, these values are best suited to Queensland, with other states and territories lacking equivalent criteria.

Concern about environmental values of river relocations are still emerging. Relocation channels were previously constructed to transfer water from one area to another, with limited concern for the river's natural values. Now, river relocation channels are planned with a consideration of regional planforms and characteristics of the natural channel, including the high interannual flow variability of Australian rivers [105].

9. Conclusions

This paper introduces the characteristics and challenges of a poorly described class of human impact on streams: river relocation channels. The term 'river diversion' has typically been ambiguous, often used for several types of engineering approaches. We suggest that 'river relocation' more accurately describes the permanent or temporary relocation of a river channel into a new course. The new course can be lined or unlined, and cut into bedrock or alluvium. A river relocation channel that does not correctly mimic natural channel characteristics can have a profound impact on the overall performance and success of the river relocation.

Traditionally, relocated channels were designed to carry large floods, but at a minimum construction cost. This means that river relocation channels were typically constructed as short, narrow, and steep as possible. The common result is excessive erosion or sedimentation in the new channel, and hanging tributaries. This has secondary consequences, including headward erosion into the upstream reach, disruption to the sediment flow regime into the downstream reach, loss of vegetation, poor water quality, loss of biodiversity, and in some cases, river channel collapse.

Rivers will continue to be relocated for infrastructure projects, flood protection, and mining operations. From an engineering perspective, it is now increasingly important to be able to design and build a permanent river relocation channel that, for the least cost, eventually has the morphology, vegetation, and dynamics of up and downstream reaches of stream. Ideally, relocated river channels should eventually be indistinguishable from the natural counterparts up and downstream. This will only be possible where managers have good understanding of the geomorphology of the river system, and the mechanisms that control major channel adjustment, such as flooding, vegetation, and sediment supply. Overall, the presence of natural features and geomorphic stability will facilitate the long-term recovery of the river relocation. Improved understanding of these natural features will allow for the identification of a natural state and projected behaviour over time. Recent analyses have identified thresholds of stream power for certain river types that have led to improved design. Poor-performing relocated channels can be a major long-term liability to companies, and once relinquished, to governments. Finally, we need to remember that river relocations will be there for millennia and need to be designed accordingly.

Author Contributions: A.F. conceived and conducted the literature review; I.D.R. supervised the project; A.F. and I.D.R. and R.H. wrote the paper; I.D.R. and R.H. provided ongoing reviews to paper drafts.

Acknowledgments: The authors would like to thank BHP-Billiton and the George Lansell Memorial Scholarship (the University of Melbourne) for financial assistance for this project. The authors wish to acknowledge the cartographic services of Chandra Jayasuriya for the production of Figure 5.

Conflicts of Interest: The authors declare no conflict of interest. The founding sponsors had no role in the design of the study; in the collection, analyses, or interpretation of data; in the writing of the manuscript, and in the decision to publish the results.

References

1. Rosenberg, D.M.; McCully, P.; Pringle, C.M. Global-Scale Environmental Effects of Hydrological Alterations: Introduction. *Bioscience* **2000**, *50*, 746–751. [CrossRef]
2. Tharme, R.E. A Global Perspective on Environmental Flow Assessment: Emerging trends in the development and application of environmental flow methodologies for rivers. *River Res. Appl.* **2003**, *19*, 397–441. [CrossRef]
3. Halim, Y. The impact of human alterations of the hydrological cycle on ocean margins. *Ocean Margin Process. Glob. Chang.* **1991**, *9*, 301–327.
4. Keller, E.A. Channelization: Environmental, geomorphic and engineering aspects. In *Geomorphology and Engineering*; Coates, D.R., Ed.; Dowden, Hutchinson and Ross Inc.: London, UK, 1976; pp. 115–140.
5. Brookes, A. Downstream morphological consequences of river channelization in England and Wales. *Geogr. J.* **1985**, *151*, 57–62. [CrossRef]
6. Brookes, A. Response of Aquatic Vegetation to Sedimentation Downstream from River Channelisation Works in England and Wales. *Biol. Conserv.* **1986**, *38*, 351–367. [CrossRef]
7. Brandt, S.A. Classification of geomorphological effects downstream of dams. *Cantena* **2000**, *40*, 375–401. [CrossRef]
8. Kingsford, R.T. Ecological impacts of dams, water diversions and river management on floodplain wetlands in Australia. *Aust. Ecol.* **2000**, *25*, 109–127. [CrossRef]
9. Fantin-Cruz, I.; Pedrollo, O.; Girard, P.; Zeilhofer, P.; Hamilton, S.K. Changes in river water quality caused by a diversion hydropower dam bordering the Pantanal floodplain. *Hydrobiologia* **2016**, *768*, 223–238. [CrossRef]
10. Poff, N.L.; Schmidt, J.C. How dams can go with the flow. *Science* **2016**, *353*, 1099–1100. [CrossRef] [PubMed]
11. Petts, G.E. Time-scales for ecological change in regulated rivers. In *Regulated Streams*; Craig, J.F., Kemper, J.B., Eds.; Plenum: London, UK, 1987; pp. 257–266. ISBN 0306426749.
12. Darby, S.; Sear, D. *River Restoration: Managing the Uncertainty in Restoring Physical Habitat*; John Wiley and Sons, Ltd.: Chichester, UK, 2008; pp. 1–315.
13. Webber, M.; Crow-Miller, B.; Rogers, S. The South-North Water Transfer Project: Remaking the geography of China. *Reg. Stud.* **2017**, *51*, 370–382. [CrossRef]
14. Hooke, J.M. Human impacts on fluvial systems in the Mediterranean region. *Geomorphology* **2006**, *79*, 311–335. [CrossRef]
15. Chatelliers, M.C.D.; Reygrobellet, J.L. Interactions between geomorphological processes, benthic and hyporheic communities: First results on a by-passed canal of the french upper Rhône river. *Regul. Rivers Res. Manag.* **1990**, *5*, 139–158. [CrossRef]
16. Department of Natural Resources and Mines. Guideline: Works that Interfere with Water in a Watercourse—Watercourse Diversions. September, 2014. State of Queensland. pp. 1–43. Available online: http://www.dnrm.qld.gov.au/__data/assets/pdf_file/0015/212424/guideline-watercourse-diversions.pdf (accessed on 28 September 2018).
17. Erskine, W.D. Channel response to large-scale ricer training works: Hunter River, Australia. *Regul. Rivers Res. Manag.* **1992**, *7*, 261–278. [CrossRef]
18. Soar, P.J.; Thorne, C.R. *Channel Restoration Design for Meandering Rivers*; U.S Army Corps of Engineers: Vicksburg, MS, USA, 2011; pp. 1–454.
19. Yevjevich, V. Water diversions and Interbasin transfers. *Water Int.* **2001**, *26*, 342–348. [CrossRef]
20. McEwan, A.L. The Failure of and Remedials to a River diversion for an opencast mine in the Witbank Coalfields of South Africa. Mine Water and Environment. In Proceedings of the 1999 IMWA Congress, Sevilla, Spain, 13–17 September 1999; pp. 79–85.

21. Water Act. Available online: http://www.austlii.edu.au/cgibin/download.cgi/au/legis/cth/consol_act/wa200783 (accessed on 26 September 2018).
22. Kemp, D.; Bond, C.J.; Franks, D.M.; Cote, C. Mining, water and human rights: Making the connection. *J. Clean. Prod.* **2010**, *18*, 1553–1562. [CrossRef]
23. Minerals Council of Australia. Water Accounting Framework for the Minerals Industry, User Guide, Version 1.2. Available online: http://www.minerals.org.au/focus/sustainable_development/water_accounting (accessed on 26 September 2018).
24. Mays, L.W. A very brief history of hydraulic technology during antiquity. *Environ. Fluid Mech.* **2008**, *8*, 471–484. [CrossRef]
25. Cronin, L. *A Journey through Ancient Kingdoms and Natural Wonders; World Heritage Sites of Australia, New Zealand and South East Asia*; Reed Books: Sydney, Australia, 1995; p. 170. ISBN 9780730104889.
26. Ratnayaka, D.D.; Brandt, M.J.; Johnson, K.M. *Twort's Water Supply*, 6th ed.; Elsevier Ltd.: Oxford, UK, 2009; p. 744. ISBN 978-0-7506-6843-9.
27. Petts, J. *Handbook of Environmental Impact Assessment: Volume 2: Impact and Limitations*; John Wiley and Sons, Ltd.: Chichester, UK, 2009; pp. 1–464. ISBN 9781444311495.
28. Seddon, C. Diversion of a River Valley. Manual of River Restoration Techniques. 2013, pp. 1–5. Available online: http://www.therrc.co.uk/MOT/Final_Versions_%28Secure%29/11.1_Sugar_Brook.pdf (accessed on 26 September 2018).
29. Holmes, P.J.; Meadows, M. *Southern African Geomorphology: Recent Trends and New Directions*; Sun Media Bloemfontein: Bloemfontein, South Africa, 2012; p. 431. ISBN 978-1-920382-02-5.
30. Atkinson, S.; Markham, A.; Rafty, M.; Heyting, M. Pilbara Creek Diversions: Resilience gained through increased ore recovery and an integrated approach to design. In Proceedings of the AUSIMM Iron Ore Conference, Perth, Australia, 24–26 July 2017; pp. 1–20.
31. Hinton, J.J.; Veiga, M.M.; Veiga, A.T.C. Clean artisanal gold mining: A utopian approach. *J. Clean. Prod.* **2003**, *11*, 99–115. [CrossRef]
32. Byizigiro, R.V.; Raab, T.; Maurer, T. Small-scale open cast mining: An important research field for anthropogenic geomorphology. *Die Erde—J. Geogr. Soc. Berl.* **2015**, *146*, 213–231. [CrossRef]
33. Mawer, J.C. Morphology of the Te Tumu Cut Under the Potential Re-Diversion of the Kaituna River. Ph.D. Thesis, University of Waikato, Hamilton, New Zealand, 2012; pp. 1–158.
34. Ward, G.H.; Irlbeck, M.J.; Montagna, P.A. Experimental River Diversion for Marsh Enhancement. *Estuaries* **2002**, *25*, 1416–1425. [CrossRef]
35. Wang, H.; Steyer, G.D.; Couvillion, B.R.; Beck, H.J.; Rybczyk, J.M.; Rivera-Monroy, V.H.; Krauss, K.W.; Visser, J.M. Predicting landscape effects of Mississippi River diversions on soil organic carbon sequestration. *Ecosphere* **2017**, *8*, 1–15. [CrossRef]
36. Fisher, K.; Ramsbottom, D. *River Diversions: A Design Guide*; Thomas Telford Publishing: London, UK, 2001; p. 176. ISBN 9780727729590.
37. Eakin, H.M. *The Porcupine Gold Placer District, Alaska*; U.S Geological Survey Bulletin: Washington, DC, USA, 1919; pp. 1–29.
38. National Research Council of the National Academies. The Coeur d' Alene System. In *Superfund and Mining Megasites: Lessons from the Coeur d' Alene River Basin*; National Academies Press: Washington, DC, USA, 2005; pp. 47–484.
39. Rocca, A. *Shasta Lake: Boomtowns and the Building of the Shasta Dam*; Arcadia Publishing: Dover, NH, USA, 2002; p. 64. ISBN 0-7385-2076-4.
40. Gilvear, D.; Perfect, C. Clay Lined River. Manual of River Restoration Techniques. 2013, pp. 1–7. Available online: http://www.therrc.co.uk/sites/default/files/projects/p449.pdf (accessed on 27 September 2018).
41. The River Restoration Center. River Restoration News. 2000. Available online: http://www.therrc.co.uk/site/default/files/newsletters/issue5.pdf (accessed on 28 September 2018).
42. Risbridger, T. Terminal 5 Twin Rivers Diversion- Critical Sub-Project in Airport's T5 expansion Programme. *Water Treat Supply*. 2007. Available online: http://www.waterprojectsonline.com/case_studies/2007/EAHeathrowT5RiverDiversion2007.pdf (accessed on 28 September 2018).
43. Pennington, M.; Spence, J. Piled foundations for the M25 motorway near Heathrow Airport. In *Advances in Transportation Geotechnics*; Ed, E., Yu, H.S., Glenn, M.D., Andrew, D., Nick, T., Eds.; CRC Press: Nottingham, UK, 2008; pp. 453–457. ISBN 978-0-415-47590-7.

44. Schultze, M.; Pokrandt, K.H.; Hille, W. Pit lakes of the Central German lignite mining district: Creation, morphometry and water quality aspects. *Limnol.—Ecol. Manag. Inland Waters* **2010**, *40*, 148–155. [CrossRef]
45. Vandvik, V.; Måren, I.E.; Ndangalasi, H.J.; Taplin, J.; Mbago, F.; Lovett, J.C. Back to Africa: Monitoring post-hydropower restoration to facilitate reintroduction of an extinct-in-the-wild-amphibian. *Ecosphere* **2014**, *5*, 1–17. [CrossRef]
46. Harvey, M.; Pearcey, M.; Price, K.; Devkota, B. Geomorphic, hydraulic and sediment transport modelling for mine related channel realignment—Case Study: Caves Creek, Pilbara, Western Australia. In *Hydrology and Water Resources Symposium 2014*; Engineers Australia: Queensland, Australia, 2014; pp. 259–266.
47. McArthur River Mining. Memorandum. September 2007, Edition 6. 2007, pp. 1–8. Available online: http://www.mcarthurrivermine.com.au/en/publications/MemorandumMagazine/MemoranduM_issue6(1).pdf (accessed on 27 September 2018).
48. Hallgath, A.; Pearson, B.A. Watercourse diversions relinquishment in Queensland—A risk-based approach to monitoring. In *Mine Closure 2016*; Fourie, A.B., Tibbett, M., Eds.; Australian Center for Geomechanics: Perth, Australia, 2016; pp. 447–460.
49. Ulan Coal Glencore. Goulburn River Diversion Erosion and Sediment Control Plan. ULN SD PLN 0130. 2017, pp. 1–38. Available online: http://www.ulancoal.com.au/en/environment/EnvironmentManagementPlan/Goulburn%20River%20Diversion%20Erosion%20Sediment%20Control%20Plan.pdf (accessed on 28 September 2018).
50. Spark, H.; Temple, L. Morwell River Diversion. ACAA.06. Ashton Coal. Appendix 1. Bowmans Creek Diversion Response to Submissions. 2010, pp. 1–7. Available online: http://www.ashtoncoal.com.au/icms_docs/254841_response-to-submissions-part-1.pdf (accessed on 28 September 2018).
51. Victorian Heritage Database. Thomson River Diversion Tunnel Site. 2018; pp. 1–8. Available online: http://vhd.heritagecouncil.vic.gov.au/places/10756 (accessed on 27 September 2018).
52. Monchy, P. Kaituna River Re-diversion and wetland creation project. In *Assessment of Environmental Effects-Summary*; Bay of Plenty Regional Council: Whakatane, New Zealand, 2014; pp. 1–11.
53. Lawrence, S.; Davies, P. Learning about landscape: Archaeology of water management in colonial Victoria. *Aust. Archeol.* **2012**, *74*, 47–54. [CrossRef]
54. Cosgrove, D. An elemental division: Water control and engineered landscape. In *Water, Engineering and Landscape*; Cosgrove, D., Petts, G., Eds.; Belhaven Press: London, UK, 1990; pp. 1–11.
55. Mason, L.; Unger, C.; Lederwasch, A.; Razian, H.; Wynne, L.; Giurco, D. *Adapting to Climate Risks and Extreme Weather: A Guide for Mining and Mineral Industry Professionals*; National Climate Change Adaptation Research Facility: Gold Coast, Australia, 2013; p. 76. ISBN 978-1-925039-24-5.
56. Laurence, D.; Stamford, C.; Barry, S.; Oosting, N.; Andryszczk, H.; Fawsett, M.; Jones, D. *A Guide to Leading Practice Sustainable Development in Mining*; Australian Government Department of Resources and Tourism: Canberra, Australia, 2011; pp. 1–204. ISBN 978-1-92-1812-49-1.
57. Hardie, R.; Nat, M.; Hall, D.; Tilleard, J. Design of a river for a modified environment. In *Hydro 2000: Interactive Hydrology*; Institute of Engineers: Barton, ACT, Australia, 2000; pp. 1024–1029. ISBN 0957824114.
58. Treadwell, S.; Dettrick, D. *Morwell River Diversion Collapse Environmental Risk Assessment*; Sinclair Knight Merz: Sydney, Australia, 2012; pp. 1–105.
59. EPA Victoria. Yallourn Mine—Community Information. Publication 1494; August 2012. Available online: http://www.epa.vic.gov.au/~/media/Publications/1494.pdf (accessed on 27 September 2018).
60. Song, M. $109m bill for Yallourn. In *Latrobe Valley Express*; Latrobe Valley Express Partnership: Morwell, Australia, 2012.
61. CLP Holdings. Media Release: CLP Holdings Limited announces 2013 Annual Results. 27 February 2014, pp. 1–2. Available online: https://www.clpgroup.com/en/Media-Resources-site/Current%20Releases/20140227_Eng.pdf (accessed on 27 September 2018).
62. Victorian Heritage Register. Thomson River Diversion Tunnel Site. 2017. Available online: http://vhd.heritagecouncil.vic.gov.au/places/10756 (accessed on 27 September 2018).
63. Hartman, H.L.; Mutmansky, J.M. *Introductory Mining Engineering*, 2nd ed.; John Wiley and Sons: Hoboken, NJ, USA, 2002; pp. 1–570. ISBN 0-471-34851-1.
64. Welsh, L.; Travers, I. Thomson River Fishway. In *Heritage Impact Statement*; Extent Heritage Pty. Ltd.: Sydney, Australia, 2017; pp. 1–24.

65. Ulan Coal Glencore. Goulburn River Diversion Remediation Plan. ULN SD PLN 0054. 2016, pp. 1–120. Available online: http://www.ulancoal.com.au/en/environment/EnvironmentManagementPlan/Goulburn%20River%20Diversion%20Remediation%20Plan.pdf (accessed on 28 September 2018).
66. URS. *Goulburn River Diversion—Long Term Stability Strategy*; Ulan Coal Mines Limited.: Sydney, Australia, 2009.
67. Eco Logical Australia Pty Limited. *Goulburn River Diversion Rehabilitation Works—Vegetation Assessment and Revegetation Recommendations*; Ulan Coal Mines Limited.: Sydney, Australia, 2014.
68. Ulan Coal Glencore. Plan for Goulburn River Diversion Erosion and Sediment Control Plan. *ULN SD PLN 0130*; Environment and Community Manager, 2017; pp. 1–38. Available online: http://www.ulancoal.com.au/en/environment/EnvironmentManagementPlan/Erosion-and-Sediment-Control-Plan.pdf (accessed on 28 September 2018).
69. White, K.; Hardie, R. Maintenance of Geomorphic Processes in Bowen Basin River Diversions. ACARP Project C8030. 2000. Available online: https://trove.nla.gov.au/version/43316977 (accessed on 28 September 2018).
70. White, K.; Lucas, R.; Hardie, R.; Moar, D.; Blackham, D. Criteria for Functioning River Landscape Units in the Mining and Post Mining Landscapes. Report prepared for ACARP Project C20017. 2014. Available online: https://www.acarp.com.au/abstracts.aspx?repId=C20017 (accessed on 28 September 2018).
71. Department of Natural Resources and Mines. *Guideline: Works that Interfere with Water in a Watercourse: Watercourse Diversions*; State of Queensland: Queensland, Australia, 2014; pp. 1–43.
72. White, K.; Hardie, R.; Lucas, R.; Merritt, J.; Kirsch, B. The evolution of watercourse diversion design in central Queensland coal mines. In Proceedings of the 7th Australian Stream Management Conference, At Townsville, Australia, 27–30 July 2014; Vietz, G., Rutherfurd, I.D., Hughes, R., Eds.; University of Melbourne: Victoria, Australia, 2014; pp. 338–348, ISBN 978 0 7340 5038 0.
73. CEEQUAL. The Twin Rivers Diversion Scheme. 1 January 2005. Available online: http://www.ceequal.com/case-studies/the-twin-rivers-diversion-scheme/ (accessed on 27 September 2018).
74. Palmer, D.; Wilbraham, P. Heathrow Terminal 5: Twin rivers diversion. *Civ. Eng.* **2008**, *161*, 25–29. [CrossRef]
75. Purohit, A.A.; Vaidya, M.M.; Kudale, M.D. Application of unstructured mesh modelling in evolving SGE of an airport at the confluence of multiple rivers in a macro tidal region. *World Acad. Sci. Eng. Technol. Int. J. Environ. Chem. Ecol. Geol. Geophys. Eng.* **2017**, *11*, 241–255.
76. Clark, P.B.; Palmer, D.K.; Joyce, P.; Pether, R.V. Environmental aspects of the hydraulic design for the twin rivers diversion for Terminal 5, Heathrow Airport. In *Environmental Hydraulics and Sustainable Water Management*; Lee, J.H.W., Lam, K.M., Eds.; Taylor & Francis: Abingdon, UK, 2004; pp. 847–852. ISBN 0415365465.
77. AMEC. *Heathrow's North-West Runway, Assessment of Flood Risk. AMEC Environment and Infrastructure*; Heathrow Airport Limited.: Heathrow, UK, 2014; pp. 1–52.
78. Lister, B. Heathrow Terminal 5: Enhancing environmental sustainability. *Civ. Eng.* **2008**, *161*, 21–24. [CrossRef]
79. Goodhue, N.D. Hydrodynamic and Water Quality Modelling of the Lower Kaituna River and Maketu Estuary. Ph.D. Thesis, University of Waikato, Hamilton, New Zealand, 2007; pp. 1–165.
80. Tortell, P. Public Involvement in the Search for a Solution to the Problems of the Maketu Estuary. In Proceedings of the 1985 Australasian Conference on Coastal and Ocean Engineering, Institute of Civil Engineers, Adelaide, Australia, 4–8 December 1985; pp. 256–269.
81. Wallace, P. Fords Cut. History and Bank Erosion Assessment. Report prepared for Environment Bay of Plenty. 2008; pp. 1–23. Available online: https://www.boprc.govt.nz/media/32659/pwallace-090803-fordscuthistory.pdf (accessed on 27 September 2018).
82. Boothway, D. Room for the River—A New Zealand Perspective. In *Hydrology and Water Resources Symposium 2014*; Engineers Australia: Queensland, Australia, 2014; pp. 670–677. ISBN 9781922107190.
83. Tuckey, B. *Kaituna River to Maketu Estuary Re-Diversion: Model Calibration and Initial Hydrodynamic Impact Assessment*; DHI Water and Environment Limited.: Auckland, New Zealand, 2009; pp. 1–63.
84. Domijan, N. The hydrodynamic and estuarine physics of Maketu Estuary. Ph.D. Thesis, University of Waikato, Hamilton, New Zealand, 2000.
85. Everitt, S.; Monchy, P. *Kaituna River Re-Diversion and Wetland Creation Project: Scoping Report for Assessment of Environmental Effects*; Bay of Plenty Regional Council Environmental Publication: Whakatāne, New Zealand, 2013; pp. 1–29.

86. Hardie, R.; Lucas, R. Bowen Basin River Diversions Design and Rehabilitation Criteria. Report prepared for ACARP Project C9068, Fisher Stewart, 2002. Available online: https://www.acarp.com.au/abstracts.aspx?repId=C8030_C9068 (accessed on 28 September 2018).
87. Canadian Environmental Assessment Agency. *Rainy River Project Environmental Assessment Report*; Canadian Environmental Assessment Agency: Ottawa, ON, Canada, 2015; pp. 1–231. ISBN 978-1-100-25429-6.
88. Schoenheinz, D.; Grünewald, U.; Koch, H. Aspects of Integrated Water Resources Management in River Basins Influenced by Mining Activities in Lower Lusatia. Global Change: Challenges for regional Water Resources. *Die Erde—J. Geogr. Soc. Berl.* **2011**, *142*, 163–186.
89. CH2M Hill. A second Runway for Gatwick Appendix A12. In *Water Flood Risk*; 2014; pp. 1–234. Available online: https://assets.publishing.service.gov.uk/government/uploads/system/uploads/attachment_data/file/368993/20_Appendix_A12_Water_Flood_Risk.pdf (accessed on 28 September 2018).
90. Cordeiro, N.J.; Lovett, J.C.; Mulungu, E.; Maina, G.G.; Gerstle, J.H. Initial trends of bird assemblages before and after river diversion in an endemic-rich African forest. *Biodivers. Conserv.* **2006**, *15*, 971–983. [CrossRef]
91. Australian Government. *A Guide to Leading Practice Sustainable Development Program for the Mining Industry*; Department of Resources Energy and Tourism: Canberra, Australian, 2008; pp. 1–112. ISBN 6642726051.
92. Palmer, M.A.; Menninger, H.L.; Bernhardt, E. River restoration, habitat heterogeneity and biodiversity: A failure of theory or practice. *Freshw. Biol.* **2009**, *55*, 1–18. [CrossRef]
93. Naylor, A.E. Improvement of the runway at Manchester Airport. *Proc. Inst. Civ. Eng.* **1984**, *76*, 221–236. [CrossRef]
94. Bebbington, A.; Williams, M. Water and Mining conflicts in Peru. *Mt. Res. Dev.* **2008**, *28*, 190–195. [CrossRef]
95. Erskine, W.D. *Review of the McArthur River Mine Open Cut Project Public Environmental Report*; The University of Newcastle: Newcastle, Australia, 2009; pp. 1–19. Available online: https://ntepa.nt.gov.au/__data/assets/pdf_file/0003/288084/rpt54_app3.pdf (accessed on 26 September 2018).
96. Lucas, R.; Crerar, J.; Hardie, R.; Merritt, J.; Kirsch, B. Isaac river cumulative impact assessment of mining developments. *Min. Technol.* **2009**, *118*, 142–151. [CrossRef]
97. Brierley, G.; Fryirs, K.; Outhet, D.; Massey, C. Application of the River Styles framework as a basis for river management in New South Wales, Australia. *Appl. Geogr.* **2002**, *22*, 91–122. [CrossRef]
98. Hallgath, A.; Pearson, B. Watercourse diversions relinquishment in Queensland—A risk based approach to monitoring. In Proceedings of the 11th International Conference on Mine Closure; Australian Centre for Geomechanics: Perth, Australian, 2016; pp. 447–460.
99. Nawrot, J.; O'Leary, W.G.; Malone, P. Stream restoration: Past practices—Current benchmarks. In Proceedings of the America Society of Mining and Reclamation, Pittsburgh, PA, USA, 4–10 June 2010; pp. 652–671.
100. US Department of the Interior. Proposed mining and reclamation plan. In *Environmental Impact Statement OSM-EIS-9*; North Rochelle Mine: Campbell County, WY, USA, 1982; pp. 1–154.
101. McCullough, C.; Schultz, M. Riverine Flow-Through of Mine Pit Lakes: Improving both Mine Pit Lake and River Water Quality Values? In Proceedings of the 10th International Conference on Acid Rock Drainage and IMWA Annual Conference, Santiago, Chile, 21–24 April 2015; pp. 1–10.
102. Grace, D.; McLeary, M.; Davies, G. Drainage principles and design for mine closure in the Pilbara. In *Hydrology and Water Resources Symposium 2014*; Engineers Australia: Queensland, Australia, 2014; pp. 94–100.
103. Heikkinen, P.; Noras, P.; Salminen, R. Environmental Techniques for the Extractive Industries. In *Mine Closure Handbook*; Outokumpu: Helsinki, Finland, 2008; ISBN 978.952-217-055-2.
104. Blanchette, M.L.; Lund, M.A.; Stoney, R.; Short, D.; Harkin, C. Bio-physical closure criteria without reference sites: Realistic targets in modified rivers. In *IMWA 2016—Mining Meets Water—Conflicts and Solutions*; Wolkersdorfer, C., Sartz, L., Sillanpää, M., Häkkinen, A., Eds.; Technische Universität Bergakademie: Freeiberg, Germany, 2016; pp. 586–592.
105. McMahon, T.A.; Finlayson, B.L. Droughts and anti-droughts: The low flow hydrology of Australian rivers. *Freshw. Biol.* **2003**, *48*, 1147–1160. [CrossRef]

water

MDPI

Review

Urban Hydroinformatics: Past, Present and Future

C. Makropoulos [1,2,3] and D. A. Savić [1,4,*]

1 KWR, Water Research Institute, Groningenhaven 7, 3433 PE Nieuwegein, The Netherlands; Christos.Makropoulos@kwrwater.nl
2 Department of Water Resources and Environmental Engineering, School of Civil Engineering, National Technical University of Athens, Iroon Politechniou 5, 157 80 Zografou, Athens, Greece
3 Department of Civil and Environmental Engineering, Norwegian University of Science and Technology, 7491 Trondheim, Norway
4 Centre for Water Systems, University of Exeter, Exeter EX44QF, UK
* Correspondence: dragan.savic@kwrwater.nl

Received: 20 May 2019; Accepted: 10 September 2019; Published: 20 September 2019

Abstract: Hydroinformatics, as an interdisciplinary domain that blurs boundaries between water science, data science and computer science, is constantly evolving and reinventing itself. At the heart of this evolution, lies a continuous process of critical (self) appraisal of the discipline's past, present and potential for further evolution, that creates a positive feedback loop between legacy, reality and aspirations. The power of this process is attested by the successful story of hydroinformatics thus far, which has arguably been able to mobilize wide ranging research and development and get the water sector more in tune with the digital revolution of the past 30 years. In this context, this paper attempts to trace the evolution of the discipline, from its computational hydraulics origins to its present focus on the complete socio-technical system, by providing at the same time, a functional framework to improve the understanding and highlight the links between different strands of the state-of-art hydroinformatic research and innovation. Building on this state-of-art landscape, the paper then attempts to provide an overview of key developments that are coming up, on the discipline's horizon, focusing on developments relevant to urban water management, while at the same time, highlighting important legal, ethical and technical challenges that need to be addressed to ensure that the brightest aspects of this potential future are realized. Despite obvious limitations imposed by a single paper's ability to report on such a diverse and dynamic field, it is hoped that this work contributes to a better understanding of both the current state of hydroinformatics and to a shared vision on the most exciting prospects for the future evolution of the discipline and the water sector it serves.

Keywords: hydroinformatics; smart cities; smart utilities; resilience; distributed systems; data; analytics; decision support; sociotechnical system; ethics; digital water

1. Introduction

1.1. Hydroinformatics—An Evolving Story

The water cycle is a system characterized by inherent complexity, variation, and uncertainty due to interlinked social, natural and engineered subsystems. Hydroinformatics, as a scientific study of this complex system takes a deliberately interdisciplinary, sociotechnical approach [1], blurring the boundaries between water science, data science and computer science. Despite having its origins in computational hydraulics [2], it, however, does not only concern itself with modelling and decision support, as is often incorrectly assumed. The modern field of hydroinformatics also embraces the social dimension of water cycle management, e.g., social needs, concerns and consequences (including equity, data privacy, ethics, legal issues, etc.). Therefore, hydroinformatics should be viewed as having a horizontal role in integrating water sciences (i.e., hydrological, hydraulic and environmental), data sciences (statistics, stochastics, data driven analytics), computer science and information and communication technologies (ICT) and society [3]. This also positions hydroinformatics as a cross-cutting field of study that underpins the transition of water authorities and utilities from reactive to proactive by leveraging technological advances to achieve to the so-called Water 4.0 state (also named Digital Water or Water Informatics) delivering sustainable and resilient water management.

As a dynamic field of research, hydroinformatics has evolved from the days of hydraulic/hydrologic modelling to an academic discipline with a thriving community of scientists, engineers and practitioners (organized around two professional organizations—the International Association for Hydro-Environment Engineering and Research, IAHR, and the International Water Association, IWA), with its own Journal [4], specialist groups and biannual international conferences. However, the discipline's network is not restricted to these institutions. It has grown around the world building strong communities and high-profile scientific journals, such as the International Environmental Modelling and Software Society (iEMSs) and their Journal [5] as well as the Consortium of Universities for the Advancement of Hydrologic Science, Inc. (CUAHSI) in the US and their Hydroinformatics Conferences. The discipline and its community run and contribute to educating new generations of hydroinformaticians through a number of professional and university degree courses offered all around the world.

Although it is beyond the scope of this paper to delve into the depths of hydroinformatics philosophy and approaches, the discipline can be thought of as a continuous process of developing and using water data, models and tools, to understand the environment, to engage all stakeholders, and help make decisions that improve society. This is a highly iterative process (Figure 1), because, as also stated in Vojinović and Abbott [6], "*hydroinformatics integrates knowledges from the social and technical domains to create so-called conjunctive knowledges, that are concerned with an understanding of how technical interventions have social consequences and how the resulting social changes in turn generate new technical developments*". This evolving nature of hydroinformatics can also be viewed through the lens of changing communities attending the biannual Hydroinformatics conferences and consequently the transformation in the research focus over a period of 25 years. While the early years attracted mostly practitioners from the mature fields of computational hydraulics and hydrology and those involved in early applications of artificial intelligence methods, the later years' conferences can be viewed as a meeting place of a community of communities, encompassing various multi-disciplinary areas. This widening of disciplinary communities resulted in changes to the scope of the work presented at conferences, for example, from purely technical approaches to managing demand for water to socio-technical approaches where customer engagement is sought through, not only technical means, but also by combining behavioral and data science. Further examples of the changes include the proliferation of real-time modelling and decision methods due to increasing computing power and the availability of data through citizen science and ubiquities sensing. Together, with the drive to open science outputs to a wider audience (via open-source tools and data), to hybridize modelling systems (via integration of physical and data-driven models), and to better visualize data, processes and decisions (via serious gaming, virtual/augmented reality), the community is well-positioned to help humanity address a range of high-impact future real-world water challenges.

1.2. Aim of This Paper

Hydroinformatics has considerable advances to show across the entire water cycle, however it would be beyond the scope of this paper to include a review of all contributions in the field, thus the focus is limited to urban water issues and perspectives. This is because as urbanization continues to accelerate concentrating ever increasing demands for water services in cities and megacities around the world [7], and as urban water infrastructure is ageing and related investments are lagging behind [8], it is argued that the urban environment urgently needs smarter solutions based on hydroinformatics more than any other domain.

The current state of the art in (urban) hydroinformatics is mapped, proposing a narrative that connects several elements and strands of work together into a coherent whole. This narrative necessarily leaves aspects of hydroinformatics out, and where applicable, references to additional review work is added to assist the reader. Specifically, the paper highlights three main pieces of the hydroinformatics puzzle: Data, analytics and decision support (the last one in both its formal planning/design and societal/communication/engagement sense) in an effort to suggest a way of thinking about the domain and to point towards a promising future.

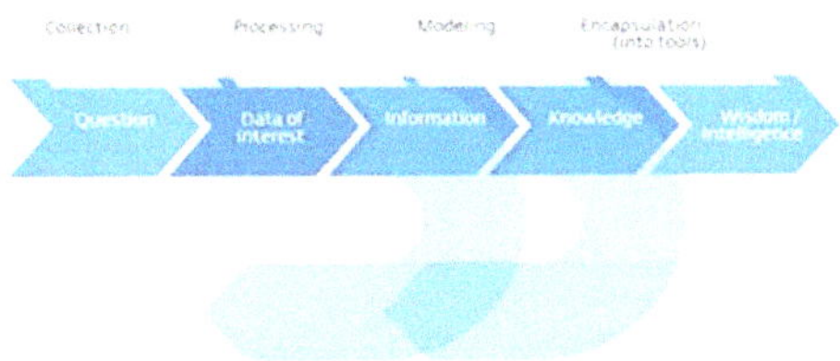

Figure 1. The circular nature of hydroinformatics.

2. From Theory to Practice

Water systems and services are highly complex [9] as they are tasked to balance water resources with demands through complex interconnected infrastructure. As such, decision making about these systems and services (at strategic, tactical and operational scales) need to be taken within a continuously changing landscape where water quality and quantity are uncertain [10]. These systems are also influenced by climatic changes and human practices water demand patterns are shifting as urbanization continues [11], influencing demands [12] as standards of living rise [13]. Lastly,, environmental legislation and customer expectations are also shifting and with them [14,15], the thresholds against which the water sector's performance is measured also change. This dynamic decision landscape is further complicated by aging infrastructure [16] and the advent of new (disruptive) technologies and concepts.

Figure 2 presents an overview of some of the main technologies and concepts that have emerged in the past few years and are influencing both research and practice in the urban water management field and hydroinformatics specifically. In this necessarily brief and elliptical sketch, new real-time information coming from smart sensors, including smart meters, also in the context of IoT developments, stored and managed through (often cloud-based) information platforms [17,18], allow for the remote monitoring and control of new more distributed interventions in the urban water cycle integrated into (and extending the useful life of) existing centralised systems and networks. This is possible due to, also, new analytics that are developed to exploit and extract value from this new information in view of design, tactical and operational decisions (from locating new technologies, to rehabilitating piped networks to understanding and managing water demands [19]). Part of the value in this improved understanding of subsystem functions is in being able to develop and calibrate whole cycle (socio-technical) system models. They are now increasingly being applied to improve the understanding of the interplays between centralised and decentralised systems as well as the interaction between infrastructure and the end users. These new, more inclusive modelling approaches underpin a more engaging approach to decision support in the form of serious games (SG), and augmented/virtual reality (AR/VR) environments, challenging and disrupting the very way decisions are made in the water sector [20]. The latest developments in artificial intelligence (AI) and machine learning (ML) have already shown that AI/ML enabled software systems can beat human players in complex games, such as chess or Go [21]. Through reinforcement learning, these systems can learn by playing games, which can be a guiding light to developing decision-support systems capable of assisting human water system operators in performing complex operational, tactical or strategic tasks. Similarly, robotic technologies and AI, which have been making great strides in the manufacturing and consumer industries, are starting to find their way to water management, e.g., underground asset inspection [22]. Lastly, the authors argue that with these data, tools and models at hand, the sector is now developing more sophisticated ways of stress-testing new and existing infrastructure, developing new methodological approaches around resilience [23]. In the remaining part of this section, a brief overview of some key literature on the subjects highlighted above is provided and an outline of their current state of art is discussed.

Figure 2. A shifting landscape for hydroinformatics research and practice.

2.1. *New Real Time Information*

The rapid developments in ICT, leveraged through advances in hydroinformatics, have created the basis for a phenomenal increase in the types and amounts of water-related data collected and analyzed, following the trend (and to some extent hype) of the so-called Big Data currently evident in numerous other fields and sectors [24]. Although the volume of water data currently collected by the sector is certainly unprecedented, attributed to an increasing deployment of dedicated sensors of various types, the data in the water sector cannot really be considered big, at least not yet. Water data are often structured data and do not usually include the main types of unstructured data (such audio, images, video, and unstructured text) that account for 95% of big data at the global scale [24]. A notable (and promising) exception is when crowdsourcing is also taken into account as a means of supplementing data obtained from more traditional sources [25]. The arrival of big data is also coinciding with a strong movement by individuals, learned societies and governments to open data for the benefit of individuals and society in general. The availability and use of open data—that anyone can access, use or share—can also increase opportunities for the collaboration and engagement of stakeholders, particularly in cities. The rise of the 'Smart City' concept, where ICT (and IoT) are used to enhance a city's livability, workability and sustainability, is another factor that impacts on the use of big data in urban water management [26]. The developments in this (growing) nexus between water and ICT (often termed digital water, Water 4.0 or water informatics), allow water companies to now be able to monitor in (near) real time their entire supply and value chain, from the sources to the consumers' tap and then 'downstream' to the wastewater plant. Smart sensors and smart meters (e.g., [27]) are becoming ubiquitous allowing for a substantial increase in coverage (e.g., [28]), resolution (e.g., [29]) and diversity (e.g., [30]) of water-related information, including water quality [30–32], which has long been the most difficult water characteristic to reliably monitor remotely. Interestingly, new water related information is not only collected by smart sensors and devices. It is also increasingly collected by the citizens/water users themselves. For example, the paper-based water quality sensor and smartphone that was used in Sicard et al. [33], or work by Farnham et al. [34] on using citizen-based water quality monitoring for combined sewer overflows.

2.2. New Distributed Infrastructure Deployment

The increasing availability of information (and remote-control capability) allows the sector to seriously consider and gain confidence in re-engineering its water management practices [35]. This can be achieved also by deploying in large scales more promising, distributed alternatives to water service provision (from treatment to separation and from reuse to drainage, see for example Larsen et al. [36]) that have hitherto been reserved for research/pilot environments. Although a review of these technological developments falls outside the scope of this paper, it is argued that their advent is both enabled by new advances in hydroinformatics (in both the hardware and software sense) and enables interesting hydroinformatic developments in the analytics, modelling and decision contexts. An example of this interplay is evident, for example, in the case of distributed water reuse technologies termed sewer mining [37]. Here, novel treatment solutions emerged, that required advanced monitoring and control systems to become deployable in remote locations [38]. This in turn, led to a need for modelling and optimisation tools, able to support the optimal location of sewer mining units in large sewerage networks [39]. The availability of the sewer mining technology as an intervention option, then meant that integrated models had to include them as options for decision makers [37]. This positive feedback is typical of the way hydroinformatics evolves in a dialectic relationship between the discipline and the water sector.

2.3. New Analytics

To make sense of this increasing amount of information, research and practice have made significant progress towards better analytics, including but not limited to those: (i) Capable of extracting valuable information from the data (from smart alerts to customized advice for water users); (ii) performing better stochastic simulations to improve the ability to produce longer timeseries (based on observations) for long-term scenario development and stress-testing; (iii) performing advanced optimisation to identify better solutions in this information richer environment; and (iv) providing novel ways of visualizing and understanding the decision tradeoffs within complex decision spaces. Examples of these new analytics, include AI/ML analytics for proactive management of water distribution systems (including burst detection) demonstrated in UK case studies [40,41], asset deterioration assessment [42], as well as the use of deep learning techniques for defining novel control strategies that are more robust against cyber-attacks of water distribution systems [43]. Examples also include recent work on using smart meter readings to parametrise residential water demand models [44] as well as the methods and tools developed to investigate the properties of these timeseries at fine timescales [29]. Based on this growing body of work, we are now in a position to assess for the first time if smart meters are effective in water demand management (see for example the review by Sønderlund et al. [27] based on 21 relevant reports and publications) or at least pinpoint the additional information needed to make this transition, including the information content, granularity, frequency and method of delivery etc.

However, getting better historical data is only part of the story. Additional work in stochastics is enabling hydroinformatics to develop simulated timeseries that explicitly represent each process of interest with any distribution model and hence conserve all of the characteristics of historical datasets (e.g., [45]). These longer timeseries can be used to drive hydroinformatic models of complex hydro-systems to better account for relevant uncertainties. However, this substantially increases the (time) burden for optimisation. Recent attention to 'optimisation on a budget' [46] shows how surrogate strategies can be employed to allow for less evaluations of expensive objective functions in evolutionary optimisation. Other authors have also focused on the challenging problem of optimal design under uncertainty and developed optimisation algorithms that exploit the concept of 'real options' [47], thus introducing flexibility into the long-term design for water systems [48]. Although an overview of the developments in optimisation is outside the scope of this paper, this is one of the most prolific fields in hydroinformatics to date. The interested reader is pointed towards an overview of this dynamic field, with a focus on water distribution networks, included in Mala-Jetmarova et al. [49] and in Maier et al. [50] for a more general overview of optimisation in water resources in general. Lastly, it is worth pointing out that developing new algorithms does not necessarily lead to better understanding or decision making. Recent attention to analytics for advanced visualisation of decision spaces suggest that developing visual analytics to explore the decision space in multi-objective (e.g., [51]) or in multi-stakeholder problems [52] is both important and necessary.

2.4. New Whole Water Cycle Socio-Technical System Models

The industry's interest in exploring new options for infrastructure provision (incl. new more distributed options discussed above) is driven in part by the process of aging infrastructure and the resulting investment gap [16]. The interest has also prompted the development and application of whole (socio-technical) system models [53] that attempt a more direct investigation of the interplay between centralized and distributed infrastructure solutions. Furthermore, the focus is also shifting towards the (often ignored) interplay between infrastructure and users (as also argued persuasively in the context of socio-hydrology by Sivapalan [54]). This integration is currently being delivered (mostly) around three axes:

- Integration between centralised and decentralised solutions and (often also) between water infrastructure and urban fabric growth in a common (whole system) modelling environment. Indicative work in this context includes the Aquacycle model [55], the Urban Water Optioneering Tool (UWOT, see Rozos and Makropoulos [56]), UVQ [57] as well as the Dance4Water model [58], to name but a few. For an overview of key models as well as a discussion on the degree of integration, the reader is referred to Bach et al. [59]. These more integrated models, sometimes termed metabolism models (e.g., [60]) are increasingly being used to evaluate alternative pathways for the evolution of water systems under uncertainty, opening up the possibility of looking at a much wider palette of options than was possible with more traditional hydraulic-only models.
- Integration between natural and engineered infrastructure systems and user interactions. This is a growing area of work, which also typically includes the explicit modelling of additional flows (e.g., the nexus between water, energy and material flows within an urban environment). Although approaches to this integration vary widely, these are based primarily on: (i) System dynamics (SD) and/or Bayesian belief networks (BBN); and (ii) agent-based models. Recent examples of the former types include Sahin et al. [61], Baki et al. [62] and Chhipi-Shrestha et al. [63]. In this context, Zomorodian et al. [64] provide an overview of SD applications for water management, while Sušnik et al. [65] provide a comparison between SD and BBN models for water management. Recent examples of the latter type include work by Kanta and Zechman [66], Berglund [67] and Koutiva and Makropoulos [68]. The power of these modelling approaches is that they enable the explicit integration of the socio-economic system into the modelling framework, which is especially important when looking into policy and end-user driven interventions, such as water demand management, water markets, innovation uptake etc.

- Integration between the physical and cyber layer of water systems. This attempt on modelling integration represents a recent development, consistent with the move towards conceptualising water systems as a cyber-physical infrastructure. This conceptualisation, advocated already 10 years ago by Edward A. Lee [69] for a range of infrastructures, is currently being operationalised in the form of integrated simulation environments for the cyber and physical layers of a water system and their interactions [70–72]. Although this work is still not rolled out in an operational sense within water companies, it is argued that it will become more important in the next few years, as part of a risk management approach for both cyber and physical risks.

It is important to note here that in support to these more integrative explorations, the hydroinformatics community has been developing and demonstrating: (a) Integrated modelling frameworks [73,74]; (b) models as services, often based on open source solutions [75]; and (c) cloud-based modelling systems [76,77], sometimes coupling both local model components and remote web services [78] in an effort to reduce the overhead required to create an integrated model in the first place and make their explorative power more accessible to the water research and practitioner communities.

2.5. New forms of Interactive and Immersive Decision Making

The multi-faceted, multi-discipline and increasingly more inclusive multi-stakeholder nature of water management considerations (and environmental management in general [79]) have given rise to new ways of setting the questions, visualizing potential results and experiencing system performance under different stresses. These ways include Serious Games [80], augmented/virtual (or mixed) reality (AR, VR, MR) and their combinations that enable a different level of immersive, playful experience of problems, options and decisions that can be used in various contexts, including operational, strategic and stakeholder collaborative decision-making. The basic idea of these (relatively new) approaches is that practical water and environmental challenges (and options to address them) can be better understood through a more direct experiential approach. These game-based learning approaches improve critical thinking, creative problem solving and teamwork [79]. They also allow stakeholders to experiment with decisions and outcomes in a safe and fun environment.

Work by several authors is currently finding its way into practical applications, engaging water stakeholders in collaborative decision making for such diverse fields as urban flood management [52], water resources management [81,82] and integrated asset management [83]. At the same time, augmented reality applications (including applications in handheld devices and smartphones) have begun to be actively used in infrastructure inspection and rehabilitations (see for example the Vidente application reported in Schall et al. [84]). The significant potential for this technology is especially evident in cases where infrastructure is underground as in the case of water distribution and sewerage networks. These applications typically superimpose data from GIS systems (such asset databases) or even data from simulations on real world views. This linking of spatial/georeferenced information directly on the real-world entities that they characterize, greatly facilitates the use of relevant data during field work (e.g., asset rehabilitation, water quality monitoring). As such, it ensures increased efficiency in maintenance activities, as well as increased understanding and learning in educational field trips and field-oriented stakeholder engagement processes (e.g., stakeholder visits in innovation demonstration case studies). An example of the latter is students participating in the EcoMOBILE project [85], who used an augmented reality application, as part of a field trip to an ecologically important lake. The virtual information was overlaid on the physical lake including hotspots—guiding students in collecting water quality measurements—but also increasing their understanding of underlying processes. It could be argued that such an increased (and more importantly shared) understanding between stakeholders, makes for a good basis for more inclusive, consensus-driven decision making.

2.6. New Design Concepts and Strategies

The availability of new ubiquitous data, advanced analytics and more integrated modeling frameworks is allowing the sector to perform more realistic stress-tests of water infrastructure (in its physical and cyber-physical sense) to help improve its performance under uncertainty. This activity is currently pushing the discipline's methodological boundaries into developing and applying novel design concepts driven to a large extent by cities worldwide demanding realistic risk management under uncertainty within a context of limited new investments (see for example the 100 Resilient Cities network supported by the Rockefeller Foundation [86]). These efforts are, recently, centered mostly around the challenging concept of resilience and the development of methods, metrics and tools to assess the resilience of urban water systems. Notable examples include models and tools developed by Irwin et al. [87], Butler et al. [88], Klise et al. [89], Makropoulos et al. [8], Kong et al. [90] as well as Sweetapple et al. [91]). Although a discussion on resilience per se is outside the scope of this paper, we note that this growing body of work, focusing on the highly interdisciplinary and multi-stakeholder context of resilience [92] is an important manifestation of the sociotechnical nature of hydroinformatics. The need to understand resilience emphasizes the role of hydroinformatics as an interface between science and policy, between water systems and urban processes as well as between technology, society and the environment.

3. Sky Is (Not) the Limit

This overview of some of the most exciting developments in hydroinformatics today, may give the impression that most of the important tasks are behind us. This, however, could not be further from the truth. As the discipline is, by definition, linked to and influenced by developments in the dynamically evolving IT sector, with every new development come new challenges and also new opportunities. Although the details of what can happen next are by virtue of this dynamic evolution, hard to predict, some of the most important trends are already visible. In an effort to summarise these future trends, four activity lines towards a hydroinformatics roadmap have been proposed below:

3.1. Tapping into the New Data Landscape

The proliferation of smart systems (including developments in the smart city and more generally the IoT arena) mean that data become more ubiquitous—although work on novel water quality sensors is still needed (see ideas on using graphene for heavy metal detection [93]). However, as more data from different sources become available the issue of standardization becomes vital. This is because standardization allows the pulling together and combined exploitation of data coming from different sources and different data providers, both within a utility but also potentially across multiple utilities, reaching the critical mass of data required to categorize water data as big data and, in turn, unlock the true potential of big data analytics. As such, data standardization, in terms, for example, of metadata, standardized markup languages (like the Open Geospatial Consortium's (OGC) WaterML [94], controlled vocabularies and ontologies [95–97] inevitably play a key role in bringing information and analytics together. Due to their importance in an IoT and related telecommunications contexts, the most successful of these standardization efforts will probably not be initiated within the water domain per se, but rather within smart city, smart home and smart industry contexts, growing towards water, energy and other utility sectors. A case in point is the work by the European Telecommunications Standards Institute (ETSI) and its Smart Appliance REFerence (SAREF) ontology [98], which is currently being expanded [99] towards energy and water, with obvious implications for smart water meters, smart(er) water consuming devices and domestic water demand forecasting and management. Another important development in this field, worth highlighting is FIWARE [100], a curated framework of open source platform components that aims to accelerate the development of smart solutions, including transport, energy, as well as more integrative smart city solutions. FIWARE has already been used to develop interesting examples of interoperability for smart agricultural water management [101] and is now expanding [102] also towards urban water management at different scales. Data quality control and validation (potentially in a distributed way, closer to the data collection itself, see for example developments in edge analytics [103]) and improvement of data access (including data sharing and open data [104]) is also expected to be at the heart of the next steps in hydroinformatics.

With this critical milestone completed, the industry may be able to exploit new developments that allow the industry to get new insights out of large, heterogeneous databases and leverage progress on AI, such as deep learning [105], from the ICT sector, to extract information, develop more accurate forecasts and offer customized services to end users. New opportunities afforded by leveraging the power of AI on larger (and more real time) water datasets, include discovering new causal relationships from data already collected to improve predictive ability, e.g., in infrastructure maintenance, water demand management or emergency response. It may also allow for progress into data assimilation techniques that couple models to field data in real time. Field data from different sources and with different uncertainties is expected to be used in combination with models, thus greatly increasing current abilities for pro-active management of water systems. This new data may also increasingly come from the customer/citizen side, where data crowd-sourcing tools will play an increasing role in collecting real time information [25] as well as in gauging public opinion towards water relevant issues (e.g., water reuse attitudes mined from micro-blogs [106]). These (significantly increased) data streams may range from data collected by smartphone embedded sensors, to information posted on social media, to data collected by, soon to be available, autonomous vehicles—cross referenced and linked to open environmental data, utility sensors and remote sensed information from new satellite networks (like NASA's Surface Water and Ocean Topography (SWOT) mission scheduled to start by 2021 [107].

3.2. Getting More Out of Existing Models

This activity line, is expected to provide the sector with more advanced optimization (including smart model calibration under uncertainty and noise), new ways of model integration (with databases and other models) as well as with real time data (including IoT sensors) to form digital twins of utilities. The concept of digital twins, where the data from the IoT sensors are seamlessly linked with asset management information and both support and are supported by models of the system's operations, recalibrated and updated in real time, across the complete value chain from water resources to customers, is expected to become possible in the near future. This ambition, of a complete integrated digital picture of a water utility may appear far-fetched at this time, but is a future in the making, judging from the interest and investment already underway in forward looking cities, such as Amsterdam [108] and its water utility (Waternet). Necessarily, this process shifts online much of the computing infrastructure for water utilities, with cloud computing for water services and software-as-service becoming the norm. This trend, however, is not without its challenges as is discussed in the following sections.

3.3. Planning for More Resilient (Cyber-Physical) Systems and Services

Armed with new data and models, the sector may also work more on model integration and higher abstraction level modelling/model coupling, where whole system strategic models—potentially linked to digital twins—can be used as real-time control, forecasting and scenario planning tools in a collaborative and inclusive way.

This direct coupling between the physical system and related infrastructure and the controlling cyber layer (from sensors to models to actuators) is expected to afford new opportunities for increased efficiency of water infrastructures throughout their lifetime, from design to building to operating. It would allow, for example, their real time control, with data from multiple sensors being continuously integrated within living models of the physical environment and the infrastructure. Furthermore, it would enable moving significant parts of these calculations to the edge [103], enabling precise and pro-active actuation of pumps, valves, sluice gates, for applications, such as flood forecasting and control [109,110], managing combined sewer overflows [111] and urban water management in general [112].

In this context of ever increasing integration between the physical and the cyber sides of water infrastructure, a growing focus on cyber-physical systems risk assessment and threat modelling (e.g., [71,72]), is expected to become more central in water company preoccupations. Cyber-physical modelling can help the sector manage emerging cyber-physical risks, especially in the context of digital twins. In the same vein, it is suggested that work on modelling cascading effects between water systems and other infrastructures may also move from the research environment [113] to the operational environment of the sector. The move may also involve other water and crisis management stakeholders at national and international levels.

3.4. Training, Engaging and Communicating

Lastly, significant advances in rethinking the way decisions are made (from the strategic to the operational) are expected. These changes in decision-making will be catalyzed through technologies that allow for more immersive and playful experiences of the decision landscape, such as Serious Games coupled with AR and VR (or mixed reality) applications and environments. The disruptive potential of such a technology shift cannot be overstated, potentially influencing everything, from immersive scenarios planning, including crisis management training, to pipe rehabilitation, innovation uptake and water education. This last point brings us, however, face to face with an important challenge: What is the form of education and indeed the skillsets required by new hydroinformaticians to be able to benefit from, engage with and ultimately help evolve this dynamic field? Popescu et al. [114] have already correctly identified this challenge some time ago, when they suggested that hydroinformaticians need to master a subject matter that is *"increasing far more rapidly than the ability of engineering curricula to cover it"*. Indeed, as if water science was not demanding enough, the domain experts also need to be fluent in data science (from statistics to machine learning) and computer science (from information theory to hands-on software development and user interfaces design). They also need to engage with topics ranging from decision theory to social science to ethics and philosophy of science. Popescu et al. [114] argued that flexibility is key here, delivered through modular design and blended forms of learning with face to face courses supplemented with online courses allowing participants to invest in deepening their knowledge in diverse areas in a more customized pace. Clearly these requirements point towards hydroinformatics as a postgraduate rather than an undergraduate course. Actually, Abbott et al., [115] used the term participant rather than student explicitly to highlight a prerequisite of solid undergraduate education in relevant fields and indeed hands-on experience before embarking in such a multi-disciplinary course. They also persuasively argued that the educational challenge posed even after this prerequisite is met, suggests another important subject for future hydroinformatics research, that is, research into the educational and training aspects of the domain. In that context, hydroinformatics may benefit from the emergence of the more immersive and playful approaches and technologies discussed above, not the least due to the active (experiential) engagement (in view, for example, of rapid developments of natural user interfaces [116]) and hazard-free, learning by doing aspects that these approaches afford. This promise, however, implies an important, additional and often neglected prerequisite: As Richert et al. [117] would argue tomorrow's hydroinformatics academics need themselves the technological competencies to allow them to both design and create these immersive environments and the training in digital coaching and joint problem solving in virtual worlds to be able to use them in meaningful and educationally productive ways. It is suggested that this prerequisite can only be delivered through new multidisciplinary forms of collaboration around education per se, both within universities and between universities and research centres and technology providers for an interesting example of emerging forms of multi disciplinarity in education see for example: [118].

4. Some Words of Caution

Although these developments can have enormous societal and technological benefits, they also raise security, privacy, legal, and ethical concerns [25].

The increased dependency of water utilities on ICT to carry out their mission and functions, as well as the tendency to provide interoperability and connect these traditionally closed systems to the Internet, opens them up to, as yet unheard of, cyber threats. A case in point is Maroochy Water Services in Australia, probably the most well-known cyber-attack in the water sector, where over a three-month period in 2000 a disgruntled former contractor took control of over 150 sewage pumping stations and released one million litres of untreated sewage into the environment [119]. Furthermore, the prospect of a large number of smart water meters being installed at customer homes, thus connecting them to the utility ICT systems, raises also a possibility of the wider water infrastructure becoming vulnerable to scalable network-borne attacks.

By the very nature of smart systems, customers adopting them share detailed information about their water usage with the utility, which is then used to better assess the demand and manage the entire system. This information sharing potentially exposes customers to privacy invasions with the main concern being the limited control over personal data by an individual, which can result in a range of negative or unintended consequences. Legal considerations relating to privacy and data protection with respect to services or applications created using customer water usage data (particularly valuable when combined with personal data), has been given insufficient attention in the literature [120]. It is, therefore, positive that the new EU General Data Protection Regulation (GDPR) [121] provides a framework for data protection and privacy for citizens. The regulation deals with the risks of accidental or unlawful destruction, loss, alteration, unauthorized disclosure of, or access to, personal data transmitted, stored or otherwise processed. The regulation's application will inevitably open up new questions and challenges which will need to be addressed, but it is important that this conversation is progressing.

Last, but certainly not least, smart systems as surveillance-enabled technologies as well as AI-based decision making, raise issues of privacy, fundamental rights, ethics and responsibility in technological innovation [122]. The need for rethinking, spelling out and agreeing upon the ethical principles on which these technologies is expected to be based [123] has never been more pressing. This is a challenge, not only for technology (and the safeguards it needs to put in place) but perhaps more importantly for ethics and the humanities that need to pick up the challenge and update their theories, methods, vocabulary and technology to make sense of and proactively manage the potential implications to society from a pace of technological development never seen before.

5. Conclusions: A Bright Future with Some Caveats

This study has presented a summary of the dynamic evolution of hydroinformatics, as a discipline at the interface between water science, data science, computer science and technology on the one hand and society on the other. In so doing, the authors have highlighted exciting advances in new real-time information; new analytics developed to extract value from this new information; novel whole cycle (socio-technical) system models that are calibrated on these new datasets; new more immersive approaches to decision support; more sophisticated ways of stress-testing new and existing cyber-physical infrastructure to improve its resilience. Four activity lines of research have also been proposed, coming up on the horizon (tapping into the new data landscape; getting more out of existing models; planning for more resilient systems and services; training, engaging and communicating). The authors suggest that these activity lines support a virtuous cycle towards more resilient water systems and services. It is further argued that their confluence can drastically change both the form and function of water services and the infrastructure that provide these services in the not too distant future—for the better—provided that important challenges around privacy, fundamental rights, ethics and responsibility in technological innovation are seriously and urgently addressed.

Author Contributions: Conceptualization, C.M. and D.A.S.; methodology, C.M.; investigation, C.M. and D.A.S.; resources, C.M. and D.A.S.; data curation, C.M. and D.A.S.; writing—original draft preparation, C.M.; writing—review and editing, C.M. and D.A.S.; supervision, D.A.S.

Funding: This research received no external funding.

Conflicts of Interest: The authors declare no conflict of interest.

References

1. Vojinovic, Z.; Abbott, M.B. *Flood Risk and Social Justice: From Quantitative to Qualitative Flood Risk Assessment and Mitigation*; IWA Publishing: London, UK, 2012; ISBN 9781843393870.
2. Abbott, M. *Hydroinformatics: Information Technology and the Aquatic Environment*; Avebury Technical: Aldershot, UK, 1991; ISBN 1856288323.
3. Holz, K.P.; Cunge, J.; Lehfeldt, R.; Savic, D. Hydroinformatics Vision 2011. In *Advances in Hydroinformatics*; Springer: Singapore, 2014; pp. 545–560.
4. Journal of Hydroinformatics. Available online: https://iwaponline.com/jh (accessed on 16 September 2019).
5. Environmental Modelling & Software. Available online: https://www.journals.elsevier.com/environmental-modelling-and-software (accessed on 16 September 2019).
6. Vojinovic, Z.; Abbott, M. Twenty-Five Years of Hydroinformatics. *Water* **2017**, *9*, 59. [CrossRef]
7. Li, E.; Endter-Wada, J.; Li, S. Characterizing and Contextualizing the Water Challenges of Megacities. *JAWRA J. Am. Water Resour. Assoc.* **2015**, *51*, 589–613. [CrossRef]
8. Makropoulos, C.; Nikolopoulos, D.; Palmen, L.; Kools, S.; Segrave, A.; Vries, D.; Koop, S.; van Alphen, H.J.; Vonk, E.; van Thienen, P.; et al. A resilience assessment method for urban water systems. *Urban Water J.* **2018**, *15*, 316–328. [CrossRef]
9. Pahl-Wostl, C. Transitions towards adaptive management of water facing climate and global change. *Water Resour. Manag.* **2006**, *21*, 49–62. [CrossRef]
10. Hounslow, A. *Water Quality Data: Analysis and Interpretation*; eBook; CRC Press: Boca Raton, FL, USA, 2018; ISBN 9781351404907.
11. UN The 2018 Revision of World Urbanization Prospects. Available online: https://population.un.org/wup/ (accessed on 28 February 2019).
12. Bouziotas, D.; Rozos, E.; Makropoulos, C. Water and the city: Exploring links between urban growth and water demand management. *J. Hydroinform.* **2015**, *17*, 176–192. [CrossRef]
13. Makropoulos, C.K.; Memon, F.A.; Shirley-Smith, C.; Butler, D. Futures: An exploration of scenarios for sustainable urban water management. *Water Policy* **2008**, *10*, 345–373. [CrossRef]
14. Brown, R.R.; Keath, N.; Wong, T.H.F. Urban water management in cities: Historical, current and future regimes. *Water Sci. Technol.* **2009**, *59*, 847–855. [CrossRef]
15. Rygaard, M.; Binning, P.J.; Albrechtsen, H.J. Increasing urban water self-sufficiency: New era, new challenges. *J. Environ. Manag.* **2011**, *92*, 185–194. [CrossRef]
16. Selvakumar, A.; Tafuri, A.N. Rehabilitation of Aging Water Infrastructure Systems: Key Challenges and Issues. *J. Infrastruct. Syst.* **2012**, *18*, 202–209. [CrossRef]
17. Wan, Z.; Hong, Y.; Khan, S.; Gourley, J.; Flamig, Z.; Kirschbaum, D.; Tang, G. A cloud-based global flood disaster community cyber-infrastructure: Development and demonstration. *Environ. Model. Softw.* **2014**, *58*, 86–94. [CrossRef]
18. Mounce, S.R.; Pedraza, C.; Jackson, T.; Linford, P.; Boxall, J.B. Cloud Based Machine Learning Approaches for Leakage Assessment and Management in Smart Water Networks. *Procedia Eng.* **2015**, *119*, 43–52. [CrossRef]
19. Eggimann, S.; Mutzner, L.; Wani, O.; Schneider, M.Y.; Spuhler, D.; Moy de Vitry, M.; Beutler, P.; Maurer, M. The Potential of Knowing More: A Review of Data-Driven Urban Water Management. *Environ. Sci. Technol.* **2017**, *51*, 2538–2553. [CrossRef] [PubMed]
20. Aubert, A.H.; Bauer, R.; Lienert, J. A review of water-related serious games to specify use in environmental Multi-Criteria Decision Analysis. *Environ. Model. Softw.* **2018**, *105*, 64–78. [CrossRef]
21. Gibney, E. Google AI algorithm masters ancient game of Go. *Nature* **2016**, *529*, 445–446. [CrossRef] [PubMed]
22. van Thienen, P.; Beuken, R.; Vertommen, I.; Slaats, N. Perspective and Preconditions for the Development and Use of Autonomous Inspection Robots in Water Mains. Available online: https://www.kwrwater.nl/en/projecten/perspective-and-preconditions-for-the-development-and-use-of-autonomous-inspection-robots-in-water-mains/ (accessed on 16 September 2019).
23. Rodina, L. Defining "water resilience": Debates, concepts, approaches, and gaps. *Wiley Interdiscip. Rev. Water* **2019**, *6*, e1334. [CrossRef]
24. Gandomi, A.; Haider, M. Beyond the hype: Big data concepts, methods, and analytics. *Int. J. Inf. Manag.* **2015**, *35*, 137–144. [CrossRef]

25. Zheng, F.; Tao, R.; Maier, H.R.; See, L.; Savic, D.; Zhang, T.; Chen, Q.; Assumpção, T.H.; Yang, P.; Heidari, B.; et al. Crowdsourcing Methods for Data Collection in Geophysics: State of the Art, Issues, and Future Directions. *Rev. Geophys.* **2018**, *56*, 698–740. [CrossRef]
26. Savic, D. A Smart City without Smart Water is Only a Pipe Dream! In Proceedings of the the Keynote Presentation Delivered at the 37th IAHR World Congress, Kuala Lumpur, Malaysia, 13–18 August 2017.
27. Sønderlund, A.L.; Smith, J.R.; Hutton, C.J.; Kapelan, Z.; Savic, D. Effectiveness of Smart Meter-Based Consumption Feedback in Curbing Household Water Use: Knowns and Unknowns. *J. Water Resour. Plan. Manag.* **2016**, *142*, 04016060. [CrossRef]
28. Di Mauro, A.; Di Nardo, A.; Bernini, R.; Sanfilippo, L.; Paleari, O.; Savic, D.; Santonastaso, G.F.; Cocco, M.; Cousin, P.; Wouters, H.; et al. Smart Water Network Monitoring using innovative On-line Sensors. *Geophys. Res. Abstr.* **2018**, *20*, 16024.
29. Kossieris, P.; Makropoulos, C. Exploring the Statistical and Distributional Properties of Residential Water Demand at Fine Time Scales. *Water* **2018**, *10*, 1481. [CrossRef]
30. Doulamis, N.; Voulodimos, A.; Doulamis, A.; Bimpas, M.; Angeli, A.; Bakalos, N.; Giusti, A.; Philimis, P.; Varriale, A.; Ausili, A.; et al. Waterspy: A high sensitivity, portable photonic device for pervasive water quality analysis. *Sensors* **2019**, *19*. [CrossRef]
31. Gholizadeh, M.; Melesse, A.; Reddi, L. A comprehensive review on water quality parameters estimation using remote sensing techniques. *Sensors* **2016**, *16*, 1298. [CrossRef] [PubMed]
32. Ponce Romero, J.; Hallett, S.; Jude, S. Leveraging Big Data Tools and Technologies: Addressing the Challenges of the Water Quality Sector. *Sustainability* **2017**, *9*, 2160. [CrossRef]
33. Sicard, C.; Glen, C.; Aubie, B.; Wallace, D.; Jahanshahi-Anbuhi, S.; Pennings, K.; Daigger, G.T.; Pelton, R.; Brennan, J.D.; Filipe, C.D.M. Tools for water quality monitoring and mapping using paper-based sensors and cell phones. *Water Res.* **2015**, *70*, 360–369. [CrossRef] [PubMed]
34. Farnham, D.; Gibson, R.; Hsueh, D.; McGillis, W.; Culligan, P.; Zain, N.; Buchanan, R. Citizen science-based water quality monitoring: Constructing a large database to characterize the impacts of combined sewer overflow in New York City. *Sci. Total Environ.* **2017**, *580*, 168–177. [CrossRef] [PubMed]
35. Nguyen, K.A.; Stewart, R.A.; Zhang, H.; Sahin, O.; Siriwardene, N. Re-engineering traditional urban water management practices with smart metering and informatics. *Environ. Model. Softw.* **2018**, *101*, 256–267. [CrossRef]
36. Larsen, T.A.; Hoffmann, S.; Luthi, C.; Truffer, B.; Maurer, M. Emerging solutions to the water challenges of an urbanizing world. *Science* **2016**, *352*, 928–933. [CrossRef]
37. Makropoulos, C.; Rozos, E.; Tsoukalas, I.; Plevri, A.; Karakatsanis, G.; Karagiannidis, L.; Makri, E.; Lioumis, C.; Noutsopoulos, C.; Mamais, D.; et al. Sewer-mining: A water reuse option supporting circular economy, public service provision and entrepreneurship. *J. Environ. Manag.* **2018**, *216*, 285–298. [CrossRef] [PubMed]
38. Karagiannidis, L.; Vrettopoulos, M.; Amditis, A.; Makri, E.; Gkonos, N. A CPS-enabled architecture for sewer mining systems. In Proceedings of the 2016 International Workshop on Cyber-Physical Systems for Smart Water Networks (CySWater), Vienna, Austria, 11 April 2016; IEEE: New York, NY, USA, 2016; pp. 1–6.
39. Psarrou, E.; Tsoukalas, I.; Makropoulos, C. A Monte-Carlo-Based Method for the Optimal Placement and Operation Scheduling of Sewer Mining Units in Urban Wastewater Networks. *Water* **2018**, *10*, 200. [CrossRef]
40. Machell, J.; Mounce, S.R.; Farley, B.; Boxall, J.B. Online data processing for proactive UK water distribution network operation. *Drink. Water Eng. Sci.* **2014**, *7*, 23–33. [CrossRef]
41. Romano, M.; Kapelan, Z.; Savić, D.A. Automated Detection of Pipe Bursts and Other Events in Water Distribution Systems. *J. Water Resour. Plan. Manag.* **2014**, *140*, 457–467. [CrossRef]
42. Berardi, L.; Giustolisi, O.; Kapelan, Z.; Savic, D.A. Development of pipe deterioration models for water distribution systems using EPR. *J. Hydroinform.* **2008**, *10*, 113–126. [CrossRef]
43. Taormina, R.; Galelli, S. Deep-Learning Approach to the Detection and Localization of Cyber-Physical Attacks on Water Distribution Systems. *J. Water Resour. Plan. Manag.* **2018**, *144*, 04018065. [CrossRef]
44. Creaco, E.; Kossieris, P.; Vamvakeridou-Lyroudia, L.; Makropoulos, C.; Kapelan, Z.; Savic, D. Parameterizing residential water demand pulse models through smart meter readings. *Environ. Model. Softw.* **2016**, *80*, 33–40. [CrossRef]
45. Tsoukalas, I.; Efstratiadis, A.; Makropoulos, C. Stochastic Periodic Autoregressive to Anything (SPARTA): Modeling and Simulation of Cyclostationary Processes with Arbitrary Marginal Distributions. *Water Resour. Res.* **2018**, *54*, 161–185. [CrossRef]

46. Tsoukalas, I.; Makropoulos, C. Multiobjective optimisation on a budget: Exploring surrogate modelling for robust multi-reservoir rules generation under hydrological uncertainty. *Environ. Model. Softw.* **2015**, *69*, 396–413. [CrossRef]
47. Marques, J.; Cunha, M.; Savić, D.A. Using real options for an eco-friendly design of water distribution systems. *J. Hydroinform.* **2015**, *17*, 20–35. [CrossRef]
48. Pellegrino, R.; Costantino, N.; Giustolisi, O. Flexible investment planning for water distribution networks. *J. Hydroinform.* **2018**, *20*, 18–33. [CrossRef]
49. Mala-Jetmarova, H.; Sultanova, N.; Savic, D. Lost in Optimisation of Water Distribution Systems? A Literature Review of System Design. *Water* **2018**, *10*, 307. [CrossRef]
50. Maier, H.R.; Kapelan, Z.; Kasprzyk, J.; Kollat, J.; Matott, L.S.; Cunha, M.C.; Dandy, G.C.; Gibbs, M.S.; Keedwell, E.; Marchi, A.; et al. Evolutionary algorithms and other metaheuristics in water resources: Current status, research challenges and future directions. *Environ. Model. Softw.* **2014**, *62*, 271–299. [CrossRef]
51. Fu, G.; Kapelan, Z.; Kasprzyk, J.; Reed, P. Optimal Design of Water Distribution Systems Using Many-Objective Visual Analytics. *J. Water Resour. Plan. Manag.* **2012**, *139*, 624–633. [CrossRef]
52. Khoury, M.; Gibson, M.J.; Savic, D.; Chen, A.S.; Vamvakeridou-Lyroudia, L.; Langford, H.; Wigley, S. A Serious Game Designed to Explore and Understand the Complexities of Flood Mitigation Options in Urban–Rural Catchments. *Water* **2018**, *10*, 1885. [CrossRef]
53. Abbott, M. On definitions. *J. Hydroinform.* **2002**, *4*, 1–27. [CrossRef]
54. Sivapalan, M. Debates-Perspectives on socio-hydrology: Changing water systems and the "tyranny of small problems"-Socio-hydrology. *Water Resour. Res.* **2015**, *51*, 4795–4805. [CrossRef]
55. Mitchell, V.G.; Mein, R.G.; McMahon, T.A. Modelling the urban water cycle. *Environ. Model. Softw.* **2001**, *16*, 615–629. [CrossRef]
56. Rozos, E.; Makropoulos, C. Source to tap urban water cycle modelling. *Environ. Model. Softw.* **2013**, *41*, 139–150. [CrossRef]
57. Mitchell, V.; Diaper, C.; Gray, S.; Rahilly, M. UVQ: Modelling the Movement of Water and Contaminants through the Total Urban Water Cycle. In Proceedings of the 28th International Hydrology and Water Resources Symposium: About Water; Symposium Proceedings, Novotel Northbeach, Wollongong, NSW, Australia, 10–13 November 2003; Available online: https://search.informit.com.au/documentSummary;dn=410125625788126;res=IELENG (accessed on 16 September 2019).
58. Rauch, W.; Urich, C.; Bach, P.M.; Rogers, B.C.; de Haan, F.J.; Brown, R.R.; Mair, M.; McCarthy, D.T.; Kleidorfer, M.; Sitzenfrei, R.; et al. Modelling transitions in urban water systems. *Water Res.* **2017**, *126*, 501–514. [CrossRef]
59. Bach, P.M.; Rauch, W.; Mikkelsen, P.S.; McCarthy, D.T.; Deletic, A. A critical review of integrated urban water modelling—Urban drainage and beyond. *Environ. Model. Softw.* **2014**, *54*, 88–107. [CrossRef]
60. Behzadian, K.; Kapelan, Z. Advantages of integrated and sustainability based assessment for metabolism based strategic planning of urban water systems. *Sci. Total Environ.* **2015**, *527–528*, 220–231. [CrossRef]
61. Sahin, O.; Siems, R.S.; Stewart, R.A.; Porter, M.G. Paradigm shift to enhanced water supply planning through augmented grids, scarcity pricing and adaptive factory water: A system dynamics approach. *Environ. Model. Softw.* **2016**, *75*, 348–361. [CrossRef]
62. Baki, S.; Rozos, E.; Makropoulos, C. Designing water demand management schemes using a socio-technical modelling approach. *Sci. Total Environ.* **2018**, *622*, 1590–1602. [CrossRef]
63. Chhipi-Shrestha, G.; Hewage, K.; Sadiq, R. Water–energy–carbon nexus modeling for urban water systems: System dynamics approach. *J. Water Resour. Plan. Manag.* **2017**, *143*, 04017016. [CrossRef]
64. Zomorodian, M.; Lai, S.H.; Homayounfar, M.; Ibrahim, S.; Fatemi, S.E.; El-Shafie, A. The state-of-the-art system dynamics application in integrated water resources modeling. *J. Environ. Manag.* **2018**, *227*, 294–304. [CrossRef]
65. Sušnik, J.; Molina, J.-L.; Vamvakeridou-Lyroudia, L.S.; Savić, D.A.; Kapelan, Z. Comparative Analysis of System Dynamics and Object-Oriented Bayesian Networks Modelling for Water Systems Management. *Water Resour. Manag.* **2013**, *27*, 819–841. [CrossRef]
66. Kanta, L.; Zechman, E. Complex Adaptive Systems Framework to Assess Supply-Side and Demand-Side Management for Urban Water Resources. *J. Water Resour. Plan. Manag.* **2014**, *140*, 75–85. [CrossRef]
67. Berglund, E.Z. Using Agent-Based Modeling for Water Resources Planning and Management. *J. Water Resour. Plan. Manag.* **2015**, *141*, 04015025. [CrossRef]

68. Koutiva, I.; Makropoulos, C. Modelling domestic water demand: An agent based approach. *Environ. Model. Softw.* **2016**, *79*, 35–54. [CrossRef]
69. Lee, E.A. Cyber Physical Systems: Design Challenges. In *Proceedings of the 2008 11th IEEE International Symposium on Object and Component-Oriented Real-Time Distributed Computing (ISORC)*; IEEE: New York, NY, USA, 2008; pp. 363–369.
70. Lin, J.; Sedigh, S.; Miller, A. Towards Integrated Simulation of Cyber-Physical Systems: A Case Study on Intelligent Water Distribution. In *Proceedings of the 2009 Eighth IEEE International Conference on Dependable, Autonomic and Secure Computing*; IEEE: New York, NY, USA, 2009; pp. 690–695.
71. Taormina, R.; Galelli, S.; Tippenhauer, N.O.; Salomons, E.; Ostfeld, A. Characterizing Cyber-Physical Attacks on Water Distribution Systems. *J. Water Resour. Plan. Manag.* **2017**, *143*, 04017009. [CrossRef]
72. Nikolopoulos, D.; Makropoulos, C.; Kalogeras, D.; Monokrousou, K.; Tsoukalas, I. Developing a Stress-Testing Platform for Cyber-Physical Water Infrastructure. In *Proceedings of the 2018 International Workshop on Cyber-physical Systems for Smart Water Networks (CySWater)*; IEEE: New York, NY, USA, 2018; pp. 9–11.
73. David, O.; Ascough, J.C.; Lloyd, W.; Green, T.R.; Rojas, K.W.; Leavesley, G.H.; Ahuja, L.R. A software engineering perspective on environmental modeling framework design: The Object Modeling System. *Environ. Model. Softw.* **2013**, *39*, 201–213. [CrossRef]
74. Wang, Y.; Chen, A.S.; Fu, G.; Djordjević, S.; Zhang, C.; Savić, D.A. An integrated framework for high-resolution urban flood modelling considering multiple information sources and urban features. *Environ. Model. Softw.* **2018**, *107*, 85–95. [CrossRef]
75. Swain, N.R.; Latu, K.; Christensen, S.D.; Jones, N.L.; Nelson, E.J.; Ames, D.P.; Williams, G.P. A review of open source software solutions for developing water resources web applications. *Environ. Model. Softw.* **2015**, *67*, 108–117. [CrossRef]
76. Vitolo, C.; Elkhatib, Y.; Reusser, D.; Macleod, C.J.A.; Buytaert, W. Web technologies for environmental Big Data. *Environ. Model. Softw.* **2015**, *63*, 185–198. [CrossRef]
77. Qiao, X.; Li, Z.; Ames, D.P.; Nelson, E.J.; Swain, N.R. Simplifying the deployment of OGC web processing services (WPS) for environmental modelling—Introducing Tethys WPS Server. *Environ. Model. Softw.* **2019**, *115*, 38–50. [CrossRef]
78. Gao, F.; Yue, P.; Zhang, C.; Wang, M. Coupling components and services for integrated environmental modelling. *Environ. Model. Softw.* **2019**, *118*, 14–22. [CrossRef]
79. Madani, K.; Pierce, T.W.; Mirchi, A. Serious games on environmental management. *Sustain. Cities Soc.* **2017**, *29*, 1–11. [CrossRef]
80. Savic, D.A.; Morley, M.S.; Khoury, M. Serious gaming for water systems planning and management. *Water* **2016**, *8*, 456. [CrossRef]
81. Medema, W.; Furber, A.; Adamowski, J.; Zhou, Q.; Mayer, I. Exploring the Potential Impact of Serious Games on Social Learning and Stakeholder Collaborations for Transboundary Watershed Management of the St. Lawrence River Basin. *Water* **2016**, *8*, 175. [CrossRef]
82. Sušnik, J.; Chew, C.; Domingo, X.; Mereu, S.; Trabucco, A.; Evans, B.; Vamvakeridou-Lyroudia, L.; Savić, D.; Laspidou, C.; Brouwer, F. Multi-Stakeholder Development of a Serious Game to Explore the Water-Energy-Food-Land-Climate Nexus: The SIM4NEXUS Approach. *Water* **2018**, *10*, 139. [CrossRef]
83. Van der Wal, M.M.; de Kraker, J.; Kroeze, C.; Kirschner, P.A.; Valkering, P. Can computer models be used for social learning? A serious game in water management. *Environ. Model. Softw.* **2016**, *75*, 119–132. [CrossRef]
84. Schall, G.; Zollmann, S.; Reitmayr, G. Smart Vidente: Advances in mobile augmented reality for interactive visualization of underground infrastructure. *Pers. Ubiquitous Comput.* **2013**, *17*, 1533–1549. [CrossRef]
85. Kamarainen, A.M.; Metcalf, S.; Grotzer, T.; Browne, A.; Mazzuca, D.; Tutwiler, M.S.; Dede, C. EcoMOBILE: Integrating augmented reality and probeware with environmental education field trips. *Comput. Educ.* **2013**, *68*, 545–556. [CrossRef]
86. Rockefeller Foundation 100 Resilient Cities. Available online: https://www.rockefellerfoundation.org/our-work/initiatives/100-resilient-cities (accessed on 28 February 2019).
87. Irwin, S.; Schardong, A.; Simonovic, S.; Nirupama, N. ResilSIM—A Decision Support Tool for Estimating Resilience of Urban Systems. *Water* **2016**, *8*, 377. [CrossRef]
88. Butler, D.; Ward, S.; Sweetapple, C.; Astaraie-Imani, M.; Diao, K.; Farmani, R.; Fu, G. Reliable, resilient and sustainable water management: The Safe & SuRe approach. *Glob. Chall.* **2017**, *1*, 63–77.

89. Klise, K.A.; Bynum, M.; Moriarty, D.; Murray, R. A software framework for assessing the resilience of drinking water systems to disasters with an example earthquake case study. *Environ. Model. Softw.* **2017**, *95*, 420–431. [CrossRef]
90. Kong, J.; Simonovic, S.P.; Zhang, C. Sequential Hazards Resilience of Interdependent Infrastructure System: A Case Study of Greater Toronto Area Energy Infrastructure System. *Risk Anal.* **2018**. [CrossRef]
91. Sweetapple, C.; Diao, K.; Farmani, R. A Tool for Global Resilience Analysis of Water Distribution Systems. In Proceedings of the WDSA/CCWI Joint Conference, Kingston, Canada, 23–25 July 2018.
92. Dunn, G.; Brown, R.R.; Bos, J.J.; Bakker, K. The role of science-policy interface in sustainable urban water transitions: Lessons from Rotterdam. *Environ. Sci. Policy* **2017**, *73*, 71–79. [CrossRef]
93. Chang, J.; Zhou, G.; Christensen, E.R.; Heideman, R.; Chen, J. Graphene-based sensors for detection of heavy metals in water: A review. *Anal. Bioanal. Chem.* **2014**, *406*, 3957–3975. [CrossRef]
94. OGC OGC®WaterML. Available online: https://www.opengeospatial.org/standards/waterml (accessed on 16 September 2019).
95. Horsburgh, J.S.; Tarboton, D.G.; Piasecki, M.; Maidment, D.R.; Zaslavsky, I.; Valentine, D.; Whitenack, T. An integrated system for publishing environmental observations data. *Environ. Model. Softw.* **2009**, *24*, 879–888. [CrossRef]
96. Daniele, L.; den Hartog, F.; Roes, J. Created in Close Interaction with the Industry: The Smart Appliances REFerence (SAREF) Ontology. In *International Workshop Formal Ontologies Meet Industries, August 2015*; Springer: Cham, Switzerland, 2015; pp. 100–112.
97. Howell, S.; Rezgui, Y.; Beach, T. Integrating building and urban semantics to empower smart water solutions. *Autom. Constr.* **2017**, *81*, 434–448. [CrossRef]
98. ETSI Smart Appliances. Available online: https://www.etsi.org/technologies/smart-appliances?jjj=1568651031056 (accessed on 16 September 2019).
99. Digital Single Market ETSI Releases Three New SAREF Ontology Specifications for Smart Cities, Industry 4.0 and Smart Agriculture. Available online: https://ec.europa.eu/digital-single-market/en/news/etsi-releases-three-new-saref-ontology-specifications-smart-cities-industry-40-and-smart (accessed on 16 September 2019).
100. FIWARE: THE OPEN SOURCE PLATFORM FOR OUR SMART DIGITAL FUTURE. Available online: https://www.fiware.org/ (accessed on 16 September 2019).
101. López-Riquelme, J.A.; Pavón-Pulido, N.; Navarro-Hellín, H.; Soto-Valles, F.; Torres-Sánchez, R. A software architecture based on FIWARE cloud for Precision Agriculture. *Agric. Water Manag.* **2017**, *183*, 123–135. [CrossRef]
102. FIWARE for the Next Generation Internet Services for the WATER Sector. Available online: https://cordis.europa.eu/project/rcn/223257/factsheet/en?WT.mc_id=RSS-Feed&WT.rss_f=project&WT.rss_a=223257&WT.rss_ev=a (accessed on 16 September 2019).
103. Kartakis, S.; Yu, W.; Akhavan, R.; McCann, J.A. Adaptive Edge Analytics for Distributed Networked Control of Water Systems. In *Proceedings of the 2016 IEEE First International Conference on Internet-of-Things Design and Implementation (IoTDI)*; IEEE: New York, NY, USA, 2016; pp. 72–82.
104. Conner, L.G.; Ames, D.P.; Gill, R.A. HydroServer Lite as an open source solution for archiving and sharing environmental data for independent university labs. *Ecol. Inform.* **2013**, *18*, 171–177. [CrossRef]
105. LeCun, Y.; Bengio, Y.; Hinton, G. Deep learning. *Nature* **2015**, *521*, 436–444. [CrossRef]
106. Fu, H.; Li, Z.; Liu, Z.; Wang, Z. Research on big data digging of hot topics about recycled water use on micro-blog based on particle swarm optimization. *Sustainability* **2018**, *10*, 2488. [CrossRef]
107. SURFACE WATER AND OCEAN TOPOGRAPHY. Available online: https://swot.jpl.nasa.gov/ (accessed on 16 September 2019).
108. DI020: Fundament Digital City <<Digitaal Wegennet Amsterdam & Voorzieningen als Gedragsregels en Openbaar Register>>>. Available online: https://amsterdamsmartcity.com/projects/digitaal-wegennet-amsterdam (accessed on 16 September 2019).
109. Bachmann, D.; Eilander, D.; de Leeuw, A.; de Bruijn, K.; Diermanse, F.; Weerts, A.; Beckers, J. Prototypes of risk-based flood forecasting systems in the Netherlands and Italy. *E3S Web Conf.* **2016**, *7*, 18018. [CrossRef]
110. Lund, N.S.V.; Falk, A.K.V.; Borup, M.; Madsen, H.; Steen Mikkelsen, P. Model predictive control of urban drainage systems: A review and perspective towards smart real-time water management. *Crit. Rev. Environ. Sci. Technol.* **2018**, *48*, 279–339. [CrossRef]

111. Garofalo, G.; Giordano, A.; Piro, P.; Spezzano, G.; Vinci, A. A distributed real-time approach for mitigating CSO and flooding in urban drainage systems. *J. Netw. Comput. Appl.* **2017**, *78*, 30–42. [CrossRef]
112. Sun, C.; Cembrano, G.; Puig, V.; Meseguer, J. Cyber-Physical Systems for Real-Time Management in the Urban Water Cycle. In *Proceedings of the 2018 International Workshop on Cyber-physical Systems for Smart Water Networks (CySWater)*; IEEE: New York, NY, USA, 2018; pp. 5–8.
113. Hasan, S.; Foliente, G. Modeling infrastructure system interdependencies and socioeconomic impacts of failure in extreme events: Emerging R & D challenges. *Nat. Hazards* **2015**, *78*, 2143–2168.
114. Popescu, I.; Jonoski, A.; Bhattacharya, B. Experiences from online and classroom education in hydroinformatics. *Hydrol. Earth Syst. Sci.* **2012**, *16*, 3935–3944. [CrossRef]
115. Abbott, M.B.; Solomatine, D.; Minns, A.W.; Verwey, A.; Van Nievelt, W. Education and training in hydroinformatics. *J. Hydraul. Res.* **1994**, *32*, 203–214. [CrossRef]
116. Wigdor, D.; Wixon, D. *Brave NUI World: Designing Natural User Interfaces for Touch and Gesture*, 1st ed.; Morgan Kaufmann Publishers Inc.: San Francisco, CA, USA, 2011; ISBN1 0123822319. ISBN2 9780123822314.
117. Richert, A.; Plumanns, L.; Gross, K.; Schuster, K.; Jeschke, S. Learning 4.0: Virtual Immersive Engineering Education. *Digit. Univ. Int. Best Pract. Appl.* **2015**, *2*, 51–66.
118. NTNU Art & Technology Task Force NTNU ARTEC. Available online: https://www.ntnu.edu/artec (accessed on 16 September 2019).
119. Slay, J.; Miller, M. Lessons Learned from the Maroochy Water Breach. In *Critical Infrastructure Protection*; Springer US: Boston, MA, USA, 2007; pp. 73–82.
120. Stewart, R.A.; Nguyen, K.; Beal, C.; Zhang, H.; Sahin, O.; Bertone, E.; Vieira, A.S.; Castelletti, A.; Cominola, A.; Giuliani, M.; et al. Integrated intelligent water-energy metering systems and informatics: Visioning a digital multi-utility service provider. *Environ. Model. Softw.* **2018**, *105*, 94–117. [CrossRef]
121. EU General Data Protection Regulation. Available online: https://eugdpr.org/ (accessed on 16 September 2019).
122. Galdon-Clavell, G. (Not so) smart cities?: The drivers, impact and risks of surveillanceenabled smart environments. *Sci. Public Policy* **2013**, *40*, 717–723. [CrossRef]
123. Kitchin, R. The ethics of smart cities and urban science. *Philos. Trans. R. Soc. A Math. Phys. Eng. Sci.* **2016**, *374*, 20160115. [CrossRef]

water

MDPI

Review

Water, Population Growth and Contagious Diseases †

Maurizio Iaccarino

Institute of Genetics and Biophysics "A. Buzzati–Traverso", CNR, 111 Via Pietro Castellino, 80131 Naples, Italy; iaccarin@igb.cnr.it; Tel.: +39-0816132431

† This paper is an elaboration of the text of a lecture given at the International Congress "*Laudato si'y Grandes Ciudades*", which took place in Rio de Janeiro, Brazil, in July 2017.

Received: 19 December 2018; Accepted: 19 February 2019; Published: 22 February 2019

Abstract: Water, essential for the biology of living organisms, is also important for agriculture, for the organization of social life and for culture. In this review we discuss the interrelationship between water availability and human population size. The total population of the globe, 3–5 million people between the years 25,000 and 5000 Before Common Era (BCE), increased about 50-fold in coincidence with the development of agriculture. Later on, after the year 200 Common Era (CE), the number of people did not change appreciably and increased slowly in the period 1000 to 1500 CE. We show that the main cause of this observed slow-down in population growth was the increase in population density, which caused the appearance and spreading of infectious diseases, often due to the use of contaminated water. Population started to increase again when people learned how to use appropriate sanitation and hygienic rules. The management of water resources, including transport of water to the areas where it is needed, separation and depuration of wastewater and production of freshwater by desalination, have become increasingly important. The population level is today very high and will continue to grow, thus causing a further increase in the density of people and an increased risk of contagious diseases. Therefore, more water for sanitation will be needed all over the world.

Keywords: freshwater needs; population density; epidemics; sanitation; global water requirements

1. Introduction

Water covers 70% of the surface of planet Earth. It is a thin layer, representing only 0.2% of the mass of our planet and it is found in a liquid or solid state or as a vapor. Most of the water (96.5%) is in the oceans, 1.7% is in the glaciers, 0.015% is in rivers, lakes and soil moisture; 1.6% is located underground, mainly in the aquifers. Humans use freshwater, namely water containing less than 0.5 parts per thousand of dissolved salts. It is estimated that the water volume of rivers is 2100 km^3 and that lakes and aquifers contain 91,000 km^3 and 10,530,000 km^3 of water, respectively. Water is used mainly (70–75%) for irrigation; 10–12% of it is used for direct human purposes (sanitation and drinking) and 15% for industrial uses (cooling, cleaning, processing, generating steam power) [1–3].

Freshwater comes from the water that precipitates on the ground: part of it replenishes rivers and lakes, part of it infiltrates the ground and stops where it finds an impermeable layer (for example clay), thus forming the aquifers [4]. The aquifers provide almost 20% of the freshwater used by humans but are not useful when, in 55% of the cases, they contain too much salt. In some cases, aquifers are contaminated with harmful chemical compounds; this may be a serious problem since it is almost impossible to decontaminate an aquifer, which as a consequence must be abandoned or the water extracted should be purified. The amount of water present in aquifers can vary with the seasons and sometimes a massive extraction leads to their exhaustion. Aquifers can cross one or more States and hence sometimes the use of water by a State can impoverish the portion of the aquifer that is beyond its boundary; in this case their management requires international collaboration and intergovernmental agreements. Indeed, in 2009 the UN approved a resolution that regulates relations between States for their use. The signatories of this resolution committed themselves to: (1) make a fair and reasonable use of cross-border aquifers located in their territory and, in cooperation with the other States involved, prepare a long-term plan of use; (2) avoid polluting the aquifers in their territory to prevent damage to neighbouring States. An inventory, compiled in 2015 by the International Groundwater Resources Assessment Centre, contains a list of 592 cross-border aquifers.

Water moves continuously from the oceans to rivers and back to the oceans. In 1715 Halley proposed the hydrological cycle: water evaporates from the seas and from the surface of the emerged lands and it then precipitates in the form of rain on the Earth's surface (including both the mainland and the oceans); the rain that falls on Earth flows on the ground and then, through the rivers, it reaches the sea [5]. The salinity of the oceans comes in part from erosion and transport from the mainland of salts dissolved in the water of the rivers but salts may also come from minerals dissolved from the bottom of the oceans.

Human beings need to drink 2 litres per day and have an autonomy of only 5 days. Water is essential for the life of cells because it is a solvent for macromolecules and participates in many biochemical reactions. The cell membrane is impermeable to water but special proteins forming the aquaporin channels [6] permit the transport of water, thus allowing the regulation of its intracellular concentration in specific cells [7]. This suggests that water may have different functions in different cells of the body and that the requirement for drinking water might be of crucial importance only for specific cells. Water is an important component of the fluids that surround the cells: it facilitates the movements of joints; it is essential for the uptake and digestion of nutrients, for their transport to all parts of the body and for the excretion of waste through urines. In human beings it is needed for the regulation of body temperature by means of sweating. The human body contains between 45% and 65% of water; this value decreases rapidly after death.

The salt concentration in the body of multicellular organisms, including plants, cannot increase over certain levels. Therefore, although living species presumably originated in the oceans, today they cannot withstand their high saline concentration (an average of 35 g NaCl per liter, as compared to 9 g per litre in the blood of human beings). Perhaps, when living beings evolved the salt concentration of the seas was lower than that observed today [8,9].

Salt toxicity might be due to an inhibition of water uptake through the membrane of some cells. In a certain sense it is analogous to the "reverse osmosis" effect: when salt is added to one side of a porous membrane it will cause water to move to the side where salt is added in order to dilute it and to achieve an equal salt concentration on either side of the membrane. Recently, evidence has been reported for osmotic homeostasis mediated by ion transport proteins and aquaporins in the gill of a fish (genus Cyprinodon) after freshwater and seawater acclimation [10]. Recent studies address the problem of salinity stress tolerance in plants in view of difficulties that may arise consequent to climate change and/or lack of water for irrigation [11].

In this review we discuss the relationship between water availability and the expansion of humans on Earth.

2. Water Availability and Human Population Growth

The necessity to find freshwater has been of utmost importance for humanity and has influenced the individual way of life and the organization of society. The earliest human settlements have been found predominantly near the borders of rivers or lakes. Parts of skeletons and remains of hunting tools were found in Ethiopia at the banks of some rivers; their dating indicates that they are 2–3 million years old and are the oldest among those today known. The rivers were used for drinking and refreshing; moreover, they were also places where food could be obtained through fishing and hunting. According to Herodotus, the fish were dried in the sun, grinded in a mortar, reduced to flour and then used to make buns or pies [12]. Hunting was facilitated by the fact that the rivers were also used for drinking by animals and hence the hunters would find it easier to catch them.

Our remote ancestors were hunter-gatherers, namely they obtained most of their food by killing wild animals or by collecting the fruits of wild plants. They lived near rivers or lakes to use water and moved to better places when the local resources became insufficient. The global population size did not increase appreciably, perhaps because of the high mortality rate but also because for nomadic groups it is difficult to take care of babies. A drastic increase in population growth was only possible following the change in the human lifestyle associated with the appearance of practices for food production. Agriculture started about 11,000 years ago, although in some parts of the world an increase in population size, probably correlated to an increased food production, started already 20,000 years ago [13].

Most anthropologists believe that agriculture was developed independently in different parts of the planet and after some time its progress became more rapid because knowledge and tools were brought to other places where the development of agricultural practices was already starting [14,15]. Three elements were crucial and they mutually reinforced each other: (1) plant domestication (namely the selection of plants suitable for human use); (2) animal domestication (namely the use of animals, like cattle and sheep, which helped working in the fields but also were useful for the production of milk and meat); and (3) abundant irrigation (which is essential to increase the fertility of the soil). Progress in these fields was probably not achieved through study and planning but by chance observation of the advantages obtained.

Plant domestication started with the use of specific plants that were found to be useful from a nutritional point of view. Later on people learned how to propagate them for the following year and in this way increased their production. Archaeologists find that the equivalent of modern ceramics was associated with the use of plants and this suggests that our ancestors discovered that cooking helped eating and digestion of plant products. Ceramic artefacts appear with the Neolithic, probably because in the Palaeolithic all food available was immediately consumed [16]. Studies with modern techniques show that plants typical of a certain region were brought and used somewhere else, even at a long distance, thus suggesting that their use was appreciated. Of course, due to climatic reasons, it was more efficient to move specific plants in an East-West-East direction than in a North-South-North direction, as indicated by the analysis of plant distribution and its comparison to their site of origin. It is likely that in many parts of the world the development of agriculture was slower because of the lack of useful plant species. With time, preference was given to plants that for empirical reasons were found to be more useful. Today we see that only a dozen plant species account for more than 80% of the annual crop on the Earth: five grains (wheat, corn, rice, barley and sorghum), one legume (soy), three tubers (potatoes, manioc and sweet potatoes), two sugar plants (sugar cane and sugar beet) and a fruit plant (banana). Cereals provide more than half of the calories consumed by the world's population [15]. Most people are fed with food produced in a farm and, if the current trend continues, within a short time the latest groups of hunter-gatherers will convert to agriculture, thus ending millions of years of history.

Animal domestication has been an important factor in the promotion of population growth because it was useful to increase agricultural productivity. Domestication implies to use animals that accept to live with human beings. Not all animal species can be domesticated: it appears that it is possible with animals that live in a herd and recognize a dominant member and a fixed hierarchy for each member of the herd. Apparently, they accept domestication when they recognize a human being as the leader of the herd and transmit this behaviour to the pets that are born in captivity [14]. Domestic animals provide meat, milk, wool and leather; their manure is very useful as a field fertilizer and sometimes it is used, after desiccation, to be burned to produce heat. They are also a source of energy, because they pull the ploughs and help moving agricultural machinery; thus, animals make it possible to overturn soils that would otherwise be left untreated and therefore contribute to increase the efficiency of farming. For a long time, they have been the only land transport available. Of course, they need food and therefore consume part of the energy they provide; an analysis of the energy produced and consumed by domestic animals has been discussed [17]. Meat from domestic animals replaced game as the primary source of protein and therefore the importance of hunting decreased.

Irrigation promotes an increase in agricultural productivity; on the other hand, the amount of water needed increases when the soil becomes potentially more productive as a consequence of plant and animal domestication. Thus, it was more efficient to use the new agricultural techniques in those parts of the planet where water was abundant and available constantly during the months of the ripening season of the plants. Agriculture started independently in different parts of the world: one of them was Mesopotamia, mainly because of the frequent flooding of the Tiger and Euphrates rivers and from there it expanded to the nearby zone, called the Fertile Crescent. It then reached Egypt, which was blessed by the river Nile. The floods of this river cover the two river banks with abundant water containing a nutrient-rich silt. They arrive once per year (due to the equatorial rains that fall upstream) and their arrival can be predicted with precision. The Egyptians built canals and reservoirs to improve the use of water, which was then assigned to specific farms. The farmers became sedentary and built huts (on the ground or on palisades) and stables for animals. They soon realized the importance of water availability for agricultural production and for this reason they developed structures to improve irrigation opportunities. This social system needed the collective work of many people and thus led to the necessity of a central authority, which was easily accepted. If a farm is used only for the cultivation of plants used for human nutrition one hectare of land can support the life of many more people (10 to 100 times) as compared to a piece of virgin land used by hunter-gatherers. On the other hand, sedentarism is associated with a decrease in mortality of infants as well as of adults, thus causing an increase in population size. As a consequence, more food is needed, thus favouring the groups that improve agricultural productivity. But of course more water per hectare is needed to support a higher level of productivity.

It should be noted that farmers live in a small portion of their cultivated land and prefer to live on the side of their property which is close to places inhabited by other farmers; therefore, when they were not working, their actual density was higher than that calculated per km^2 of their farm and they often found it useful to build structures for collective use. This societal structure is an opportunity for the development of cultural relationships and for information exchanges.

Agriculture caused an increase in population size (see below) and induced new forms of social interactions. At the same time, as described in the following chapter, this facilitated different types of contaminations, as for example human contacts with excrements of animal or human origin that cause transmission of pathogens through complex pathways. In the absence of sanitation and sewerage facilities that isolate faecal material from the environment, pathogenic microorganisms can spread into fields and ambient waters and thus cause the appearance of many epidemics [18]. Contamination with faecal sources has been shown to be harmful in recreational water bodies [19] or in child nutrition [20].

3. Contagious Diseases Limit the Rate of Population Growth

The size of the total population of the planet in pre-modern times is difficult to determine; the estimates reported by experts are extrapolations from archaeological findings and only few of them quote confidence intervals. In the absence of a straightforward means to assess the error of such estimates, a rough idea of expert consensus can be gained by comparing values given in independent publications. More recently the statistical analysis of sequence variations of genomes of contemporary humans has given information on the number of individuals present in the past [21,22] but these data are restricted to specific geographic areas.

Data reported by different authors [23,24] and quoted by Kremer [25] indicate that in the years 25,000 to 5000 BCE the total population of the planet was rather constant, at a level of 3 to 5 million people; it then started to increase (50 million people in the year 1000 BCE; 190 million people in the year 200 CE). However, in the following years the global number of people increased only slightly, to 265 million in the year 1000 CE and only 350 million people in the year 1400 CE. After the year

1500 the population size started to increase steadily and reached today, according to the United Nations Population Division, the number of 7550 million people [26].

The reason for the slow growth rate that occurred after the year 200 CE has seldom been discussed in the literature [27]. We propose here that the slow population growth was a consequence of the increase in population density, which in turn facilitated the spreading of contagious diseases; in many cases the epidemics were a consequence of the use of contaminated water [28]. Only when humanity understood the importance of sanitation and personal hygiene the mortality rate decreased and the number of people started to increase again.

With the advent of agriculture, many factors caused an increase in the transmission of infectious diseases: (1) the substantial increase in population density, which facilitates contagion; (2) the sharing of the same surroundings for sleeping, often used together with domestic animals; (3) the ignorance of the mechanisms of contagion and the non-observance of hygienic rules; (4) the accumulation of microorganisms in food stored for long time before eating; (5) the accumulation of manure, of animal or human origin and the lack of precautions in its use (microorganisms can make up to 60% of the dry mass of the faeces); (6) the use of contaminated water; (7) the increase in number of the insects acting as vectors of an infective agent; (8) travel to other regions, due for example to the trade of products, thus contributing to the spreading of diseases. The increase in morbidity rate connected with some of the agriculture-associated practices was already noted long time ago: even Herodotus writes "wheat is cultivated with manure and therefore the life of those who eat it is short" [29]. In conclusion, we stress the fact that the increased frequency of contagious diseases [28] was due to the lifestyle of farmers and to the increase in population density. An important factor in the spreading of contagious diseases was the high amounts of water needed for crops cultivation: this may be used for irrigation but it is dangerous if used for drinking.

Sometimes an infectious disease starts occurring and spreading in an animal species and then the microorganism causing it adapts to human beings (for example this was the case of tuberculosis, that originally appeared in bovines). The promiscuity of animals and humans facilitates a change in target species and the possibility of infecting humans as well: in fact, when an infection is successful, the pathogenic microorganism will multiply enormously (even many billion times), thus increasing the possibility of the chance appearance of few microorganisms that are able to infect also humans (and therefore multiply in humans and spread the disease among them). Moreover, the danger to become infected by a microorganism originated in animals increases substantially when farmers grow them in crowded places, like in the case of poultry or pigs breeding: an infectious agent spreads among the animals and therefore it increases its chances to propagate to humans. Moreover, poultry and livestock farms may be infected with a virus coming from the environment /for example an influenza virus from wild birds) and develop new strains that are able to infect humans and spread through respiratory droplets [30,31].

In conclusion, agriculture permitted an increase in the population size of the planet; but high population density, the proximity with domestic animals, the lack of personal hygiene and the use of contaminated water caused the appearance of many diseases that limited population growth. We wish to stress that population density is essential for human contagion but in some cases contaminated water is a deadly instrument for the spreading of the disease. We have the historical documentation of many epidemics that caused the death of millions of people [32]. The severity of epidemics is confirmed by the description of genetic diseases that confer some resistance to a specific infectious disease: in fact, the frequency of genotypes conferring resistance to a specific disease increases among the survivors to an epidemic [33].

Typhoid fever, caused by *Salmonella typhi*, is spread by contamination of food or drinking water with the faeces of an infected person or by the contact with flying insects feeding on faeces [34]. According to Thucydides, a plague, probably typhoid fever, killed in 430 BCE 25% of the population of Athens, thus ending the golden age of Pericles and the dominance of Athens in the Greek world [35,36].

The plague of Justinian, which first emerged during the reign of the Emperor of the Byzantine Empire Justinian, caused Europe's population to drop by around 50% between the 6th and 8th centuries CE. Its effect is clearly reflected in the data reported by Kremer and quoted above [25]: the number of people indicated in the year 400 CE is 190 million and increased only to 200 million in the year 600 CE.

The Black Death pandemic of the 14th century, caused by the bacterium *Yersinia pestis*, reduced the world population from an estimated 450 million in 1340 to between 350 and 375 million in 1400 (a drop of about 20%).

Many deadly smallpox epidemics are described in the literature [37]. They were described in India, Egypt and China as early as 1500 years BCE and several centuries later in Europe (smallpox was eradicated in 1977).

Gastroenteritis (also called diarrheal disease) may be caused by a virus or a bacterium or a parasite; it is usually caused by food or water contaminated by faeces but sometimes it comes directly from contacts with an infected person [38,39]. Gastroenteritis infections cause diarrhoea and have been deadly during history. Still today 2 to 5 billion cases of infectious diarrhoea occur per year, mainly in poor areas, where sanitation is not given enough care. Although their lethality decreased substantially, they may cause almost 1 million deaths per year, mainly in areas of greatest population growth and among young children [40,41].

Cholera, caused by the bacterium *Vibrio cholerae*, spreads mostly by the use of unsafe water but also by food contaminated with human faeces containing the bacteria [42]. Historical descriptions of a dysentery resembling cholera are found as early as the 5th century BCE. Humans are the only animals affected and still today 3 to 5 million people worldwide become sick with cholera, which causes 30,000 to 130,000 deaths per year. During cholera infection the aquaporin water transporter is down regulated, probably in an attempt of the host to counteract the abundant secretion of water with diarrhoea [43]. Mutations causing cystic fibrosis are rather frequent in humans, possibly because, when in the heterozygote state, they cause resistance to cholera [44]. In fact, the cystic fibrosis transmembrane conductance regulator has been suggested to activate a specific aquaporin in airway epithelial cells [45].

Malaria is caused by the protozoon *Plasmodium*, transmitted by a mosquito [46]. *Plasmodium* probably existed already long time ago (even 100,000 years) but its population size increased 10,000 years ago with the advent of agriculture and the development of human settlements. In some cases, farmers realized that it was better to live on top of the hills, where mosquitos are rare but they had to go downhill to work in the farms where marshes favoured the growth of infectious mosquitos and in this way the danger to contract malaria increased. Mutations in different human genes have been selected because, under certain conditions, cause resistance to malaria [47].

In the case of other diseases, the contagion takes place through direct contact with sick people (and of course the chances of this event increase when humans live in crowded places). Smallpox is caused by a virus, usually transmitted through droplets coming from the oral, nasal or pharyngeal mucosa of an infected person. Tuberculosis, caused by the bacterium *Mycobacterium tuberculosis*, is usually transmitted in droplets coming from an infected person.

Several other diseases are directly or indirectly caused by unsafe water: for example, amoebiasis, cryptosporidiosis, dengue, hepatitis A, giardiasis, legionellosis and so on [48].

In conclusion, many people in the past died because of infection by different microorganisms; one factor for diseases spreading was the high density of people; another was the use of contaminated material, in most cases water. For a long time, people believed that the use of water to clean the body was dangerous and as a consequence they did not wash themselves: they used perfumes to cover the unpleasant odours and removed fleas manually. A slow process led to the understanding that something could be done to prevent the spreading of diseases. One way to reduce contagion was to leave crowded towns and move to isolated places: in the Decameron (written in the year 1353) Boccaccio describes a group of young people that decide to leave Florence, where in the year 1348 a plague was killing many people and move to the country to avoid the danger of contagion. Later on, it was realized that the use of out of town thermal baths was not only useful for the convalescence of sick people but also to protect the health of the accompanying persons. Population growth started again and it was due mainly to a decline in mortality from infectious diseases [49], due to a better understanding of the danger to use contaminated water and of the importance of personal hygiene and of sanitary structures. Thus, water for sanitation has become important but it is required in amounts much higher than the 2 litres per day per person needed for drinking.

The frequency of waterborne diseases is higher in countries where water distribution and wastewater treatment are not appropriate; inadequate sanitation is considered by the World Health Organization responsible for 4% of all deaths worldwide. Of course, an epidemic originating in a specific country increases the probability to spread the disease to well organized countries and for this reason it is in the interest of all countries to prevent the occurrence of waterborne diseases; WHO provides an appropriate forum where to discuss this issue.

Prevention of the spreading of waterborne diseases requires an appropriate management of the water used by humans. At the same time the reduction in mortality rate consequent to an efficient prevention causes an increase in population size, thus leading to new necessities in water management.

4. Management of Water Resources

The primary source of freshwater for the use of human beings comes from the water that runs off after rain and feeds rivers, lakes and aquifers. However, freshwater is unevenly distributed on the planet: about three-quarters of the annual precipitations occur in specific areas and about 80% of the available freshwater is located in few basins, like in the Great Lakes of North America, in Africa and in the Baikal lake of Siberia; or in the five major fluvial systems: Amazon, Ganges with Brahmaputra, Congo, Yangtze and Orinoco. Large parts of the globe do not have enough water for human use and in many cases transport of water from nearby zones is required. In a growing number of cases freshwater is produced by desalination (see below).

The development of agricultural practices allowed the feeding of more people per hectare, thus inducing an increase in population size. Human beings soon realized that it was useful to move water to zones were land fertility could be increased through irrigation. Appropriate canals were constructed already a long time ago: remains of them, dating perhaps from the 6th millennium BCE, can be seen in Iran; others, built at least three millennia BCE, are found in India or China; and others, built one or two millennia BCE, have been found in Africa, departing from the Niger River. With the discovery of the New World, the Spanish *conquistadores* were astonished at the sight of a network of canals covering, on the Andes of Peru, an area of about 700 km^2; these canals were dug into the rock with rudimentary stone tools and without the help of animals.

Many canals were built in modern times and today the irrigated areas of the planet are very large: the 2008 estimate was about 3 million km^2, almost 70% of which are located in Asia, 15% in America and the rest distributed between Europe, Africa and Oceania. The following are few examples of modern structures built for the transfer of water. The aqueduct of Arizona is an open-air canal, 540 km long, constructed to provide irrigation of 400,000 hectares of land in Arizona. The aqueduct of California is a complex system of canals, tunnels and pipelines having a total length of about 1100 km. In Libya, the Great Man-Made River was built to transport water extracted from more than 1300 wells located in the South of the country in order to distribute it to the cities of Tripoli, Sirte and Benghazi. The transport of water required the assembly of large pipelines, for a total length of 3000 km. Unfortunately, the pipelines were damaged in 2011 during a war conflict and for this reason the Libyan cities are today facing a water shortage. This is a good example of the tight relationship between water needs and security [50]. Extensive use of water for irrigation has caused many cases of aquifers exhaustions, river depletion (for example the Yellow River, the Indus, the Colorado, the Nile) and a serious decrease in the volume of some lakes, as in the case of the Aral Sea [51]. Sometimes water sharing becomes a problem because of climate change or because two or more States do not agree in the sharing of the water of a lake or of a river or of an aquifer. A recent publication dealing with these problems is the World Water Development Report 2019 [52].

The management and supply of water for irrigation is facilitated by the use of dams, mostly built in modern times. In 1997 the World Commission on Dams estimated that 800,000 dams exist in the world and contribute to 12–16% of the global agricultural production.

As stressed above, agriculture also changed the nomadic habits of people, who became sedentary and started new activities, like the production of artisan tools and their trading: this led to a social division of labour and a social stratification, thus generating the appearance of towns. Today the management of water resources for the inhabitants of a city is a complex problem. It implies to provide enough water for drinking and for personal hygiene; to manage the disposal of wastewater and to use structures that avoid the spreading of waterborne diseases; at the same time, it is important to provide enough water for irrigation. In this respect, it is interesting to reflect on what happened during the development and expansion of ancient Rome. This city was founded in the 8th century BCE (the year 753, according to the legend) when it was inhabited by very few people, perhaps a few thousand; at the time of maximum expansion it approached (or perhaps exceeded) one million people. This remarkable increase was also permitted by the efficient management of the freshwater used and by an accurate wastewater disposal. The city of Rome grew from small settlements, mainly located on the hills called *Palatinus*, *Capitolinus* and *Aventinus*, which were close to a ford on the river Tiber. The inhabitants used the river for procuring water, for fishing, for navigation and to exchange goods with the Etruscans that lived on the other bank of the river. They also used water coming from few springs that were located on the top of the hills and disposed of their waste material downhill. The Romans were open to immigration and therefore the number of inhabitants increased rapidly. As a consequence, they needed: (1) more space; (2) more water; and (3) a waste disposal structure. They began to settle in the space located between the hills and increased the provision of water by taking it from springs located far from Rome and transported through aqueducts built for this purpose. They soon realized that drinking water should not be mixed with wastewater and, to cope with the problems related to the needs of waste disposal, they started building a sewer. This was originally built around the year 600 BCE as an open-air canal that drained the sewage to the river Tiber; later on, the Romans covered the canal and turned it into a sewer system for the city, called the "Greatest Sewer" (*Cloaca Maxima*). In the city of Rome, therefore, we observe the first example of an accurate separation between clean water and wastewater (today many countries have built sewer networks and wastewater treatment facilities that have reduced the incidence of waterborne diseases). At the time of maximum expansion, the city of Rome utilized 11 aqueducts that carried 700,000 m^3 of water per day. At the same time, about 600 aqueducts were built throughout the Roman Empire (in Italy, Spain, Germany, France and North Africa) and contributed to the foundation or expansion, of many cities. The main reason to build the aqueducts out of Rome was to increase the efficiency of irrigation and thus to produce food to be imported (and cope with the problem of lack of water for irrigation). In fact, cities need land to grow vegetables and water to irrigate it. But cities have very little space for agriculture and therefore they need to import food. When they do it, they indirectly import the water that has been used to produce it, namely they import water-intensive commodities. The "water footprint" concept has been introduced by Hoekstra and collaborators. It refers to the total volume of freshwater used to produce a specific good, under standard conditions [53].

As discussed above, a long time ago people changed their nomadic habits, became sedentary and established themselves near the farms where they worked. Recently however, almost a century ago, a drastic social change took place: with the advent of the industrial revolution and the increased productivity in agriculture (the so-called green revolution), job opportunities in the country-side decreased and people were attracted by better salaries in the cities or by a higher standard of living and centralized services. The social, economic and environmental problems associated with a predominantly urbanized population are considerably different from those of the rural population of the past. Moreover, the population density is higher and therefore the spreading of contagious diseases and of antibiotic resistant pathogens becomes more efficient; therefore, the quantity of water necessary for personal hygiene and sanitation becomes higher [54]. At the same time antibiotic resistant infections are becoming more frequent and their danger is discussed within WHO among the States member of this organization [55].

Large metropolitan areas encounter special problems that are not present in small or medium size cities. A megacity is usually defined as a metropolitan area with a total population in excess of ten million people. The new megacities experienced a recent and rapid urbanization, which sometimes modifies the territory and alters the pre-existing water fluxes, thus creating the necessity to adapt the infrastructures for distribution. Moreover, the administrative boundaries of a megacity may not coincide with those previously existing in the urbanized area and therefore the upgrading of the networks for water delivery and the infrastructures for sewage disposal is more difficult to manage. Finally, in megacities the percentage of impervious soil is high as compared to that of villages and therefore the sewerage system should be able to combine the management of the usual type of urban waste with the abundant rain water that occasionally comes from the streets when it rains. In conclusion, in large cities water is distributed through complex and expensive structures; moreover, it is important to establish an accurate system of sanitation performed under the surveillance of a central authority.

Water supply policies and regulations are under the responsibility of the highest Authority of the country. In the ancient Rome the top officer, named "*curator aquarum*" reported directly to the Emperor. Today the Member States of the European Union conform their decisions to the European Directive number 60/2000. Water and sanitation policy in the USA is under the responsibility of the Environmental Protection Agency, which reports directly to the President. In most other countries the responsibility is entrusted directly to different ministries, as the Ministry of Environment or the Ministry of Health, of Public Works and so forth. In general, the political authorities publish each year a report on the diseases presumably caused by contaminated water and share this information with the World Health Organization. The publication of these data contributes to raise the awareness of the public opinion and thus it may induce the governments to increase the prevention of diseases caused by contaminated water.

Water supply for personal use should in principle reach all private houses, 24 h per day, at constant pressure. It should be clean, non-toxic and free from pathogenic microorganisms. For this reason, before distribution, the level of a detailed list of substances is analysed, as well as the presence of specific microorganisms; the analyses are repeated at predetermined intervals of time and numerous samples are taken at different points of the distribution chain. The results of the analyses are communicated to the authority entrusted for surveillance and are publicized. When a specific parameter increases over a certain value, predetermined precautions are enforced. It should be pointed out that taste cannot be determined through chemical and microbiological analyses and therefore water distributed by water companies, although safe, sometimes is not pleasant to drink. Furthermore, the parameters analysed are not necessarily complete, since new advances in technology may introduce into the environment new substances that may be toxic [56]. For this reason, the existing rules are sometimes revised and the analysis of one or more parameters is added.

Water has become very important from an economic point of view but the economic value of something that is essential for human survival and for human dignity is difficult to define. Moreover, lack of water and of sanitation structures favours the appearance of contagious diseases that later on spread to all sectors of the population: it is therefore in the interest of society to make available to everybody enough water for sanitation. One principle that is widely accepted is that the price should be affordable by all citizens, including the very poor. Moreover, when there is a shortage, water should be distributed with impartiality.

Wastewater management for appropriate sanitation is more important today as compared to the past: in fact, our hunter-gatherer ancestors produced very little pollution per person as compared to what happens today and water contamination is today more frequent because of the presence of domestic animals [57]. Wastewater sanitation is today insufficient in many parts of the world and proper management of sewer material should become a priority. This is important to preserve the quality of freshwater necessary for humanity: in fact, today, as the percentage of the total available freshwater required for human purposes is increasing (see below), it is becoming more important to preserve the water that is not used. The effluent of the wastewater treatment plants sometimes is reused for different purposes, as recovery of nutrients or irrigation or even drinking. Wastewater treatment is usually adapted to the type of material received by the depurator plant. After separation of solid material, a prolonged aeration is used to allow microbial digestion of organic substances and of nitrates, followed by disinfection to kill pathogenic bacteria. Membrane filtration may be used to remove some impurities. Thus, the wastewater treatment plants contain complex apparatus of different types. Wastewater treatment is more complex in a city as compared to that required for a small village [58]. Sewage treatment plants receive all types of hazardous waste from households, hospitals and industries. Combined sewers require much larger and more expensive treatment facilities as compared to plants serving small, well defined areas and therefore it is sometimes more practical and less expensive to use different waste disposal plants that are specialized for the substances produced in different portions of the city. In fact, there are potentially thousands of components of sludge that remain untested or undetected and are disposed of from modern society and that have been proven to be hazardous to both human and ecological health. Residents living in certain zones of the city sometimes show an increased risk for certain respiratory, gastrointestinal and other diseases [59]. Although correlation does not imply causation, these observations may lead to conclude that precaution is necessary.

Water desalination is widely used in different parts of the world [60]. According to the International Desalination Association, in 2015 there were more than 18,000 plants worldwide, mainly located in arid areas, providing water for 300 million people. Most of these plants use membranes to separate water (usually seawater) from the dissolved salts; energy is needed to apply pressure on one side of the membrane. Less energy would be required if the permeability of membranes to water is selectively enhanced and some private companies are beginning to try to use analogues of the aquaporin proteins to increase the efficiency of the process [61,62].

In conclusion, the management of water availability has been a major factor in the development of humanity. Therefore, it is crucial to be prepared to make water available for a growing population and at the same time take care of appropriate sanitation.

5. Future Needs

We describe an intricate relationship between water availability, food production, increase in the world population, contagious diseases.

In principle a human being needs only 2 L of water per day for drinking. Hunter-gatherers needed very little extra water but today the situation is drastically changed and human beings need enormous quantities of water: in the year 2000, according to the Environmental Outlook published by OECD [63], 2384 km^3 of water were consumed for irrigation; 504 km^3 were consumed for industrial purposes and 348 km^3 for domestic purposes. Note that water for irrigation is needed in specific places (where farms are present) and at specific times (when plant growth and ripening are occurring) [51], thus making this requirement more stringent. Moreover, as described in this Review, the use of domestic water is important to prevent the spreading of contagious diseases. It is worth noting that the human population increased because of agriculture production and developed where water was abundant; but today much water per person is needed to support people in areas where water is scarce. Globally, much more water will be needed in the future and the above quoted OECD document predicts that the total amount of water needed on the globe, 3236 km^3 in the year 2000, will increase to 5420 km^3 in the year 2050.

Water availability is not sufficient today [64]: in the poorest areas of the planet more than 2 billion people use on average 10 litres of water per day per person, thus generating migrations and wars.

Moreover, it would be a common interest of poor and rich countries to provide enough water for sanitation to avoid the occurrence and spreading of infectious diseases [65].

The population of the planet increased in the last centuries mainly because of a decrease in mortality [49] and therefore it is unlikely that the present trend of population increase will change in the immediate future. Thus, the freshwater requirement will increase dramatically, not only to give water in an equitable way to everybody on the planet but also because demographic studies (see Table 1 in Ref. [26]) indicate that the total amount of people on the globe will increase to about 9800 million by the year 2050: therefore, we should be prepared to give enough water to an increased population [66,67]. Moreover, the density of people in certain areas will increase, thus increasing the danger of disease contagion; as a consequence, the use of water for sanitation will become more important and more attention should be given to the prevention of some diseases (for example through vaccination or better ecological management).

The water volume of rivers is about 2100 km^3 [3]; keeping in mind that part of this water is used to feed lakes and aquifers, we should begin to think in terms of total amount of rainfall water per year and compare the human needs to the figures assigned to the different components of the water repositories.

The provision of water will have a great social, financial and political relevance; all social sectors will have to be involved and many habits will have to change.

Acknowledgments: The author warmly thanks Anna Maria Aliperti for manuscript editing.

Conflicts of Interest: The author declares no conflict of interest.

References

1. Gleick, P.H. *Water in Crisis: A Guide to the World's Fresh Water Resources*; Oxford University Press: New York, NY, USA, 1993; ISBN 9780195076288.
2. Boberg, J. *Withdrawing Liquid Assets: How Demographic Changes and Water Management Policies Affect Freshwater Resources*; Rand Labor and Population: Santa Monica, CA, USA, 2005; ISBN 0-8330-3807-9.
3. Shiklomanov, I.A. Appraisal and Assessment of World Water Resources. *Water Int.* **2009**, *25*, 11–32. [CrossRef]
4. Eckstein, G. *The International Law of Transboundary Groundwater Resources*; Taylor Francis Ltd.: Abingdon, UK, 2017; ISBN 9781138842991.
5. Bengtsson, L. *The Earth's Hydrological Cycle*; Springer: Berlin, Germany, 2014; ISBN 978-94-017-8789-5.
6. Agre, P.; Brown, D.; Nielsen, S. Aquaporin water channels: Unanswered questions and unresolved controversies. *Curr. Opin. Cell Biol.* **1995**, *7*, 472–483. [CrossRef]
7. Kapilan, R.; Vaziri, M.; Zwiazek, J.J. Regulation of aquaporins in plants under stress. *Biol. Res.* **2018**, *51*, 4. [CrossRef] [PubMed]
8. Curry, R.; Dickson, B.; Yashayaev, I. A change in the freshwater balance of the Atlantic Ocean over the past four decades. *Nature* **2003**, *426*, 826–829. [CrossRef] [PubMed]
9. Takei, Y. From Aquatic to Terrestrial Life: Evolution of the Mechanisms for Water Acquisition. *Zool. Sci.* **2015**, *32*, 1–7. [CrossRef] [PubMed]
10. Lema, S.C.; Carvalho, P.G.; Egelston, J.N.; Kelly, J.T.; McCormick, S.D. Dynamics of gene expression responses for ion transport proteins and aquaporins in the gill of a euryhaline pupfish during freshwater and high salinity acclimation. *Physiol. Biochem. Zool.* **2018**, *91*, 1148–1171. [CrossRef] [PubMed]
11. Jha, U.C.; Bohra, A.; Jha, R.; Parida, S.K. Salinity stress response and 'omics' approaches for improving salinity stress tolerance in major grain legumes. *Plant Cell Rep.* 2019. [CrossRef]
12. Herodotus. *Histories, Book 1: Clio*; SMK Books: Plano, TX, USA, 2014; p. 200, ISBN-10 1617207691.
13. Leonardi, M.; Barbujani, G.; Manica, A. An earlier revolution: Genetic and genomic analyses reveal pre-existing cultural differences leading to Neolithization. *Sci. Rep.* **2018**, *9*, 682. [CrossRef]
14. Diamond, J. *Guns, Germs and Steel: The Fates of Human Societies*; W.W. Norton: New York, NY, USA, 1997; ISBN 0-393-31755-2.

15. Hassan, F. *Water History for Our Times. IHP Essays on Water History*; UNESCO Publishing: Paris, France, 2011; Volume 2, Available online: http://unesdoc.unesco.org/images/0021/002108/210879e.pdf (accessed on 19 February 2019).
16. Bellwood, P. *First Farmers: The Origins of Agricultural Societies*; Wiley-Blackwell: Hoboken, NJ, USA, 2004; ISBN 978-0-631-20566-1.
17. Smil, V. *Energy and Civilization: A History*; MIT Press: Cambridge, MA, USA, 2017; ISBN 9780262035774.
18. Ercumen, A.; Pickering, A.J.; Kwong, L.H.; Arnold, B.F.; Parvez, S.M.; Alam, M.; Sen, D.; Islam, S.; Kullmann, C.; Chase, C.; et al. Animal faeces contribute to domestic fecal contamination: Evidence from *E. coli* measured in water, hands, food, flies and soil in Bangladesh. *Environ. Sci. Technol.* **2017**, *51*, 8725–8734. [CrossRef]
19. Schoen, M.E.; Soller, J.A.; Ashbolt, N.J. Evaluating the importance of faecal sources in human-impacted waters. *Water Res.* **2011**, *45*, 2670–2680. [CrossRef]
20. Headey, D.; Nguyen, P.; Kim, S.; Rawat, R.; Menon, P. Is Exposure to Animal Feces Harmful to Child Nutrition and Health Outcomes? A Multicountry Observational Analysis. *Am. J. Trop. Med. Hyg.* **2017**, *96*, 961–999. [CrossRef] [PubMed]
21. Schraiber, J.G.; Akey, J.M. Methods and models for unravelling human evolutionary history. *Nat. Rev. Genet.* **2015**, *16*, 727–740. [CrossRef] [PubMed]
22. Petraglia, M.; Clarkson, C.; Boivin, N.; Haslam, M.; Korisettar, R.; Chaubey, G.; Ditchfield, P.; Fuller, D.; James, H.; Jones, S.; et al. Population increase and environmental deterioration correspond with microlithic innovations in South Asia ca. 35,000 years ago. *Proc. Natl. Acad. Sci. USA* **2009**, *106*, 12261–12266. [CrossRef] [PubMed]
23. McEvedy, C.; Jones, R. *Atlas of World Population History*; Penguin: London, UK, 1978; ISBN1 0713910313, ISBN2 9780713910315.
24. Biraben, J.-N. Essai sur l'évolution du nombre des hommes. *Population* **1979**, *34*, 13–25. [CrossRef]
25. Kremer, M. Population Growth and Technological Change: One Million, B.C. to 1990. *Q. J. Econ.* **1993**, *108*, 681–716. [CrossRef]
26. World Population Prospects. Available online: https://esa.un.org/unpd/wpp/Publications/Files/WPP2017_KeyFindings.pdf (accessed on 19 February 2019).
27. Armelagos, G.J.; Goodman, A.H.; Jacobs, K.H. The origins of agriculture: Population growth during a period of declining health. *Popul. Environ.* **1991**, *13*, 9–22. [CrossRef]
28. Percival, S.L.; Yates, M.V.; Williams, D.D.; Chalmers, R.; Gray, N. *Microbiology of Waterborne Disease*; Elsevier: Amsterdam, The Netherlands, 2014; ISBN 978-0-12-415846-7.
29. Herodotus. *Histories, Book 8 (Thalia)*; Cambridge University Press: Cambridge, UK, 2008; ISBN 9780521575713.
30. Yassine, H.M.; Lee, C.W.; Saif, Y.M. Interspecies transmission of influenza a viruses between Swine and poultry. *Curr. Top. Microbiol. Immunol.* **2013**, *370*, 227–240. [CrossRef]
31. Donatelli, I.; Castrucci, M.R.; De Marco, M.A.; Delogu, M.; Webster, R.G. Human-Animal Interface: The Case for Influenza Interspecies Transmission. *Adv. Exp. Med. Biol.* **2017**, *972*, 17–33. [CrossRef]
32. Kohn, G.C. *Encyclopedia of Plague and Pestilence: From Ancient Times to the Present*; Checkmark Books: Princeton, NJ, USA, 2002; p. 454, ISBN-10 9780816048939.
33. Withrock, I.C.; Anderson, S.J.; Jefferson, M.A.; McCormack, G.R.; Mlynarczyk, G.S.A.; Nakama, A.; Lange, J.K.; Berg, C.A.; Acharya, S.; Stock, M.L.; et al. Genetic diseases conferring resistance to infectious diseases. *Genes Dis.* **2016**, *2*, 247–254. [CrossRef]
34. Crump, J.A.; Mintz, E.D. Global trends in typhoid and paratyphoid fever. *Clin. Infect. Dis.* **2010**, *50*, 241–246. [CrossRef]
35. Thucydides. *History of the Peloponnesian War*; Penguin Books: London, UK, 2004; Chapter VII; ISBN 9780142004371.
36. Littman, R.J. The Plague of Athens: Epidemiology and Paleopathology. *Mt. Sinai J. Med.* **2009**, *76*, 456–467. [CrossRef] [PubMed]
37. Hopkins, D.R. *The Greatest Killer: Smallpox in History*; University of Chicago Press: Chicago, IL, USA, 2002; ISBN 978-0-226-35168-1.
38. Lim, M.L.; Wallace, M.R. Infectious diarrhea in history. *Infect. Dis. Clin. N. Am.* **2004**, *18*, 261–274. [CrossRef] [PubMed]

39. Prüss-Ustün, A.; Bartram, J.; Clasen, T.; Colford, J.M., Jr.; Cumming, O.; Curtis, V.; Bonjour, S.; Dangour, A.D.; De France, J.; Fewtrell, L.; et al. Burden of disease from inadequate water, sanitation and hygiene in low- and middle-income settings: A retrospective analysis of data from 145 countries. *Trop. Med. Int. Health* **2014**, *19*, 894–905. [CrossRef] [PubMed]
40. Ercumen, A.; Gruber, J.S.; Colford, J.M., Jr. Water distribution system deficiencies and gastrointestinal illness: A systematic review and meta-analysis. *Environ. Health Perspect.* **2014**, *122*, 651–660. [CrossRef] [PubMed]
41. GBD Diarrhoeal Diseases collaborators. Estimates of global, regional and national morbidity, mortality and aetiologies of diarrhoeal diseases: A systematic analysis for the Global Burden of Disease Study 2015. *Lancet Infect. Dis.* **2017**, *17*, 909–948. [CrossRef]
42. Nair, G.B.; Takeda, Y. *Cholera Outbreaks*; Springer: Berlin/Heidelberg, Germany, 2014; ISBN 978-3-642-55404-9.
43. Flach, C.F.; Qadri, F.; Bhuiyan, T.R.; Alam, N.H.; Jennische, E.; Holmgren, J.; Lönnroth, I. Differential expression of intestinal membrane transporters in cholera patients. *FEBS Lett.* **2007**, *581*, 3183–3188. [CrossRef]
44. Gabriel, S.E.; Brigman, K.N.; Koller, B.H.; Boucher, R.C.; Stutts, M.J. Cystic fibrosis heterozygote resistance to cholera toxin in the cystic fibrosis mouse model. *Science* **1994**, *266*, 107–109. [CrossRef] [PubMed]
45. Schreiber, R.; Nitschke, R.; Greger, R.; Kunzelmann, K. The cystic fibrosis transmembrane conductance regulator activates aquaporin 3 in airway epithelial cells. *J. Biol. Chem.* **1999**, *274*, 11811–11816. [CrossRef] [PubMed]
46. Carter, R.; Mendis, K.N. Evolutionary and Historical Aspects of the Burden of Malaria. *Clin. Microbiol. Rev.* **2002**, *15*, 564–594. [CrossRef]
47. Kwiatkowski, D.P. How Malaria has affected the human genome and what human genetics can teach us about malaria. *Am. J. Hum. Genet.* **2005**, *77*, 171–192. [CrossRef]
48. Water Related Diseases Surveillance. Available online: http://www.unece.org/fileadmin/DAM/env/water/cadialogue/docs/Almaty_Oct2011/Presentations_Eng/11.Shinee_WRD_Eng.pdf (accessed on 19 February 2019).
49. McKeown, T. *The Modern Rise of Population*; Academic Press: Cambridge, MA, USA, 1976; ISBN-10 0124855504.
50. Feitelson, E.; Tubi, A. A main driver or an intermediate variable? Climate change, water and security in the Middle East. *Glob. Environ. Chang.* **2017**, *44*, 39–48. [CrossRef]
51. Falkenmark, M.; Lannerstad, M. Consumptive water use to feed humanity—Curing a blind spot. *Hydrol. Earth Syst. Sci.* **2005**, *9*, 15–28. [CrossRef]
52. World Water Development Report 2019. Available online: https://en.unesco.org/events/world-water-development-report-2019-official-launch (accessed on 19 February 2019).
53. Hoekstra, A.Y.; Mekonnen, M.M. The Water footprint of Humanity. *Proc. Natl. Acad. Sci. USA* **2012**, *109*, 3232–3237. [CrossRef] [PubMed]
54. Wuijts, S.; van den Berg, H.H.; Miller, J.; Abebe, L.; Sobsey, M.; Andremont, A.; Medlicott, K.O.; van Passel, M.W.; de Roda Husman, A.M. Towards a research agenda for water, sanitation and antimicrobial resistance. *J. Water Health* **2017**, *15*, 175–184. [CrossRef] [PubMed]
55. World Health Organization. Antimicrobial Resistance: An Emerging Water, Sanitation and Hygiene Issue. World Health Organization Briefing Note. WHO/FWC/WSH/14.7. Available online: https://www.who.int/water_sanitation_health/publications/antimicrobial-resistance/en/ (accessed on 19 February 2019).
56. Golea, D.M.; Upton, A.; Jarvis, P.; Moore, G.; Sutherland, S.; Parsons, S.A.; Judd, S.J. THM and HAA formation from NOM in raw and treated surface waters. *Water Res.* **2017**, *112*, 226–235. [CrossRef] [PubMed]
57. *Wastewater, The Untapped Resource*; The United Nations World Water Development Report 2017; UNESCO Digital Library: Paris, France, 2017; ISBN 978-92-3-100201-4.
58. Iaccarino, M. Como obtenir agua de buena calidad en las ciudades en desarrollo. In *Laudato si' y grandes ciudades. Propestuas de una Ecologia Integral*; Los Libros de la Catarata: Madrid, Spain, 2018; ISBN 978-84-9097-448-3.
59. Rushton, L. Health hazards and waste management. *Br. Med. Bull.* **2003**, *68*, 183–197. [CrossRef] [PubMed]
60. Kucera, J. *Desalination: Water from Water*; Wiley: Hoboken, NJ, USA, 2014; ISBN-13 978-1118208526.
61. Nielsen, C.H. Biomimetic membranes for sensor and separation applications. *Anal. Bioanal. Chem.* **2009**, *395*, 697–718. [CrossRef] [PubMed]

62. To, J.; Torres, J. Can Stabilization and Inhibition of Aquaporins Contribute to Future Development of Biomimetic Membranes? *Membranes* **2015**, *5*, 352–368. [CrossRef] [PubMed]
63. OECD. Environment: The Water Challenge: Sharing a Precious Commodity. March 2012. Available online: http://www.oecd.org/environment/environmentthewaterchallengesharingapreciouscommodity.htm (accessed on 19 February 2019).
64. Rodell, M.; Famiglietti, G.S.; Wiese, D.N.; Reager, J.T.; Beaudoing, H.K.; Landerer, F.W.; Lo, M.-H. Emerging Trends in global freshwater availability. *Nature* **2018**, *557*, 651–659. [CrossRef] [PubMed]
65. United Nations Development Programme. *Human Development Report, Beyond Scarcity: Power, Poverty and the Global Water Crisis*; UNDP: New York, NY, USA, 2006; ISBN 0-230-50058-7.
66. Vorosmarty, C.J.; Green, P.; Salisbury, J.; Lammers, R.B. Global water resources: Vulnerability from climate change and population growth. *Science* **2000**, *289*, 284–288. [CrossRef]
67. Schlosser, C.A.; Strzepek, K.; Gao, X.; Fant, C.; Blanc, E.; Paltsev, S.; Jacoby, R.; Reilly, J.; Gueneau, A. The future of global water stress: An integrated assessment. *Earths Future* **2014**, *2*, 341–361. [CrossRef]

water

MDPI

Article

Robust Assessment of Uncertain Freshwater Changes: The Case of Greece with Large Irrigation—and Climate-Driven Runoff Decrease

Georgia Destouni * and Carmen Prieto

Department of Physical Geography, Bolin Centre for Climate Research and Navarino Environmental Observatory, Stockholm University, 10691 Stockholm, Sweden; carmen.prieto@natgeo.su.se
* Correspondence: georgia.destouni@natgeo.su.se; Tel.: +46-8-164785

Received: 1 October 2018; Accepted: 8 November 2018; Published: 13 November 2018

Abstract: We develop a data-driven approach to robustly assess freshwater changes due to climate change and/or human irrigation developments by use of the overarching constraints of catchment water balance. This is applied to and tested in the high-uncertainty case of Greece for five nested catchments of different scales across the country and for freshwater changes from an early period (1930–1949) with small human influences on climate and irrigation to a recent period (1990–2009) with expected greater such influences. The results show more or less equal contributions from climatic decrease in precipitation and from human irrigation development to a considerable total decrease in runoff (*R*) over Greece. This is on average -75 ± 10 mm/year and is greatest for the Ionian catchment in the west (-119 ± 18 mm/year) and the Peloponnese catchment in the south (-91 ± 16 mm/year). For evapotranspiration (*ET*), a climate-driven decrease component and an irrigation-driven increase component have led to a net total increase of *ET* over Greece. This is on average 26 ± 7 mm/year and is greatest for the Mainland catchment (29 ± 7 mm/year) and the Aegean catchment in the east (28 ± 6 mm/year). Overall, the resulting uncertainties in the water-balance constrained estimates of *R* and *ET* changes are smaller than the input data uncertainties.

Keywords: freshwater changes; irrigation; climate change; evapotranspiration; runoff; catchments; Greece

1. Introduction

Water is a peculiar liquid [1], critical for human existence and all life as we know it. Accurately understanding and being able to predict water conditions changes is among humanity's greatest needs and challenges [2,3]. This does not only apply to the major ocean part of Earth's hydrosphere but also to the essential freshwater on land. Freshwater changes on land interact with other Earth System changes, including changes in climate and in the landscape [4,5] and in both the ecosystems [6] and the societal systems [7–9] of the latter. In recognition of these interactions and critical open questions related to them, new hydrological research directions have developed such as the Panta Rhei community initiative [10] and the fields of ecohydrology [11] and sociohydrology [12].

The freshwater system itself includes various landscape manifestations and formations (streams, lakes, wetlands, water in biota, soil water, groundwater, glaciers, permafrost), with different aspects of variability and change interacting and propagating across them [13–16]. This propagation also involves interactions with different human uses of freshwater and structures for managing these, and all these freshwater interactions prevail and are organized within (the surface and subsurface parts of) hydrological catchments of different scales.

Furthermore, each hydrological catchment of any scale is subject to overarching water balance. In this, the total area-integrated water-fluxes of precipitation (*P*) and evapotranspiration (*ET*) and the concurrent water-storage changes occurring over the catchment are balanced with each other and the total runoff flux (*R*; area-normalized total discharge) from the catchment [17]. This emergent [18] catchment water balance is then also in simultaneous alignment with (i.e., both depends on and determines) and as such couples the different local water manifestations, formations, uses and management structures and the conditions of and changes in these over each catchment [19]. However, our ability to understand, interpret and predict these conditions and changes is currently fragmented among various research fields, associated with the different (liquid, frozen) freshwater manifestations and formations (e.g., surface water hydrology, hydrogeology, hydro-meteorology, ecohydrology, limnology, cryospheric science), or the different human uses of and management structures for freshwater (e.g., water resources engineering, hydraulic engineering, agricultural water management, urban water management, sociohydrology).

On the one hand, such freshwater fragmentation may be needed for in-depth process study and understanding. On the other hand, it may lead to important knowledge gaps and uncertainties between the fragments. Decrease of these gaps and uncertainties require ability to synthesize and jointly interpret various types of data, related to different freshwater aspects, which interact and coevolve with each other and other landscape and climate changes over hydrological catchments of different scales. For such synthesis and joint interpretation of different data for assessment of overall freshwater conditions and changes in a catchment of any scale, various studies have proposed [6,20] or indicated [4,5,19] that the constraints implied by catchment water balance may decrease the combined result uncertainties compared to the uncertainties associated with different underlying data and aspects.

In this paper, we put this indication to further testing for a case of particularly large underlying uncertainties [21]: Greece and five nested catchments of different scales across it. For these catchments, we investigate freshwater conditions and their changes over the time period 1930–2009, for which relevant data are available. Using the constraints implied by overarching water balance in each catchment, we develop, apply and test a general approach to synthesize, assess and interpret the available data for freshwater changes over the study period, the possible drivers of these changes and associated input-data and result uncertainties. Through the development and specific case quantification and testing of this approach, this study also outlines and uses advancements made in recent research on such long-term, large-scale freshwater changes.

2. Materials and Methods

2.1. Greece and Its Catchments

For several reasons, Greece is a relevant and useful case example for the present investigation. The country constitutes in itself the major part of a regional catchment draining into the Ionian and Aegean parts of the eastern Mediterranean Sea (Figure 1). Through Greece, this regional catchment is more or less peninsular and its long coastline facilitates selection of various nested coastal catchments of different scales within it, for comparative multi-catchment investigation using different parts of the available data for the total regional catchment.

Furthermore, previous comparative multi-catchment studies have shown that hydro-climatic changes and associated freshwater interactions with the atmosphere are subject to particularly large uncertainties in Greece [21]. These include: (a) uncertainties due to observation data limitations in terms of both temporal extent and spatial coverage across the country; and (b) climate model uncertainties for freshwater conditions, especially for model estimations of *ET* and, as a consequence, also of *R* and other related freshwater conditions in the Greek landscape for the reference period 1961–1990 [21]. These uncertainties make Greece a particularly important and useful case for investigating a possible robust methodology for catchment-wise data synthesis and interpretation and testing if the constraints implied by catchment water balance decrease the combined uncertainty effects for freshwater conditions and their changes.

Finally, in addition to climate change, Greece has undergone considerable agricultural irrigation developments over the study period 1930–2009, which is chosen due to availability of different types of relevant data for Greece over this period. Previous studies have reported such developments to be important drivers of freshwater changes in different parts of the world [7,8,17,22–24]. The occurrence of such developments over the study period in Greece contributes to making this a relevant case for studying freshwater changes and their interactions and co-evolution with such human-driven developments as well as with climate change.

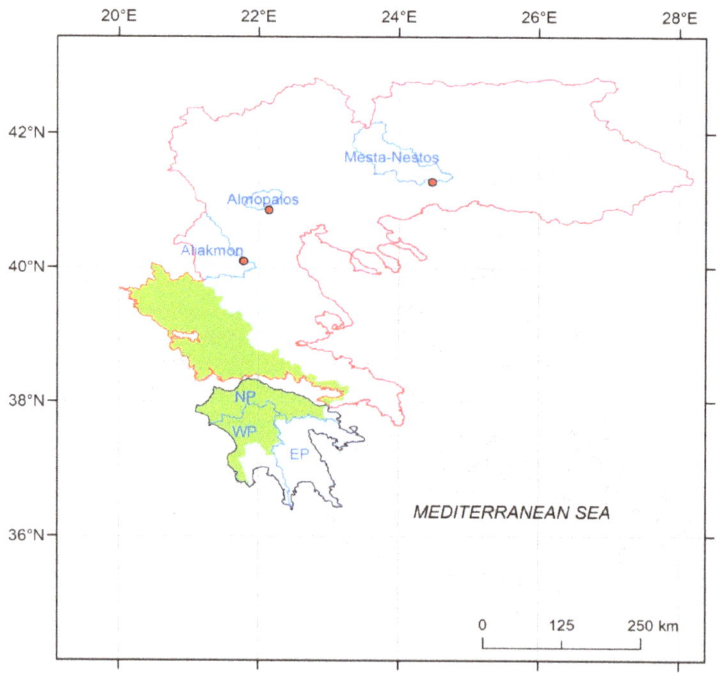

Figure 1. Location and extent of the nested study catchments. These include: the total regional catchment (coloured) and within it the catchments of: Mainland Greece (red outline) and Peloponnese (black outline); and of the Ionian Sea (green area) and the Aegean Sea (light blue area). Data-wise, the Mainland catchment includes three local catchments (blue outlines) with reasonable openly reported runoff data series. The Peloponnese catchment includes three water management districts and associated local catchments (blue outlines) of Northern Peloponnese (NP), Western Peloponnese (WP) and Eastern Peloponnese (EP) with available temporal average runoff and other freshwater data.

Overall, freshwater conditions, interactions and changes across Greece are here quantified and interpreted from the available reported data on hydro-climate and irrigated land-use and water-use for the period 1930–2009. To distinguish freshwater changes and associated uncertainties that may have been driven by the atmospheric climate change and/or the irrigation development occurring over Greece during this period, we compare long-term average conditions in relevant water-related variables between two 20-year sub-periods, 1930–1949 and 1990–2009, for five nested catchments of different scales (Figure 1). The two comparative sub-periods are chosen long enough to represent long-term average climate (and not just temporary weather) conditions and with as long time between them as possible for capturing freshwater changes from an early time with relatively small human influences on climate and irrigation to the recent time with expected much greater such influences.

With regard to the five catchments, the largest one is the total regional catchment (coloured in Figure 1; 178,984 km^2) that includes the whole of Greece and drains into the eastern part of the

Mediterranean Sea through different coastline stretches. In addition to Greece (without its numerous islands, which are not considered in this study), this total catchment also includes some parts in the north that extend into the territories of the former Yugoslav Republic of Macedonia, Bulgaria, Turkey and a minor part of Serbia. This regional catchment is further divided into: the catchment of Mainland Greece (red outline in Figure 1; 157,550 km^2) and that of Peloponnese (black outline in Figure 1; 21,434 km^2); and the catchment draining into the Ionian Sea (green area in Figure 1; 31,958 km^2) and that draining into the Aegean Sea (light blue area in Figure 1; 147,026 km^2). These catchments are identified from reported watershed boundaries and associated vectorized river networks at 15 arc second by 15 arc second resolution [25].

Data-wise, the Mainland catchment includes three local catchments (blue outlines within the Mainland catchment in Figure 1) with reasonable openly reported data series of runoff (explaining further what is considered reasonable in the data section below). Moreover, the Peloponnese catchment includes three water management districts and associated local catchments (blue outlines within that catchment in Figure 1) of Northern Peloponnese (NP), Western Peloponnese (WP) and Eastern Peloponnese (EP), for which relevant temporal average runoff and other freshwater data are available (as described further below).

2.2. Hydro-Climatic Data

In general, open access data for the study region is limited in space and time. Long-term observations of the atmospheric variables surface temperature (T) and P are readily available from global gridded databases like the CRU TS2.1 data set [26] and the more recently updated CRU TS3.10/CRU TS3.10.01 data sets at 0.5° grid-cell resolution [27]. However, for the water variables R and ET in the landscape, the spatio-temporal coverage of available regional data is much more limited.

Openly reported data in the Global Runoff Data Centre (GRDC) [28] show only three stream discharge stations in Greece with reasonable long-term annual average values for R. The reasonableness is judged by the long-term average R being smaller than the long-term average P, since equal or greater long-term average R (as found for other Greek discharge stations) implies unrealistic zero or negative long-term average $ET = P - R$. The reasons for the physically unreasonable discharge values reported for many stations in Greece are not known to us. For most of these discharge stations and the associated catchments, the annual aggregation of the R values (discharge divided by catchment area) implied by the reported discharge data in GRDC equals more or less the annual aggregation of the corresponding catchment-average P values in the CRU data; this indicates that the reported discharge data may not be independently measured but derived from catchment-average P data without accounting for the partitioning of the P water input between ET and R. At any rate, the three discharge stations with physically reasonable data and their local catchments are located in mainland Greece with the associated rivers being: Mesta-Nestos, Almopaios and Aliakmon (Figure 1). Data from these stations in GRDC [28] have a maximum temporal extent of 24 successive years in the second half of the 20th century (Table 1).

For the Peloponnese, estimates representative of current long-term average R conditions are available from reports by the Greek Water Management Authority [29] for the three water management districts and associated local catchments NP, WP and EP (Figure 1). For these, average values of P, R and ET are reported for different subcatchments, leading to the average ET/P ratios listed in Table 1.

Table 1. Data-based estimates of the ratio between long-term average evapotranspiration (ET) and long-term average precipitation (P) for the local catchments in mainland Greece and the river basin districts of Western, Northern and Eastern Peloponnese with relevant data availability. The term R represents long-term average runoff.

	Data in Mainland Greece					Data in Peloponnese					
Catch-ment (Sta-tion)	**Area [a] (km²)**	P [b] **(mm/year)**	R [c] **(mm/year)**	ET [g] **(mm/year)**	ET/P	**District**	**Area [h] (km²)**	R [i] **(mm/year)**	ET [i] **(mm/year)**	P_{eff} [m] **(mm/year)**	ET/P_{eff}
Mesta-Nestos (Teme-nos)	4948	666	283 [d]	383	0.57	Northern Peloponnese (NP)	6108	312 [j]	472 [j]	784	0.60
Almo-paios (Prof Ilias)	993	487	189 [e]	298	0.61	Western Peloponnese (WP)	7235	456 [k]	559 [k]	1015	0.55
Aliak-mon (Ila-rion)	5002	763	315 [f]	448	0.59	Eastern Peloponnese (EP)	8442	172 [l]	452 [l]	624	0.72

[a] Area calculated in ArcGIS. [b] Area-average P data over the years with runoff data from the gridded CRU TS2.1 data set [26]. [c] Temporal average over all years with available data from the GRDC [28]. [d] Monthly runoff data available for the period 1966–1989. [e] Monthly runoff data available for the period 1989–1994. [f] Monthly runoff data available for the period 1963–1987. [g] $ET = P - R$. [h] Greek Water Management Authority [29]. [i] Area-average of reported values from different subunits in management districts (as reported from the Greek Water Management Authority [29]). [j] Considered subunits: GR27 and GR28 (without the island subunit GR45) [29]. [k] Considered subunits: GR29 and GR32 [29]. [l] Considered subunits: GR30, GR31 and GR33 [29]. [m] Effective precipitation partitioning into the reported R and ET in the district: $P_{eff} = R + ET$.

2.3. Land- and Water-Use Data for Irrigation

Openly available information on irrigation rates and areas for Greece is not complete over either the whole investigated geographical space or the whole investigated time period. Two different types of such data are available: national irrigation data sets from statistics reported by the Organisation for Economic Co-operation and Development (OECD) [30,31]; and the global map of irrigated areas version 5.0 (GMIAv5.0) [32]. The national OECD data sets [30,31] contain statistics in terms of area that is actually irrigated (*Aai*) and total volume of water withdrawal for irrigation (*Iwwv*) in Greece, especially from 1990 onwards (Figure 2). For earlier times, sporadic data on total irrigated area is provided by [33]. Furthermore, the global map of irrigation areas [32] represents the situation around year 2005 with spatial resolution of 5 arc minute by 5 arc minute (see Figure 3 for Greece).

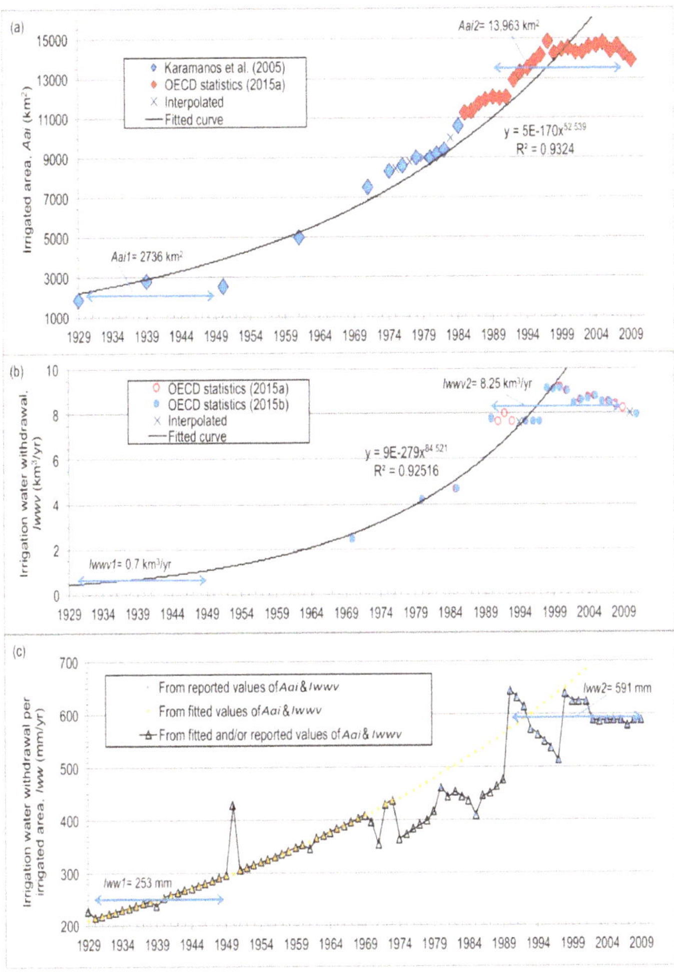

Figure 2. (**a**) Reported irrigated area, *Aai*, in Greece, shown with the best power-law fit to the reported data. (**b**) Reported annual water withdrawal for irrigation, *Iwwv*, in Greece, shown with the best power-law fit to the reported data until year 2001, after which *Iwwv* conditions stabilize. (**c**) Irrigation water withdrawal per irrigated area, *Iww* = *Iwwv*/*Aai*, estimated from the reported and fitted data on *Aai* and *Iwwv* in panels (**a**,**b**), respectively. The panels also show average variable values over the two 20-year sub-periods 1930–1949 and 1990–2009.

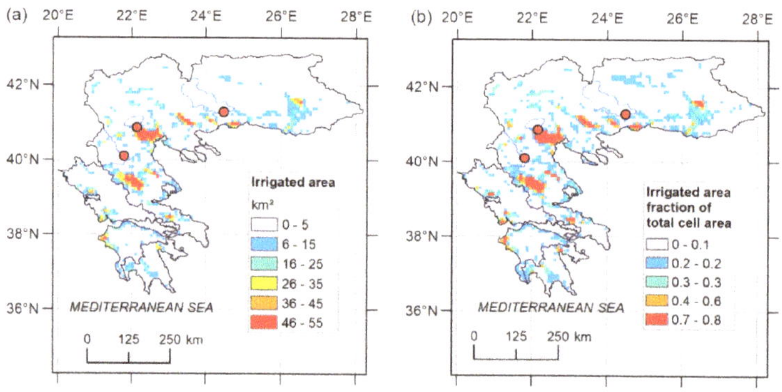

Figure 3. (**a**) Area actually irrigated within each grid cell in km^2. (**b**) Irrigated area relative to total cell area. These maps are created from the global map of irrigation areas [32] and represent conditions in the recent period 1990–2009. In these maps, the Ionian and Aegean catchments are indicated by black outlines and the three local catchments in mainland Greece with reasonable runoff data time series are indicated by blue outlines.

2.4. Assessing Freshwater Conditions and Changes

Temporal average T and P for the two comparative periods ($T1$, $P1$ for the early period 1930–1949 and $T2$, $P2$ for the recent period 1990–2009) are calculated from data on monthly P and T, as provided by the CRU TS 3.10/3.10.1 worldwide database extending over the period 1901–2009 [27]. Catchment-average values of these temporal averages are further quantified over all CRU grid cells with at least 34% of their area intersecting each considered catchment; Supplementary Figure S1 shows this CRU mask, used to extract catchment-average values of all analysed variables for each study catchment.

While period-average conditions for the atmospheric variables T and P can be calculated directly from available data series for the two comparative study periods, data on the average water fluxes in the landscape, ET and R, are not obtainable in this way for the earlier period 1930–1949. To estimate the early-period long-term average values of $ET1$ and $R1$ and the associated ET and R changes to the average values $ET2$ and $R2$ in the recent period 1990–2009, we start by evaluating the long-term average $ET2$ value for 1990–2009 from the available P and R data for this period as $ET2 = P2 - R2 = (ET2/P2) \times P2$, based on the overarching long-term average water balance in each catchment. The data-given range of currently applicable ET/P values in Table 1 is calculated on this water-balance basis, following previous studies that have found [9,34] and assumed [5,7,8] long-term catchment-average changes in water storage to be near-zero over such long time periods.

To further estimate the change in average ET between the two 20-year periods, we follow the approach of previous studies [5,7,8] in estimating this total change as:

$$\Delta ET = ET2 - ET1 \approx \Delta ETclim + \Delta ETirr. \tag{1}$$

The second equality is approximate in assuming that total ΔET involves two main change components: $\Delta ETclim$ driven by climate change and $\Delta ETirr$ driven by irrigation changes. By evaluating these components, $ET1$ can be estimated from Equation (1) as $ET1 = ET2 - \Delta ETclim - \Delta ETirr$.

Furthermore, from this quantification of $ET1$, the corresponding average $R1$ can be consistently estimated as:

$$\begin{aligned} R1 \approx P1 - ET1 = P1 - (ET2 - \Delta ETclim - \Delta ETirr) = P1 - ((P2 - R2) - \Delta ETclim - \Delta ETirr) \\ = R2 - ((\Delta P - \Delta ETclim) - \Delta ETirr), \end{aligned} \tag{2}$$

with $\Delta P = P2 - P1$ being the P change, and

$$\Delta R = R2 - R1 \approx (\Delta P - \Delta ETclim) - \Delta ETirr = \Delta Rclim + \Delta Rirr \tag{3}$$

being the total R change between the two periods. The climate-driven component of total ΔR is thus quantified as $\Delta Rclim = \Delta P - \Delta ETclim$ and the corresponding irrigation-driven component is quantified as $\Delta Rirr = -\Delta ETirr$.

To estimate the climate-driven change component $\Delta ETclim$, we further relate the period-average temperature ($T1$, $T2$) and precipitation ($P1$, $P2$) to the corresponding average potential evapotranspiration ($PET1$, $PET2$) based on the Langbein functional relationship [35] and further to corresponding ET conditions in each period ($ETclim2$, $ETclim1$) based on the Turc functional relationship [36]. From these quantifications, the climate-driven ET change can be estimated as $\Delta ETclim = ETclim2 - ETclim1 = f(PET2,P2) - f(PET1,P1)$. Alternative functional relationships for $ETclim = f(PET,P)$ may be used for this estimation but previous work has shown small differences in resulting changes of long-term catchment-average ET by use of alternative such functions [34].

Furthermore, to estimate the irrigation-driven change component $\Delta ETirr = ETirr2 - ETirr1$, we quantify the average irrigation water withdrawal per irrigated area in each period ($Iww1$, $Iww2$) based on: the data and associated best-fit function for the temporal Iww evolution (Figure 2c); the area actually irrigated in absolute terms ($Aai2$ from Figure 3a for recent period 1990–2009) and relative to total cell area (Aai_rat2 from the map in Figure 3b for the recent period); and the assumption that more or less all water used for irrigation feeds back into the average ET over each period. The latter assumption is supported by previous studies of different irrigated areas of the world [17,22,23], as well as by recent findings of ET variations correlating well with concurrent, independently determined transpiration variations [16]. As irrigation is applied precisely for feeding into transpiration, it is reasonable to assume that it largely does so, or otherwise feeds into increased evaporation, either directly from soil water, or from potentially added surface water runoff; in any case, the applied irrigation water may be expected to largely feed into the total ET of each catchment. Moreover, a wide range of different possible Iww and Aai_rat evaluation scenarios are investigated (as described further in the following uncertainty analysis section) and may be expected to also cover uncertainty effects associated with varying parts of the irrigation water use (Iww) and / or the relative irrigated catchment area (Aai_rat) feeding into ET in each catchment. In general, based on these considerations and assumptions, an irrigation-driven part of total ET in each period is estimated as $ETirr \approx Iww \times Aai\_rat$. The difference between the resulting values of $ETirr$ for each of the two periods then provides the irrigation-driven change $\Delta ETirr = ETirr2 - ETirr1$.

In the absence of available data on the spatial distribution of irrigated area in the earlier period 1930–1949, we assume that the relative grid-cell irrigated area in that period (Aai_rat1) is some fraction (αirr) of the corresponding relative irrigation area in the recent period (i.e., that $Aai\_rat1 = Aai\_rat2 \times \alpha irr$). The fraction αirr is further assumed to be more or less the same for all grid cells and is estimated as such from the total irrigated area over Greece (Figure 3a) in the early period (total Greek $Aai1$) relative to that in the recent period (total Greek $Aai2$), that is, as $\alpha irr \approx$ (total Greek $Aai1$)/(total Greek $Aai2$).

Table 2 summarizes a set of estimated input variables values, needed in the above-described calculations for each study catchment. Footnotes to Table 2 explain how these values are estimated and this evaluation is in the following referred to as the base case scenario for each catchment. This scenario is obtained using mean (or other relevant characteristic) values of the main input variables for each catchment and time period. In general, values that to some degree differ from those in the base case scenario may also be estimated for uncertain variables, depending on choices of data and assumptions made for the variable evaluation. To account for such evaluation uncertainty, we also consider possible alternative values of main uncertain variables, as explained further in the following section.

Table 2. Average variable values used in the base case scenario for each catchment and time period, 1930–1949 (*Per1*) and 1990–2009 (*Per2*). The terms *T*, *P*, *ET*, *Iww*, *Aai* and *Aai_rat* represent temperature, precipitation, evapotranspiration, irrigation water withdrawal per irrigated area, irrigated area and ratio between *Aai* and total catchment area, respectively.

Catchments	Total		Mainland		Peloponnese		Ionian		Aegean	
Total area [a] (km^2)	178,984		157,550		21,434		31,958		147,026	
Time Period	*Per1* [b]	*Per2* [b]	*Per1*	*Per2*	*Per1*	*Per2*	*Per1*	*Per2*	*Per1*	*Per2*
T [c] (°C)	12.1	12.4	11.7	12.1	15	14.8	14.4	14.4	11.6	12
P [c] (mm/year)	676	630	663	622	773	693	861	767	637	602
ET/*P* [d]		0.59		0.59		0.62		0.59		0.59
Iww [e] (mm/year)	0.692/2736 = 253	8.25/13,963 = 591	253	591	253	591	253	591	253	591
Aai (km^2)	2736 [e]	11,046 [f]	2394 [g]	9665 [f]	342 [g]	1381 [f]	547 [g]	2255 [f]	2189 [g]	8791 [f]
Aai_rat	0.015 [h]	11,046/178,984 = 0.062	0.015 [h]	9665/157,550 = 0.061	0.016 [h]	1381/21,434 = 0.064	0.017 [h]	2255/31,958 = 0.071	0.015 [h]	8791/147,026 = 0.06

[a] Spatial extent of each catchment. [b] *Per1*: 1930–1949, *Per2*: 1990–2009. [c] Spatio-temporal averages from the monthly CRU TS3.10.01 data set [27]. [d] Using the mean *ET*/*P* value from the three local catchments with data over mainland Greece and that from the Peloponnese local catchments over the Peloponnese. [e] From the national reported data on irrigated areas and irrigation water withdrawals [30,31] (Figure 2). [f] Spatial average values for each catchment with the area actually irrigated obtained from the GMIAv5.0 map dataset [32] (Figure 3). [g] Assuming that the ratio of areas actually irrigated in this catchment relative to that in the mainland Greece and the Peloponnese is more or less the same in 1930–1949 as in 1990–2009. [h] With *Aai_rat1* estimated from: $Aai\_rat1 = Aai\_rat2 \times \alpha irr$ and $\alpha irr \approx Aai1/Aai2 = 0.25$ in the base case scenario.

2.5. Uncertainty Analysis

For comparison with and uncertainty analysis of other possible variable values than those considered in the base case scenario (Table 2), we also evaluate alternative scenarios of uncertain variable values. Table 3 compares the base and the alternative scenarios and lists their variable value differences for the total regional catchment; Supplementary Table S1 does the same for the other four study catchments.

With regard to P data, the alternative scenarios outlined in Table 3 (named in the first and explained in the second column, in direct comparison with the base case) consider possible corrections for P undercatch bias [37], or for both undercatch and orographic biases [38]. Moreover, the ET/P ratio used for estimating average ET in the recent period 1990–2009 ranges from 0.55 to 0.72 among the local catchments with available data across Greece (Table 1). The base case scenario considers the mean value of ET/P, whereas two alternative ET/P scenarios consider corresponding minimum and maximum values (Table 3); the ET/P values are multiplied with P in each data grid cell to obtain grid-cell values of ET over each catchment.

Furthermore, considerable uncertainty about irrigation variables may stem from the estimation of water use (withdrawal) per irrigated area in the 1990–2009 period (*Iww2*). For example, for the total regional catchment (Table 3), *Iww2* may be as large as 747 mm/year if the recent irrigated area in its calculation is estimated from the national reported irrigated area [30] (*Aai2* = 13,963 km^2) instead of 591 mm/year in the base case scenario that estimates *Iww2* from GMIAv5.0 [32] (*Aai2* = 11,046 km^2). The resulting uncertainty range of *Iww2* is then ±78 mm/year.

With this uncertainty range given for *Iww2*, a corresponding uncertainty range of ±78 mm/year may also be expected for *Iww1*, yielding an even wider uncertainty range for the temporal change in *Iww* by considering different possible combinations of *Iww2* and *Iww1* values in the base and alternative scenarios (as outlined in Table 3). Finally, also the value of the irrigation fraction $\alpha irr = Aai1/Aai2$ differs if the recent total irrigated area (*Aai2*) is estimated from the GMIAv5.0 data [32] (11,046 km^2 for the regional catchment; base case scenario), or from the national reported data [30] (13,963 km^2 for the regional catchment; alternative scenario).

3. Results and Discussion

3.1. General Approach

A main result of this study is the general approach developed, used and tested for robustly assessing freshwater changes in terms of catchment-scale long-term average ET and R, driven by climate and landscape (here irrigation) developments from an earlier period (here 1930–1949) to recent conditions (here 1990–2009). Figure 4 illustrates schematically this principal approach, which is based on fundamental catchment water balance, aiming to constrain the result and uncertainty ranges of different possible variable evaluations. This approach provides a consistent flow-chart basis for corresponding quantitative assessments of freshwater changes in various comparative scenarios and catchment cases. Different scenarios may represent various possible variable evaluations from limited available observation data and the catchments may be of different scales and in different parts of the world.

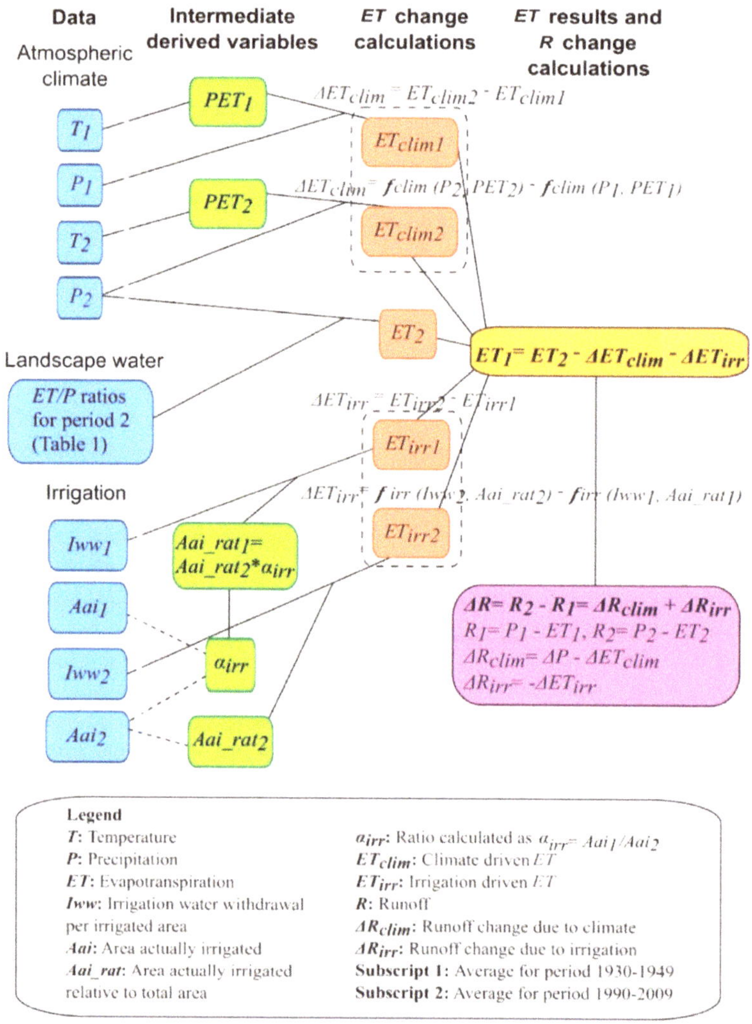

Figure 4. Schematic illustration of the approach developed and used to assess long-term average evapotranspiration (*ET1*) and runoff (*R1*) in an early 20-year period and associated changes (Δ) from this to a recent 20-year period. The sources and types of data used for the assessment in this study are exemplified for the base case scenario in the light blue boxes to the left, grouped as: atmospheric climate, landscape water and irrigation data. The variables extracted from the data are shown in the dark blue boxes. The green boxes show intermediate derived variables for estimating the different components of total *ET* change, as outlined in the red boxes and synthesized in the yellow box. The lilac box shows the final synthesis for estimating corresponding components and total change for *R*.

Table 3. The base case and alternative (Alt.) evaluation scenarios and their variations. Results are listed for the total regional catchment, with the terms *P*, *ET*, *Iww*, *αirr* and *Per* representing precipitation, evapotranspiration, irrigation water withdrawal per irrigated area, ratio *Aai1*/*Aai2* between irrigated area (*Aai*) and time period. The latter is period 1 (1930–1949) or period 2 (1990–2009).

Total Regional Catchment			
Scenario	**Variation**	***Per1***	***Per2***
Base *P*	No correction of *P* observation data (mm/year)	676	630
		−46	
Alt.1 *P*	Undercatch correction of *P* observation data (mm/year)	734	686
		−48	
Alt.2 *P*	Undercatch and orographic correction of *P* observation data (mm/year)	769	718
		−51	
Base *ET*/*P*	Mean *ET*/*P*		0.59
Alt.1 *ET*/*P*	Min *ET*/*P*		0.55
Alt.2 *ET*/*P*	Max *ET*/*P*		0.72
Base *Iww*	Mean *Iww1*, Min *Iww2* (mm/year)	253	591
		338	
Alt.1 *Iww*	Min *Iww1*, Min *Iww2* (mm/year)	175	591
		416	
Alt.2 *Iww*	Max *Iww1*, Min *Iww2* (mm/year)	331	591
		260	
Alt.3 *Iww*	Min *Iww1*, Max *Iww2* (mm/year)	175	747
		572	
Alt.4 *Iww*	Mean *Iww1*, Max *Iww2* (mm/year)	253	747
		494	
Alt.5 *Iww*	Max *Iww1*, Max *Iww2* (mm/year)	331	747
		416	
Base *αirr*	Max *αirr*	0.25	
Alt.1 *αirr*	Min *αirr*	0.20	

3.2. *Freshwater Changes*

Figure 5 shows the regional conditions and changes in the atmospheric variables *T* and *P* (top left in the general approach illustration, Figure 4), in terms of the spatial distribution of their temporal averages over the whole study period 1930–2009 (Figure 5a,b) and the changes in period-average *T* and *P* from 1930–1949 to 1990–2009 (Figure 5c,d). Spatially, long-term average *T* increases from a local range of around 7–11 °C in the north to around 15–18 °C in the south, while average *P* increases from a local range of around 460–800 mm/year in the east to around 800–1400 mm/year in the west. The corresponding changes also follow a north-south gradient for *T* and an east-west gradient for *P*. The temperature *T* increases in the cooler north and decreases in the warmer south (i.e., cool gets warmer and warm gets cooler), while *P* increases (or decreases the least) in the dry east and decreases (the most) in the wet west (i.e., dry gets wetter—or dries the least and wet gets drier—or dries the most).

Catchment-average results for *T* and *P* (Table 2, for the base case scenario) are consistent with the mapped grid-cell results (Figure 5) in showing that the most north-extending and coolest catchment (Aegean) has warmed the most (increase in average *T* of 0.4 °C) while the most southern and warmest catchment (Peloponnese) has cooled (decrease in average *T* of −0.2 °C). Moreover, while average *P* has decreased in all catchments, the most east-extending and driest catchment (Aegean) exhibits the smallest P decrease (−35 mm/year) and the most western and wettest catchment (Ionian) exhibits the largest P decrease (−94 mm/year).

Overall, the atmospheric hydro-climatic changes in T and P from 1930–1949 to 1990–2009 have driven most catchments (except the Peloponnese and Ionian) toward somewhat warmer conditions and all catchments toward drier conditions. Furthermore, these changes have decreased the variability range of average T and P among the nested catchments, from earlier ranges of 3.4 °C (from minimum 11.5 to maximum 15 °C) and 224 mm/year (from 637 to 861 mm/year) to recent ranges of 2.8 °C (from 12 to 14.8 °C) and 145 mm/year (from 622 to 767 mm/year), respectively.

With regard to irrigation, this has increased greatly in Greece in terms of all associated variables: irrigated area (Figure 2a) and amount of water used for irrigation in terms of absolute volume (Figure 2b) and per irrigated area (Figure 2c). These changes imply corresponding irrigation increases also in the different study catchments (Table 2). As a consequence of these human-driven irrigation developments, combined with the atmospheric hydro-climatic changes (Figure 5b), the long-term average values of the landscape hydro-climatic variables ET and R have also changed over Greece and in the different study catchments. Figures 6 and 7 illustrate the changes in ET and R, respectively, for all catchments. Further result details are listed in Table 4 for the total regional catchment; corresponding results for the other catchments are listed in Supplementary Table S2 for Mainland, Table S3 for Peloponnese, Table S4 for Ionian and Table S5 for Aegean.

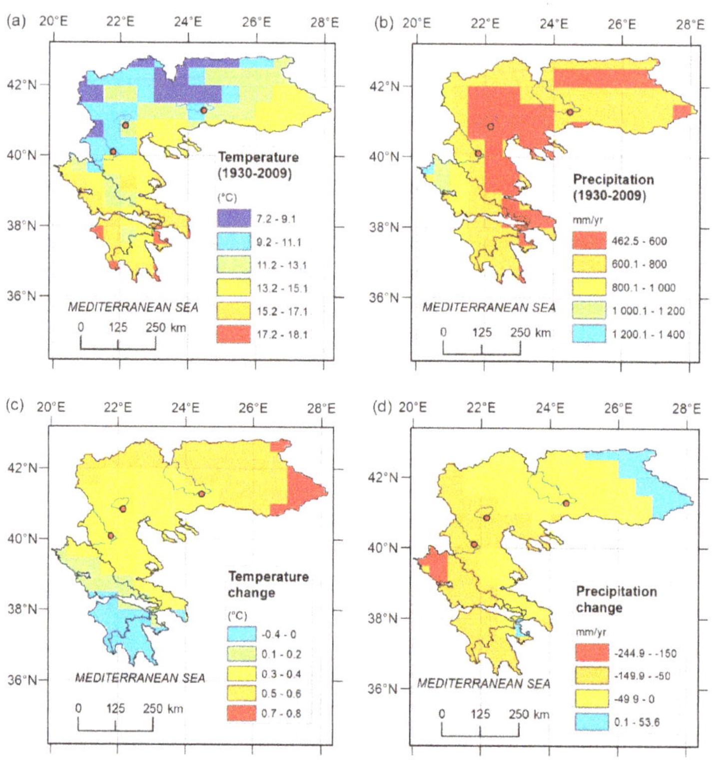

Figure 5. The spatial distribution of temporal average temperature (**a**) and precipitation (**b**) over the period 1930–2009 and corresponding period-average changes ((**c**,**d**), respectively) from the period 1930–1949 to the period 1990–2009. In these maps, the Ionian and Aegean catchments are indicated by black outlines and the three local catchments in mainland Greece with reasonable runoff data time series are indicated by blue outlines.

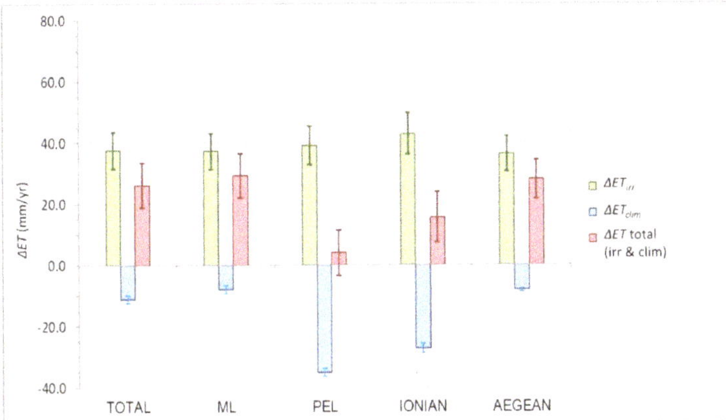

Figure 6. Evapotranspiration changes from 1930–1949 to 1990–2009 in total (ΔET) and their climate ($\Delta ETclim$) and irrigation ($\Delta ETirr$) components. Main bars show mid-range results and error bars show the range of change estimates for different evaluation scenarios (Table 3 and Supplementary Table S1). Results are shown for the total regional catchment (TOTAL) and the nested Mainland (ML), Peloponnese (PEL), Ionian (IONIAN) and Aegean (AEGEAN) catchments.

Table 4. Freshwater changes and main uncertainty estimates for the total regional catchment. The terms P, ET and R represent precipitation, evapotranspiration and runoff, respectively, while Δ stands for change and subscript *clim* and *irr* indicate climate driven and irrigation driven change, respectively. Scenario definitions (Base, Alt.) are as given and explained in Table 3.

Type of Scenario and Uncertainty	ΔP (mm/year)	$\Delta ETclim$ (mm/year)	$\Delta ETirr$ (mm/year)	ΔET (mm/year)	$\Delta Rclim$ (mm/year)	$\Delta Rirr$ (mm/year)	ΔR (mm/year)
P scenarios ± Irrigation uncertainty	−46 (Base) −48.4 (Alt. 1) −51 (Alt. 2)	−12.5 −11.5 −10		24.9 ± 6 25.9 ± 6 27.4 ± 6	−33.5 −36.9 −41		−70.9 ± 6 −74.3 ± 6 −78.4 ± 6
Irrigation (Iww) scenarios ± P uncertainty			32.6 (Base) 33.8 (Alt. 1) 31.4 (Alt. 2) 43.4 (Alt. 3) 42.2 (Alt. 4) 41 (Alt. 5)	21.3 ± 1.3 22.5 ± 1.3 20.1 ± 1.3 32.1 ± 1.3 30.9 ± 1.3 29.7 ± 1.3		−32.6 −33.8 −31.4 −43.4 −42.2 −41	−69.7 ± 3.8 −70.9 ± 3.8 −68.5 ± 3.8 −80.6 ± 3.8 −79.4 ± 3.8 −78.2 ± 3.8
All scenarios ± Total uncertainty	−48.5 ± 2.5	−11.3 ± 1.3	37.4 ± 6	26.1 ± 7.3	−37.1 ± 3.8	−37.4 ± 6	−74.6 ± 9.8

Overall, ET has increased to some greater or lesser degree, while R has decreased considerably over all catchments. For ET, the irrigation developments have driven the largest change component $\Delta ETirr$ (overall increase of around 40 mm/year), while the climate-driven change $\Delta ETclim$ is in the opposite direction (decrease) and mostly of smaller absolute magnitude (around or less than 10 mm/year for the total, Mainland and Aegean catchments). For the Ionian and Peloponnese catchments, the absolute magnitude of the decrease $\Delta ETclim$ is relatively close to that of the increase $\Delta ETirr$. As a consequence, the total net increase ΔET is relatively small in these two catchments (around 15 mm/year in the Ionian and 4 mm/year in the Peloponnese) and larger in the other catchments (around 25–30 mm/year).

The main reason for the relatively large decreases $\Delta ETclim$ of around −27 mm/year and −35 mm/year in the Ionian and the Peloponnese catchment, respectively, is that they are subject to the largest precipitation decreases ΔP (of −94 mm/year and −80 mm/year). These are also combined with ΔT conditions of zero warming (Ionian) or even cooling (of −0.2 °C in Peloponnese), which do not drive ET increase. A question to investigate in further climate-change research for this region is whether the generally large irrigation-driven ET increase ($\Delta ETirr$ component) and the additional latent heat flux that this implies from the land surface to the atmosphere have contributed to significant local cooling of the land surface, thereby counteracting local warming effects of global climate change. This has been found in other irrigated regions of the world [17,23] and such irrigation-driven surface cooling would be consistent with the relatively small increases of local surface T over Greece, including the cooling and zero warming in the Peloponnese and Ionian catchments with the largest and second largest increase components $\Delta ETirr$, respectively.

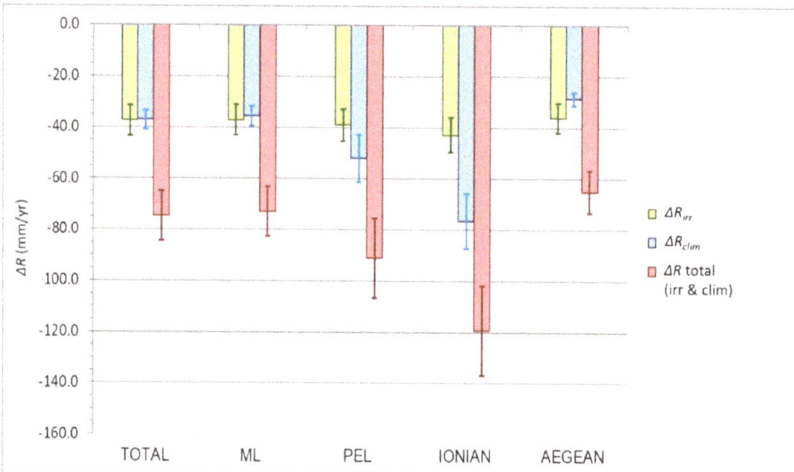

Figure 7. Runoff changes from 1930–1949 to 1990–2009 in total (ΔR) and their climate ($\Delta Rclim$) and irrigation ($\Delta Rirr$) components. Main bars show mid-range results and error bars show the range of change estimates for different evaluation scenarios (Table 3 and Supplementary Table S1). Results are shown for the total regional catchment (TOTAL) and the nested Mainland (ML), Peloponnese (PEL), Ionian (IONIAN) and Aegean (AEGEAN) catchments.

In contrast to ET, the climate and irrigation change components of ΔR are both in the same direction, thus reinforcing each other and leading to considerable decrease in R across all catchments (Figure 7). In total, ΔR is around: −75 mm/year, −73 mm/year and −65 mm/year in the three largest catchments, total regional, Mainland and Aegean, respectively; and −91.1 mm/year and −119 mm/year in the two smallest catchments, Peloponnese and Ionian, respectively. The irrigation driven change component $\Delta Rirr = -\Delta ETirr$ (Equation (3)) has similar magnitude (decrease of around −40 mm/year) across all catchments. In most catchments (total, Mainland, Aegean), the climate-driven change component $\Delta Rclim = \Delta P - \Delta ETclim$ (Equation (3)) is also a decrease of somewhat smaller magnitude (around −35 to −30 mm/year) than $\Delta Rirr$. However, in the Ionian and Peloponnese catchments, the decrease $\Delta Rclim$ is larger (around −75 mm/year and −50 mm/year, respectively) than $\Delta Rirr$. Overall, the decrease $\Delta Rclim$ is mainly due to the precipitation decrease ΔP across all catchments.

The results of net total increase in average ET and decrease in average R, which for ET masks and for R exacerbates the decrease expected from only the observed atmospheric climate change (in T and P), are consistent with corresponding results for other irrigated areas of the world. For example, in the Indian Mahanadi River catchment, draining into the Bay of Bengal, with decreased long-term average P by −60 mm/year and increased irrigation water use by 81 mm/year (from 1901–1955 to 1956–2000), the average ET increased by around 55–70 mm/year while the average R decreased by around −130–−115 mm/year [39]. Furthermore, in the case of the Aral Sea catchment (ASC), P increased by 11 mm/year and irrigation water use increased by 23 mm/year (from 1901–1950 to 1983–2002), while the average ET increased by 15 mm/year and the average R decreased by −28 mm/year [39]. In both of these irrigation cases, as also found here for Greece, the irrigation development has led to ET increase in spite of P decreasing (or to greater ET increase than the increase in P for ASC) and consequently to much greater decrease in R than the decrease in P (or to R decrease in spite of P increasing for ASC); in both cases, as in Greece, the large ET increase is also not explainable by just the observed warming (increase in average T) over each catchment. A global study of ET changes around the world's land areas has also shown statistically that irrigated areas (and areas with dam and reservoir developments for meeting the irrigation and other increased water demands) have experienced significantly greater ET increases than other, more undisturbed land areas with regard to such human developments [8].

The irrigation-driven increases in ET imply greater losses of freshwater from the irrigated catchments. For most such catchments, these water losses are not compensated by corresponding increases in observed P, since the observed P changes (even combined with the observed T changes) cannot explain the total ET increases [8,39]. From its use for helping crops to survive dry season and drought conditions [16], the irrigation water adding to increased ET in an irrigated catchment is thus lost from that catchment and goes to feed other catchments or maybe even the sea. This leaves less freshwater for other uses in the irrigated catchment, such as for households, industry, energy generation and/or ecosystems. Water managers and decision makers need to understand the involved trade-offs and make conscious, informed choices for sustainably balancing freshwater uses among societal sectors and ecosystems. Further research is also needed to support such choices by investigating and revealing the implications of different development scenarios.

3.3. Uncertainty in Freshwater Changes

The error bars in Figures 6 and 7 show the uncertainty ranges of the estimated freshwater changes ΔET and ΔR, as obtained in total from all considered uncertainty scenarios (Table 3, Supplementary Table S1). Table 4 lists further result details for the total regional catchment; corresponding results for the other catchments are listed in Supplementary Table S2 for Mainland, Table S3 for Peloponnese, Table S4 for Ionian and Table S5 for Aegean. In general, the total uncertainty range of ΔET ($\pm$6–8 mm/year across all catchments) is smaller than that of ΔR, which is also more variable among the catchments (from $\pm$8 for the Aegean up to $\pm$18 mm/year for the Peloponnese). Overall, these uncertainty ranges do not change the main results and implications discussed above based on the mean freshwater changes.

With regard to the investigated underlying uncertainties, the largest range is associated with the different scenarios of irrigation water use per irrigated area (Iww; Table 3). For the total regional catchment (Table 3), the estimated change in Iww is on average 416 mm/year and the associated uncertainty range is $\pm$156 mm/year. In comparison, the resulting uncertainty ranges for this catchment are much smaller for both ΔET and ΔR, at $\pm$7 and $\pm$10 mm/year, respectively. Similar results are obtained for this uncertainty propagation in all catchments, implying a major decrease in absolute range magnitude from that of underlying uncertainties (Table 3, Supplementary Table S1) to that of the total resulting ΔET and ΔR uncertainties (Table 4, Supplementary Tables S2–S5).

Also in relative terms, the largest underlying uncertainty range is that for the Iww change at $\pm$38%. For the decrease in R (Figure 7), the resulting relative uncertainty range is overall smaller: 13% for the total regional, Mainland and Aegean catchments (with mean ΔR of -75 mm/year, -73 mm/year and -65 mm/year, respectively); 15% for the Ionian catchment (with mean ΔR of -119 mm/year); and 17% for the Peloponnese catchment (with mean ΔR of -91 mm/year). For the increase in ET (Figure 6), the relative uncertainty range is larger and varies more among the catchments than that of ΔR. Specifically, the relative ΔET range is: $\pm$28% for the total regional, $\pm$25% for the Mainland and $\pm$23% for the Aegean catchment (with mean ΔET of 26 mm/year, 29 mm/year and 28 mm/year, respectively); $\pm$54% for the Ionian catchment (with mean ΔET of 16 mm/year); and $\pm$194% for the Peloponnese catchment (with mean ΔET of just 4 mm/year).

In general, the largest contribution to the total ΔET range (of $\pm$6–8 mm/year) stems from irrigation uncertainty ($\pm$5–6 mm/year contribution; with additional $\pm$1 mm/year stemming from climate uncertainty). For the total ΔR range (of $\pm$8–18 mm/year), the largest uncertainty contribution varies among catchments. For the Aegean, Mainland and total regional catchment, the largest contribution stems from irrigation uncertainty ($\pm$6–8 mm/year contribution to their total ΔR range of $\pm$8–10 mm/year; with additional $\pm$2–4 mm/year stemming from climate uncertainty, in which the main part is due to the ΔP uncertainty of around $\pm$2 mm/year). In contrast, for the Ionian and Peloponnese catchments, the largest contribution stems from climate uncertainty ($\pm$9–11 mm/year, in which the main part is due to the ΔP uncertainty of $\pm$7–8 mm/year, contributing to the total ΔR range of $\pm$16–18 mm/year; with $\pm$6–7 mm/year stemming from irrigation uncertainty).

Across all catchments, the scenario combination with maximum *Iww1* and minimum *Iww2* (i.e., Alt. 2 *Iww* from Table 3) yields the lower ΔET and the upper ΔR limit, while the opposite scenario combination with minimum *Iww1* and maximum *Iww2* (Alt. 3 *Iww*, Table 3) yields the upper ΔET and the lower ΔR limit. The actual values of these limits depend also on the considered ΔP scenario, with the undercatch correction (Alt 1 *P*, Table 3) and even more so the undercatch and orographic correction (Alt 2 *P*, Table 3) increasing the ΔET values and decreasing the ΔR values compared to the base case scenario. The alternative scenarios for ET/P (Supplementary Table S6) and αirr (not shown) do not significantly change the results compared to the Base case scenario.

4. Conclusions

This paper has developed a general approach for synthesizing and jointly interpreting various types of data, related to different interacting and co-evolving drivers and aspects of freshwater change, subject to various degrees of uncertainty, over hydrological catchments of different scales. The application of this approach to the high-uncertainty case of Greece and its freshwater changes from 1930–1949 to 1990–2009, based on catchment water balance, provides support for an overall decrease in the combined uncertainties of resulting catchment-scale *ET* and *R* changes from the magnitude of the various underlying uncertainties.

For Greece, the study results are robust in showing that climate-driven *P* decrease and concurrent human-driven irrigation development from the first half of the 20th century to recent time have combined in driving considerable total decrease in *R* over the country. In the three largest study catchments (total regional, Mainland and Aegean), the total *R* decrease ranges from −65 to −75 mm/year ±13%. This total *R* decrease is driven by the local irrigation development to more or less the same degree (possibly somewhat more but within the uncertainty range) as by climate change (predominantly decrease in *P*, with relatively small *T* increase over Greece and even decrease in Peloponnese). For the relatively small Ionian and Peloponnese catchments, the *R* decrease is up to −119 mm/year ±15% and −91 mm/year ±17%, respectively, due to the particularly large *P* decrease experienced in these catchments, combined with similar irrigation-driven *R* decrease as in the other catchments.

In each catchment, the irrigation-driven component of the total *R* decrease has largely fed a corresponding irrigation-driven component of *ET* increase. Recent results for water deficit propagation and partitioning over different parts of Europe [16] indicate that this *ET* increase likely represents increased transpiration, thereby contributing to support vegetation and crop status against the decreased *P* water input and the likely more frequent and/or more anomalous drought events that may be expected under such *P* decrease in Greece. Under these conditions of decreased *P* combined with relatively small *T* increase (or even *T* decrease), the atmospheric climate change drives *ET* decrease, that is, opposite change direction to the *ET* increase driven by irrigation. In net total, the irrigation-driven component dominates, such that total *ET* has increased (but less than *R* has decreased) over Greece, except in Peloponnese where the climate—and irrigation-driven components are more or less equal and total *ET* remains essentially unchanged.

Supplementary Materials: The following are available online at http://www.mdpi.com/2073-4441/10/11/1645/s1, Figure S1: Location and extent of CRU cells considered for each study catchment, Table S1: Base case and alternative (Alt.) quantification scenarios and their variations for different catchments, Table S2: Freshwater changes and main uncertainty estimates for the Mainland catchment, Table S3: Freshwater changes and main uncertainty estimates for the Peloponnese catchment, Table S4: Freshwater changes and main uncertainty estimates for the Ionian catchment, Table S5: Freshwater changes and main uncertainty estimates for the Aegean catchment, Table S6: Evapotranspiration conditions and changes showing insignificant influence of choice of ET/P scenario (defined in main Table 3) for the example of the total regional catchment.

Author Contributions: G.D. conceived the study and analysis approach and wrote the paper. C.P. performed most analyses for the Figures and Tables and contributed to the writing.

Funding: This research was funded by The Swedish Research Council Formas, grant number 2016-02045.

Conflicts of Interest: The authors declare no conflict of interest.

References

1. Pettersson, L.G.M.; Henchman, R.H.; Nilsson, A. Water—The most anomalous liquid. *Chem. Rev.* **2016**, *116*, 7459–7462. [CrossRef] [PubMed]
2. Nature Editorial. Get water governance on the global agenda. *Nature* **2016**, *540*, 169–170. [CrossRef]
3. World Economic Forum. The Global Risks Report 2018. Available online: http://www3.weforum.org/docs/WEF_GRR18_Report.pdf (accessed on 30 September 2018).
4. Bosson, E.; Sabel, U.; Gustafsson, L.G.; Sassner, M.; Destouni, G. Influences of shifts in climate, landscape and permafrost on terrestrial hydrology. *J. Geophys. Res. Atmos.* **2012**, *117*, D05120. [CrossRef]
5. Jaramillo, F.; Destouni, G. Developing water change spectra and distinguishing change drivers worldwide. *Geophys. Res. Lett.* **2014**, *41*, 8377–8386. [CrossRef]
6. Elmhagen, B.; Destouni, G.; Angerbjörn, A.; Borgström, S.; Boyd, E.; Cousins, S.A.O.; Dalén, L.; Ehrlén, J.; Ermold, M.; Hambäck, P.A.; et al. Interacting effects of change in climate, human population, land use and water use on biodiversity and ecosystem services. *Ecol. Soc.* **2015**, *20*, 23. [CrossRef]
7. Destouni, G.; Jaramillo, F.; Prieto, C. Hydroclimatic shifts driven by human water use for food and energy production. *Nat. Clim. Chang.* **2013**, *3*, 213–217. [CrossRef]
8. Jaramillo, F.; Destouni, G. Local flow regulation and irrigation raise global human water consumption and footprint. *Science* **2015**, *350*, 1248–1251. [CrossRef] [PubMed]
9. Kalantari, Z.; Ferreira, C.S.S.; Walsh, R.P.D.; Ferreira, A.J.D.; Destouni, G. Urbanization development under climate change: Hydrological responses in a peri-urban Mediterranean catchment. *Land Degrad. Dev.* **2017**, *28*, 2207–2221. [CrossRef]
10. Montanari, A.; Young, G.; Savenije, H.H.G.; Hughes, D.; Wagener, T.; Ren, L.L.; Koutsoyiannis, D.; Cudennec, C.; Toth, E.; Grimaldi, S.; et al. "Panta Rhei—Everything Flows": Change in hydrology and society—The IAHS Scientific Decade 2013–2022. *Hydrol. Sci. J.* **2013**, *58*, 1256–1275. [CrossRef]
11. Smettem, K.R.J. Welcome address for the new 'Ecohydrology' Journal. *Ecohydrology* **2008**, *1*, 1–2. [CrossRef]
12. Sivapalan, M.; Savenije, H.H.G.; Blöschl, G. Socio-hydrology: A new science of people and water. *Hydrol. Process* **2012**, *26*, 1270–1276. [CrossRef]
13. Karlsson, J.M.; Lyon, S.W.; Destouni, G. Thermokarst lake, hydrological flow and water balance indicators of permafrost change in Western Siberia. *J. Hydrol.* **2012**, *464–465*, 459–466. [CrossRef]
14. Berghuijs, W.R.; Woods, R.A.; Hrachowitz, M. A precipitation shift from snow towards rain leads to a decrease in streamflow. *Nat. Clim. Chang.* **2014**, *4*, 583–586. [CrossRef]
15. Karlsson, M.J.; Jaramillo, F.; Destouni, G. Hydro-climatic and lake change patterns in Arctic permafrost and non-permafrost areas. *J. Hydrol.* **2015**, *529*, 134–145. [CrossRef]
16. Orth, R.; Destouni, G. Drought reduces blue-water fluxes more strongly than green-water fluxes in Europe. *Nat. Commun.* **2018**, *9*, 3602. [CrossRef] [PubMed]
17. Destouni, G.; Asokan, S.M.; Jarsjö, J. Inland hydro-climatic interaction: Effects of human water use on regional climate. *Geophys. Res. Lett.* **2010**, *37*, L18402. [CrossRef]
18. De Haan, J. How emergence arises. *Ecol. Complex.* **2006**, *3*, 293–301. [CrossRef]
19. Baresel, C.; Destouni, G. Novel quantification of coupled natural and cross-sectoral water and nutrient/pollutant flows for environmental management. *Environ. Sci. Technol.* **2005**, *39*, 6182–6190. [CrossRef] [PubMed]

20. Thorslund, J.; Jarsjo, J.; Jaramillo, F.; Jawitz, J.W.; Manzoni, S.; Basu, N.B.; Chalov, S.R.; Cohen, M.J.; Creed, I.F.; Goldenberg, R.; et al. Wetlands as large-scale nature-based solutions: Status and challenges for research, engineering and management. *Ecol. Eng.* **2017**, *108 Pt B*, 489–497. [CrossRef]
21. Bring, A.; Asokan, S.M.; Jaramillo, F.; Jarsjö, J.; Levi, L.; Pietroń, J.; Prieto, C.; Rogberg, P.; Destouni, G. Implications of freshwater flux data from the CMIP5 multi-model output across a set of Northern Hemisphere drainage basins. *Earth's Future* **2015**, *3*, 206–217. [CrossRef]
22. Jarsjö, J.; Asokan, S.M.; Prieto, C.; Bring, A.; Destouni, G. Hydrological responses to climate change conditioned by historic alterations of land-use and water-use. *Hydrol. Earth Syst. Sci.* **2012**, *16*, 1335–1347. [CrossRef]
23. Asokan, S.M.; Rogberg, P.; Bring, A.; Jarsjö, J.; Destouni, G. Climate model performance and change projection for freshwater fluxes: Comparison for irrigated areas in Central and South Asia. *J. Hydrol. Reg. Stud.* **2016**, *5*, 48–65. [CrossRef]
24. Jaramillo, F.; Destouni, G. Comment on "Planetary boundaries: Guiding human development on a changing planet". *Science* **2017**, *348*, 1217. [CrossRef] [PubMed]
25. Lehner, B.; Verdin, K.; Jarvis, A. HydroSHEDS Technical Documentation. World Wildlife Fund US, Washington, D.C. Available online: http://hydrosheds.cr.usgs.gov (accessed on 24 April 2013).
26. Mitchell, T.D.; Jones, P.D. An improved method of constructing a database of monthly climate observations and associated high-resolution grids. *Int. J. Clim.* **2005**, *25*, 693–712. [CrossRef]
27. Harris, I.; Jones, P.D.; Osborn, T.J.; Lister, D.H. Updated high-resolution grids of monthly climatic observations—the CRU TS3.10 Dataset. *Int. J. Clim.* **2014**, *34*, 623–642. [CrossRef]
28. GRDC. Global Runoff Data Centre, Koblenz, Germany. Available online: http://www.bafg.de (accessed on 3 June 2013).
29. Greek Water Management Authority. *Water Management Plan for the Greek River Basin Districts (GR06)*; Greek Water Management Authority: Athens, Greece, 2013; Volume 2.
30. Organisation for Economic Co-operation and Development, Statistics. Extracted by Selecting Environmental Indicators for Agriculture, Environmental Performance of Agriculture 2013, Water Resources, Enviromental Indicator: Irrigation, Irrigated Areas (ha) and Irrigation Freshwater Withdrawals (million m^3). Available online: http://stats.oecd.org/Index.aspx# (accessed on 4 May 2015).
31. Organisation for Economic Co-Operation and Development, Statistics. Extracted by Selecting Environment\Water\Freshwater Abstractions (million m^3). Variable: Irrigation\Total freshwater for Greece. Available online: http://stats.oecd.org/Index.aspx# (accessed on 4 May 2015).
32. Siebert, S.; Henrich, V.; Frenken, K.; Burke, J. *Global Map of Irrigation Areas Version 5*; Rheinische Friedrich-Wilhelms-University: Bonn, Germany; Food and Agriculture Organization of the United Nations: Rome, Italy, 2013.
33. Karamanos, A.; Aggelides, S.; Londra, P. *Irrigation Systems Performance in Greece*; Options Méditerranéennes, Series B, No. 52; CIHEAM: Bari, Italy, 2005.
34. Jaramillo, F.; Prieto, C.; Lyon, S.W.; Destouni, G. Multimethod assessment of evapotranspiration shifts due to non-irrigated agricultural development in Sweden. *J. Hydrol.* **2013**, *484*, 55–62. [CrossRef]
35. Langbein, W.B. Annual runoff in the United States. *Geol. Surv. Circ.* **1949**, *52*. [CrossRef]
36. Turc, L. "The water balance of soils", relation between precipitation, evaporation and flow. *Ann. Agron.* **1954**, *5*, 491–569.
37. Adam, J.C.; Lettenmaier, D.P. Adjustment of global gridded precipitation for systematic bias. *J. Geophys. Res.* **2003**, *108*, 1–14. [CrossRef]
38. Adam, J.C.; Clark, E.A.; Lettenmaier, D.P.; Wood, E.F. Correction of global precipitation products for orographic effects. *J. Clim.* **2006**, *19*, 15–38. [CrossRef]
39. Asokan, S.M.; Destouni, G. Irrigation effects on hydro-climatic change: Basin-wise water balance-constrained quantification and cross-regional comparison. *Surv. Geophys.* **2014**, *35*, 879–895. [CrossRef]

water

MDPI

Review

Muddy Waters: Refining the Way forward for the "Sustainability Science" of Socio-Hydrogeology

Paul Hynds [1,*], Shane Regan [2], Luisa Andrade [3], Simon Mooney [4], Kevin O'Malley [5], Stephanie DiPelino [6] and Jean O'Dwyer [7,8,*]

1 Environmental Sustainability and Health Institute (ESHI), Dublin Institute of Technology, Dublin 7, Ireland
2 Department of Civil, Structural and Environmental Engineering, Trinity College, Dublin 2, Ireland; regans@tcd.ie
3 School of Architecture, Planning, and Environmental Policy, University College Dublin, Dublin 4, Ireland; Luisa.Andrade@ucd.ie
4 College of Business, Dublin Institute of Technology, Aungier Street, Dublin 2, Ireland; simon.mooney@ucdconnect.ie
5 Department of Psychology, University of Limerick, Limerick, Ireland; Kevin.OMalley@ul.ie
6 Department of Public Health Science, Queen's University, Kingston, ON K7L3N6, Canada; stephanie.dipelino@hotmail.com
7 School of Biological, Earth & Environmental Sciences, University College Cork, Distillery Fields, North Mall, Cork, Ireland
8 Environmental Research Institute, University College Cork, Cork, Ireland
* Correspondence: hyndsp@tcd.ie (P.H.); Jean.ODwyer@ucc.ie (J.O.)

Received: 12 July 2018; Accepted: 17 August 2018; Published: 21 August 2018

Abstract: The trouble with groundwater is that despite its critical importance to global water supplies, it frequently attracts insufficient management attention relative to more visible surface water sources, irrespective of regional climate, socioeconomic profile, and regulatory environment. To this end, the recently defined sub-discipline of "socio-hydrogeology", an extension of socio-hydrology, seeks to translate and exchange knowledge with and between non-expert end-users, in addition to involving non-expert opinion and experience in hydrogeological investigations, thus emphasising a "bottom-up" methodology. It is widely acknowledged that issues pertaining to groundwater quality, groundwater quantity, climate change, and a poor general awareness and understanding of groundwater occurrence and movement are global in their scope. Moreover, while effective communication and engagement represent the key tenet of socio-hydrogeology, the authors consider that multiple actors should be identified and incorporated using stakeholder network analysis and may include policymakers, media and communications experts, mobile technology developers, and social scientists, to appropriately convey demographically focused bi-directional information, with the hydrogeological community representing the communication keystone. Accordingly, this article aims to highlight past and current work, elucidate key areas of development within socio-hydrogeology, and offer recommendations to ensure global efficacy of this increasingly important and growing field going forward. The authors seek to assist in protecting our global groundwater resource for future generations via an improved framework for understanding the interaction between communities and hydrogeological systems.

Keywords: socio-hydrogeology; groundwater management; communication; engagement; socio-economic aspects

---

Preface:

While socio-hydrology is a well-established paradigm for the incorporation of sociological factors into water resource management, the sociological nuances associated with the subsurface environment and hydrogeological phenomena are frequently under-represented within management strategies.

In response to this, 'socio-hydrogeology' as distinct from socio-hydrology, has been argued within the literature, and represents an opportunity to guide the development and optimization of inter and multidisciplinary paradigms that focus on the subsurface environment. However, to date, there has been limited discussion contextualizing the myriad challenges facing socio-hydrogeology and it is clear that solutions need to be offered in order to move towards a cyclical paradigm that integrates both scientific and non-scientific stakeholders. As such, this paper aims to highlight past and current work, elucidate key areas of development within socio-hydrogeology, and offer recommendations to ensure global application of this increasingly important and growing field going forward. The structure of this paper (represented graphically in Figure 1) incorporates five main thematic areas, moving from the genesis of the concept of socio-hydrogeology to its global significance and the challenges and opportunities for socio-hydrogeological development in a connected, high-tech world. These three areas culminate in highlighting the need for heightened stakeholder engagement and network analysis to achieve sustainable groundwater management, which we discuss. Finally, the ideas and knowledge presented are synthesized towards an improved socio-hydrogeological paradigm that incorporates a circular socioeconomic approach that aims to put the 'socio' in socio-hydrogeology and move towards integrated groundwater resource management.

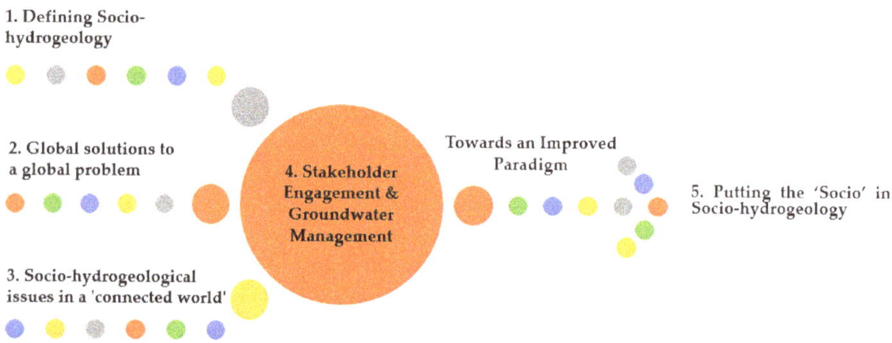

Figure 1. Overview of paper structure: We present the evolution of the definition of socio-hydrogeology, its importance in a global context, as well as challenges and solutions for socio-hydrogeology. We highlight the importance of stakeholder network analysis, culminating in a move towards a new paradigm that puts the socio in socio-hydrogeology.

1. Hydro-Sociology, Socio-Hydrology, and the Genesis of Socio-Hydrogeology

Four decades ago, Widstrand (1978) [1] recognised the need for multidisciplinary methodologies when managing interactions between people and water, and more specifically, the importance of integrating the social sciences. Shortly after, hydro-sociology, with an overarching objective of improving analysis of the social consequences of water related projects, was introduced by Falkenmark (1979) [2]. Many studies since then have addressed these concepts, albeit in the absence of concrete terms or models [3,4]. The term socio-hydrology, first coined by Sivapalan et al. (2012) [5], refers to the myriad of interactions and feedback loops between social and hydrological processes and pressures, and was introduced to the hydrological lexicon as a response to the discipline's perceived failure to adequately examine and address human-modified water sources. At its core, socio-hydrology comprises two social components: (i) absorption of people and their activities into hydrological science, and (ii) ensuring that water-related decisions take the stakeholder perspective into consideration, that is, how and why water is used [6]. Furthermore, socio-hydrology focuses on observing, understanding, and predicting future trajectories of human–water system interactions and the relationships between the two [5,7]. Socio-hydrology thereby represents an interdisciplinary field that attempts to integrate the dynamic reactions and interactions between water and people. For example, process-based models of coupled human-water systems seek to include societies and communities as internal model variables, as opposed to boundary conditions [8]. As such, increasingly accurate long-term predictions pertaining to issues including flooding and water quality may become achievable, as the socio-hydrological perspective seeks to capture the co-evolution of human-water system dynamics [5], for example, water usage, demands, migration, infrastructural development, and so on. However, as recently noted by Pande & Sivapalan (2017) [9], use of the term has been inconsistent.

The challenges facing (non-expert) communities and policymakers regarding groundwater, as opposed to surface water, resource management are quite unique; it is difficult to comprehend and consequently to garner support for the maintenance and remediation of a resource that cannot be easily seen. As a result, while pressures and approaches to the assessment and use of groundwater remain at the global scale, remediation measures have taken a socio-integrative shift.

With the continuous refinement of socio-hydrology, it was perhaps inevitable (and necessary) that a groundwater specific branch would develop. While Burke et al. (1999) [10] were perhaps the first to make the distinction between socio-hydrological and socio-hydrogeological systems and processes, the term "socio-hydrogeology" was first introduced by Re (2015) [11] in the *Hydrogeology Journal*. Re (2015) presents the Bir Al-Nas (bottom-up integrated approach for sustainable groundwater management in rural areas) approach, which seeks to integrate socio-hydrological and science-based groundwater management practices. The Bir Al-Nas (Arabic translation—"the peoples well") approach comprises a strong social component (Figure 2), including stakeholder analysis, public engagement, and socio-economic assessment, and as such, differs from many developed socio-hydrology models [12] in that it places a particular emphasis on surveying, stakeholder network analysis, and local consultation. Re (2015) refers to socio-hydrogeology as *"a way of incorporating the social dimension into hydrogeological investigations"*, similarly, Limaye (2017) [13] notes that the basis for any socio-hydrogeological intervention is effective communication. As such, and as substantiated by Re (2015), this represents one of the primary differences between socio-hydrology and socio-hydrogeology; because of widespread misunderstanding of hydrogeological principles (irrespective of location, socio-economic status, and/or geopolitical setting), higher levels of awareness nurturing via translation and communication are required. Moreover, it seems that a distinct definition of, and model for, applying socio-hydrogeology is required to address the inherent differences between hydrological and hydrogeological systems and processes.

Traditionally, models used in hydrological studies have often assumed stationarity as opposed to temporal variation. Additionally, human-induced water resources management activities are included as external variables in water cycle dynamics. However, in considering the human population's current impact on the water cycle in terms of a growing population (and subsequent demands), increasing river basin management, and climate change, it is unclear whether this approach is still appropriate. Furthermore, research directed at the evolution of water resources and society has shown that the components constituting the human–water system are changing interdependently [14]. Thus, water cycle dynamics should be approached from an integrated perspective in which humans are considered endogenous forces to (and within) the system [14]. As noted by Gorelick & Zheng (2015) [15], and specific to hydrogeology, a new generation of aquifer management approaches and models are required to compete with the new (and existing) interconnected generation of groundwater challenges including global climate change, aquifer storage/depletion, land subsidence, saline intrusion, and hydro-ecology. These global issues require global solutions, which can only be developed and fostered through the application of malleable and evolving socio-hydrogeological principles.

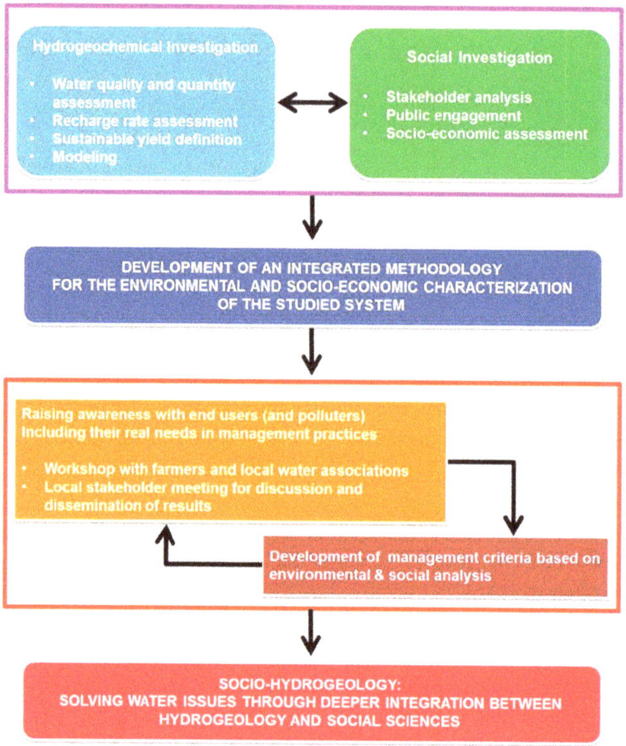

Figure 2. Schematic description of the bottom-up integrated approach for sustainable groundwater management in rural areas (Bir Al-Nas) approach for socio-hydrogeology (Re, 2015).

2. Socio-Hydrogeology: Global Solutions to a Global Problem

Current concerns regarding water scarcity and insecurity are both ubiquitous and varied, manifesting themselves in a myriad of respects worldwide [16]. This is particularly evident from a research perspective. As shown (Figure 3), the geographical distribution of published articles relating to groundwater contamination (from Scopus, the largest global abstract and citation database of peer-reviewed literature) demonstrates that groundwater issues occur across a range of socioeconomic and geo-political regions. However, to date, the majority of socio-hydrogeological applications have occurred in low income countries (Table 1), thus potentially painting a picture of socio-hydrogeology as a science relevant to the "developing" world primarily. However, this is not the case; most, if not all, of the studies employ methodological approaches that are pan-global in terms of application, and recent work by Re et al. (2015) [11] has designed a replicable model for implementing a socio-hydrogeological approach in rural areas, regardless of location. For example, the majority of the issues discussed within these articles are not unique to low income regions, nor are pressures such as population growth, climate change, the shift toward increasingly water-dependent economies and societies, and reduced groundwater availability. Rather, it is evident that human social processes are catalysing hydrogeological degradation at a global level and, therefore, must also be the agents of change and remediation.

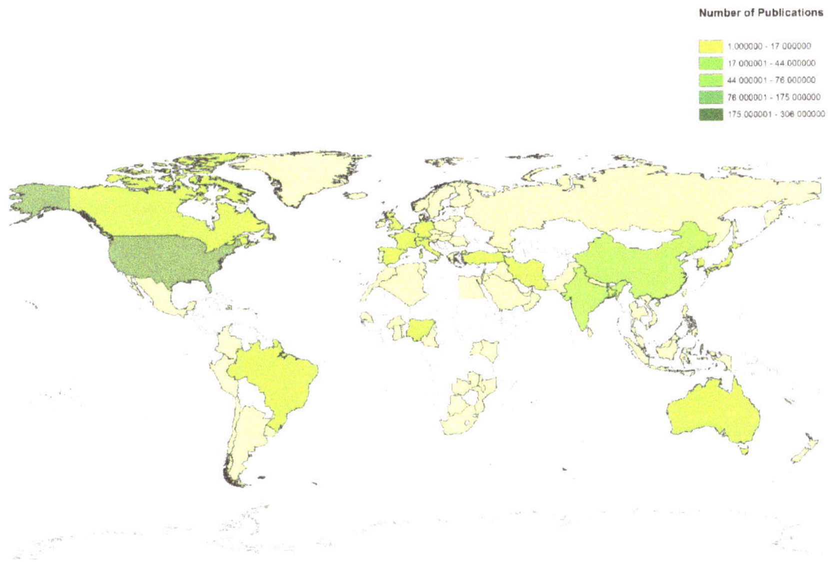

Figure 3. Geographical distribution and frequency of published articles relating to groundwater contamination (January 1975–September 2017).

Unfortunately, because of the inherent nature of groundwater occurrence and transport, the development and implementation of effective socio-hydrogeology management faces many barriers. For example, despite best efforts by the hydrogeological community, groundwater traditionally receives significantly less attention than surface water, inevitably resulting in lower levels of monitoring and management [11,17]. Moreover, as pointed out by both Re (2015) and Limaye (2017) [11,13], a global gap exists between academic research and the general population's awareness, understanding, and daily requirements in relation to groundwater; again largely due to the 'out of sight, out of mind' nature of the resource. For example, in a recent study by Hynds et al. (2013) [18], 245 private well owners from diverse hydrogeological settings in the Republic of Ireland were interviewed; findings indicate that while just 1.2% of respondents had involved an accredited hydrogeologist during the design phase of their domestic water source.

Compounding these issues, physical barriers also exist and need to be accounted for that largely do not exist for surface water; for example, access to groundwater resources may be limited; boundaries are more difficult to establish among and between users; and human–water relationships more likely occur at the micro-level, that is, individual private well-owners [19]. Accordingly, development and implementation of appropriate socio-hydrogeological practices seeks to surmount or dispense with these barriers and promote increased awareness of groundwater-related issues and needs among local communities [11,13,20]. Moreover, research relating to groundwater resources in medium- and high-income countries has shown that contamination, and particularly microbiological pollution, is a recurring problem leading to endemic and epidemic waterborne infectious diseases in these regions [21–23].

Table 1. Articles published to date (2015–2017) applying socio-hydrogeological methods.

Study	Study Year	Country/Region	Field of Study	Engagement with Socio-Hydrogeology
Re [11]	2015	Italy	Groundwater Management, Rural Development	Discussion
Re & Sacchi [24]	2017	Morocco	Groundwater Isotopes, Salinity, Coastal Aquifers	Application
Leduc et al. [25]	2017	Mediterranean	Groundwater Resources, Exploitation, Management	Introduction
Re et al. [20]	2017	Tunisia	Nitrate Contamination	Application
Rodrigues-Capítulo et al. [26]	2017	Argentina	Coastal Aquifers, Urban Development	Introduction
Limaye [13]	2017a	India	Risk Awareness and Communication	Application
Limaye [27]	2017b	India	Rural Communities, Communication, Water Management	Application
Tringali et al. [28]	2017	Tunisia	Groundwater Management	Application
López-Corona et al. [29]	2018	Mexico	Statistical Theory of Groundwater Management	Application
Re [30]	2018a	Myanmar	Water Resources Assessment	Application
Re et al. [31]	2018b	Italy	Climate Change, Future Hydrogeologists, Engagement	Discussion

We consider that researchers, hydrogeologists, and policy-makers should not view socio-hydrogeology or human–water systems as separate entities in different regions. For example, while two socio-hydrological basin processes may not be hydro(geo)logically connected, they may be joined socially or economically by basin consumers. Increased understanding of these spatial socio-hydrological connections will assist socio-hydrogeology in becoming a frequently and consistently employed discipline between countries, regions, economies, and communities. The belief that high-income countries' 'technical capabilities' to 'make several alternative solutions' may be an inappropriate assumption, given the uncertainty associated with the extent to which technology can ensure a sustainable future. As previously stated by Pahl-Wost (2002) [32], while problems were once resolved with purely 'top down' approaches (e.g., increasing treatment sophistication, changing legislation, etc.), it is increasingly apparent that those no longer suffice. As such, it is important to discuss what socio-hydrogeology and alternatives to 'top down' management look like in a high-tech, 'connected' world.

3. Socio-Hydrogeological Issues in a 'Connected World'

Communication is the most vital element for effective socio-hydrogeological applications and interventions [11,13], as lack of knowledge surrounding groundwater resources is a limiting factor for social integration. Therefore, accessibility to information tailored to non-expert audiences, appropriate local/regional translation, and face-to-face communication of research is imperative to stimulate hydro-geological awareness and education [13] and emulate the successes of hydrological awareness. However, in light of significant technological advances during recent decades, communication with the general public has become increasingly complex, demographically distinct, and challenging. There is a vast range of approaches for science communication within the general public. As such, many fundamental difficulties have and will continue to arise when attempting to insert society or sociology into the realm of hydrogeological modelling and research [25]. Compounding this, the media can also play a key role in shaping the perceptions and/or misconceptions of hydrogeological science; for example, groundwater contamination in one area may affect groundwater perceptions in another. As such, the challenges of communicating hydrogeological science are multifaceted, but must be overcome.

For example, high levels of media exposure and education, in addition to technological and intellectual developments, shape responses to socio-hydrogeological interventions, thus science communication can and does frame the way these interventions occur. For example, through media exposure and education, public opinion is currently being molded towards increased environmental awareness [33]. This is undoubtedly advantageous in terms of the development and integration of socio-hydrogeology, if it can be utilised effectively. However, despite all the inherent advantages experienced through media exposure, particularly in high economic regions, a surprising direct effect of financial security and media presence is that it decreases people's likelihood to act when there are changes in hydrological (and hydrogeological) conditions [12]. In other words, communities and individuals in financially secure regions with consistent media exposure have a higher threshold to change and require a greater stimulus to undertake socio-hydrogeological interventions. A solution to this disengagement may lie, perhaps counter-intuitively, in the evolution of handheld mobile technology (e.g., tablets and smartphones), which has allowed collaborative engagement and knowledge transfer to become an everyday reality. At present, there are approximately 20 countries worldwide with levels of smartphone utilisation above 65%; 90% of these are considered high-income regions, with a strong correlation (R^2 = 0.87) also found between per capita income and internet usage [34]. However, the rest of the "emerging world" is catching up; in 2015, a median of 54% of those surveyed reported occasional internet usage or smartphone ownership, an increase of 9% compared with 2013, with much of the increase coming from large emerging economies including Malaysia, China, and Brazil [34].

Inevitably, this will result in members of the public not only consuming scientific knowledge, but contributing their own unique ideas, views, and criticisms via blogs, podcasts, and social media. As such, while communication remains a central challenge for effective socio-hydrogeology, scope also exists for the development of citizen science strategies to move the discipline forward, particularly in medium- and high-income regions. Pragmatically, increased usage of mobile technology represents a novel data source and could assist in reducing the burden of large data requirements for socio-hydrogeological models [35]. Examples of the use of technology for safeguarding water quality can be seen all over Africa. For example, the use of mobile transmitters has made selected handpumps 'smart' by automatically sending usage data via short message service (SMS). Trials ran by Oxford University in Rwanda and Kenya suggest the technology can work and delivers promising results, offering data in four areas: (1) objective monitoring of daily water use; (2) use of monitoring data to rapidly identify and repair broken handpumps; (3) condition monitoring to predict failure prior to occurrence; and (4) non-intrusive, shallow aquifer monitoring [36]. However, further work is needed to elucidate the efficacy of this approach in varied societal and socioeconomic structures.

Furthermore, as illustrated by Re (2015) [11], many hydrogeologists spend substantial time in the field, and as such, they should be the first point of contact for well owners, farmers, and other groundwater users. However, as previously outlined, research by Hynds et al. (2013) [18] reports that only 1.2% of private well utilisers surveyed in the Republic of Ireland consulted with a hydrogeologist, with many citing lack of communication as a deterrent. The authors consider that appropriate development of interactive applications that embrace the communication of socio-hydrogeology in a highly technological world may facilitate hydrogeologists to act as mediators between theory and practice, or between the problem and the (potential) proposed solution to issues of resource quality; quantity; and above all, sustainability. However, the efficacy of socio-hydrogeological interventions are predicated on the identification of the key stakeholders; unlike surface water, groundwater is not as familiar to many people, and thus socially driven groundwater management requires carefully planned stakeholder engagement to ensure sustainable and continuous development of socio-hydrogeological paradigms that impact those most affected.

4. Socio-Hydrogeology, Stakeholder Engagement, and Groundwater Management

The value of the functions provided by freshwater ecosystems, such as rivers, wetland, and floodplains, has gained significant prominence in recent years as these "ecosystem services" represent a vast invaluable resource with respect to regional/national economies and human well-being [37].

However, the ecosystem services of groundwater, and particularly the provision of drinking water, are largely ignored; likely because of its hidden, and often complex nature, which can be difficult to manage. While groundwater resource management undoubtedly requires the technical skills of hydrogeological and engineering professionals, the spatially and temporally heterogeneous characteristics of groundwater flow and a myriad of environmental interactions actually necessitate active community and stakeholder engagement, arguably to a greater extent than surface water. However, the stakeholders for groundwater resources are not as obvious or various. As a result, at the forefront of all groundwater management strategies should be stakeholder network analysis; a process that investigates and categorizes the relationships between stakeholders [38] and identifies the key actors likely to positively influence the implementation of new management practices resulting from a hydrogeological investigation. Importantly, groundwater management and stakeholder identification must be holistic in its approach, considering environmental degradation in addition to resource assessment and engage a wide range of parties ranging from those who physically use and extract groundwater to those who manage and are affected by the benefits granted by groundwater influenced ecological systems (Figure 4). While detailed stakeholder analyses can be costly, they are also extremely valuable; a professionally facilitated process that begins with a carefully conducted stakeholder network analysis can help ensure that all interests are adequately met and that those affected by future groundwater policy have the opportunity to decide who will govern groundwater use. For example, in contrast to surface water (or hydrological) stakeholder network analysis, which frequently identifies engineers and water managers as the 'key actors' [39], hydrogeologically associated stakeholders tend to include residents and water user groups as key considerations [40]. This is one of the key differences between surface and groundwater users; for well owners, their supply is entirely personal and is typically not governed by an overarching management facility. As such, well owners are often the most valuable stakeholder in assessing quality and guiding policy and, therefore, should form a key component of 'integrated water resource management (IWRM)' planning [10,40]; an increasingly deployed approach to managing the water cycle in both high [41] and low-income countries [42]. However, groundwater is still frequently under-represented in water management plans, including IWRM and indeed, often added only as an afterthought [40].

Nevertheless, effective and sustainable management of water resources is vital for ensuring sustainable development [43,44]. However, while physical problems are well understood by the technical and scientific community, the range and complexity of socio-economic responses to these physical problems are not immediately recognised [10]. It has been reported that the failure to effectively engage stakeholders and communities in the planning and implementation of infrastructural projects of varying scales can prove costly, resulting in public controversy, and delayed or abandoned projects [38]. Previous studies have found that social, institutional, and political factors represent the primary obstacles to sustainable management of the world's groundwater resources, and typically lag behind technological and technical developments in hydrogeology [10,45]. Stakeholder and community involvement are thus crucial in the decision-making process [44] and to the successful development of any project, particularly ones involving groundwater that must engage a large number of individual users [10]. Accordingly, while water management is typically driven by top-down government approaches, participatory bottom-up approaches are now increasingly employed to involve local stakeholders, such as farmers, in the decision and planning process, and have been demonstrated to be a successful management strategy [45].

As constantly alluded to throughout the current article, sustainable current and future groundwater management strategies are confronted with momentous challenges [46]. Frameworks for sustainable groundwater management must respond to these emerging hurdles, while adhering to policy and legislative directives and the needs of communities and stakeholder bodies, thereby integrating important elements of engagement and technical insight. Moreover, stakeholders not directly using groundwater, such as river and conservation managers, must also be incorporated into framework development as they too have a voice and role in long-term resource sustainability.

Developing frameworks for management assists in avoiding situations where the public becomes engaged with groundwater only after a 'failure' occurs, such as the excessive exploitation of aquifers in countries with a high dependency on potable groundwater [46], meaning it is important to establish standards and principles for long-term monitoring and sustainable use. Social barriers to water management must thus be addressed. In recent years, social research and theory has been employed as an increasingly important factor in understanding and responding to the challenges associated with evolving a more sustainable society [39,47]. An example of this approach is the application of 'transition theory', which is generically defined as *"a gradual, continuous process of structural change within a society or culture"* [40,41], and may well provide a useful framework for socio-hydrogeology to go forward, in that it provides a basis for coherent, consistent public policy, which is not deterministic, but rather offers a range of possible pathways for change [40,41]. This social model facilitates community and stakeholder engagement by providing pathways between different levels of social structure (Figure 4), permitting transformative change [48]. The model contains three tier levels [48], as follows:

1. Technical learning and outcomes associated with implementing and refining technologies and policy instruments (first-order learning);
2. Conceptual learning associated with questioning (reconceptualising) the fundamental policy aims and objectives (second-order learning);
3. Social learning (third-order) providing the opportunity and leverage to promote regular shifts (transformation) in the sophistication of learning, from technical to conceptual. Indeed, it is suggested that without 'social learning', conceptual shifts in understanding will only occur following a crisis or persistent policy failure [42].

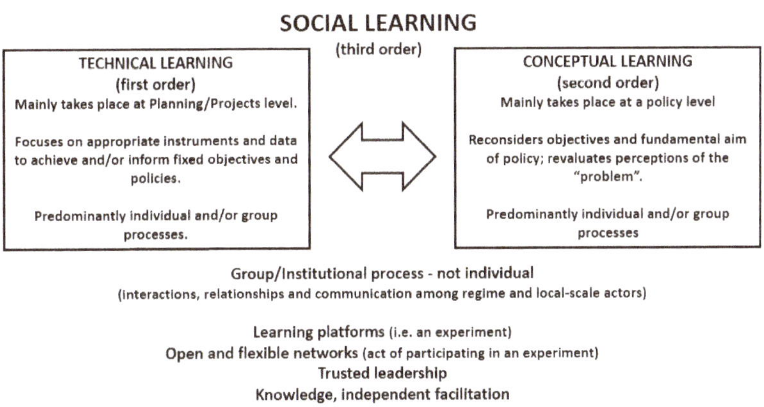

Figure 4. Social learning and transformative change, adapted from the literature [48].

This learning must be fostered by effectively communicating with stakeholders and end-users, which requires a nuanced approach if it is to foster stewardship and engagement on the part of the wider (non-expert) community [49–51]. This is one of the obvious spaces for the communication and social sciences to contribute to positive outcomes in socio-hydrogeology research. As such, a brief comment on some of the psychological factors related to this domain is warranted, before a wider discussion of the inclusion of sociological elements.

As stated earlier, one of barriers to effective communication is the combination of 'invisibility' and complexity, which can make it difficult to convey information in a way that is comprehensible. For a number of reasons, each of which must be addressed, this can make it challenging to garner enthusiasm and commitment to maintenance on the part of stakeholders. One key example is the fact that groundwater-related concerns, like many environmental issues, may be too "psychologically distant" [52] to be perceived as meaningful to end-users. For example, the comparative infrequency of waterborne infection across an individual lifespan, the ratio of instances of safe to unsafe water exposure, could be said to reduce the 'closeness in probability' experienced by those with more established access to groundwater resources. This may be further compounded by the ostensible randomness of events that pose an acute threat to groundwater sources such as extreme weather and flooding, as well as industrial or agricultural accidents. Additionally, a lack of understanding of the severity of the potential harm caused by waterborne disease such as Verotoxigenic *Escherichia coli* (VTEC) may decrease the degree to which people experience it as 'socially close'.

Although it may seem that simply presenting the reality of each of these components should result in more engagement on the part of end-users, a wide body of health and social psychological research has not only indicated that this is not the case [53], but has also enumerated a number of psychological models and socio-cognitive factors that must be taken into consideration [54].

Kasl & Cobb (1966) [55] define health behaviours as those undertaken by healthy individuals to promote continued wellbeing. A number of psychological models have been developed to explain the performance of such behaviours, such as the Health Beliefs Model [56] and the Theories of Reasoned Action and Planned Behaviour [57,58]. Although these theoretical frameworks diverge in some ways, one of the key areas of overlap, which is also of key concern to those wishing to communicate effectively on environmental issues, is that of perceived self-efficacy and control, which can cause social actors to disengage from an issue if not managed correctly [59]. Hydrogeological information may be particularly prone to reducing self-efficacy and perceived control on the part of end-users if not managed effectively. This is potentially best illustrated with a specific example.

The same features that may increase the 'psychological distance' for groundwater users, the randomness of weather and the socio-political space between end-users and policy making, may contribute to a sense of helplessness or extreme difficulty if communication is not adequately tailored to the target audience. This perceived difficulty has been shown to cause individuals to actively disengage from health behaviors [60], even so far as inhibiting physiological responses. The aim then must be that research is communicated to the wider community, rather than at the wider community in order to promote hydro-geologically related 'health behaviours' [61] and truly put the 'socio' in socio-hydrogeology. Communicating in this balanced way can help to foster the sense of shared social identity, inclusive of both the hydrogeological research community and the wider public, which will reinforce and bolster the sense of individual and group efficacy [62].

5. Putting the "Socio" in Socio-Hydrogeology

Ruddell & Wagener (2015) [63] assert that the solution to many of the 21st centuries grand challenges in hydrological education lie in the establishment of educational networks and partnerships between different strands within and beyond hydrology. To better serve and understand society, hydrogeology, distinct from hydrology, must thus draw impetus from previous and ongoing developments in social and communications science, and attempt to weave together findings from socio-hydrological investigations [11,28]. Hydrogeologists can, and indeed must, be leaders within socio-hydrogeology via advocation, mediation, translation, and promotion of best practice; they are in a unique position to advocate for appropriate groundwater management and protection, promote and develop experience at the local/catchment scale into regional or national management strategy, and assist in translating between the ideal (science) and the achievable (practise) [11,13]. However, in a rapidly changing world, it may be difficult for hydrogeologists to act as "social hydrogeologists", and more so when not operating within multi- or transdisciplinary teams.

A recent study by Hynds et al. (2018) [64] surveyed 1634 domestic wastewater treatment system (septic tank) owners from the Republic of Ireland, the majority (≈65%) of whom derived their household water supply from a private borehole. While approximately 5.5% of respondents acquired relevant information (septic tank management, etc.) from public meetings and/or face-to-face interaction, approximately 37% of respondents obtained their information from the Internet, with a further 8% acquiring it directly from social media. Moreover, and perhaps more crucially, respondents ascribed a significantly higher level of efficacy (vis. trust and accuracy) to information from the Internet (89.5%) than from personal contact (83.3%). O'Dwyer et al. (2014) [65] found that proactivity (consumption avoidance) among Irish private well users (n = 132) as a direct result of individual-level knowledge (microbiological quality of their supply) was not evident across an entire survey cohort. As such, the authors consider that the "all things to all men" socio-hydrogeological approach likely does not represent the ideal; hydrogeologists may not have the time to act as social hydrologists, and in many cases, will not possess the necessary expertise with regard to appropriate scientific translation and communication.

Thus, the authors recommend an approach that truly "puts the "socio" in socio-hydrogeology", whereby hydrogeologists utilise and translate their sectorial expertise and local experience for social, communications, and media experts, thus fostering inclusion of the social dimension in hydrogeological investigations and ensuring appropriate demographically-focused information finds its way to the appropriate audience as efficiently as possible. As such, we present (Figure 5) a prototype circular socio-economy that incorporates both a "bottom up" and "top down" dimension, but importantly, also acknowledges that there must be a symbiosis between these two through 'multi-level governance'. For example, hydrogeologists, working in cooperation with social scientists and media/communications experts, in concurrence with continuing bi-directional stakeholder interaction and traditional scientific communication, permits an open discursive dialogue that can be augmented proportionally as needs arise. This inclusion of a communication loop (Figure 5) allows for continuous engagement and knowledge exchange, which can be used to stimulate new ideas and solutions as well as to develop bespoke groundwater management strategies at the local level. Importantly, and as shown within the model, interdisciplinary within hydrogeology should be encouraged, appropriately acknowledging it as a "community-facing" scientific discipline. As such, hydrogeologists require training in the integration of social and physical sciences, thus producing effective socio-hydrogeologists, and ensuring successful realisation of the hydrogeology-sociology-local knowledge feedback and knowledge exchange loop. However, even with appropriate training, interdisciplinary collaboration, particularly between the natural and social sciences, is crucial to solving the significant challenges facing humanity [66]; socio-hydrogeology would appear to be a perfect example of this, with the inclusion of social and communications experts also likely taking pressure off the hydrogeological community in both the short- and long-term.

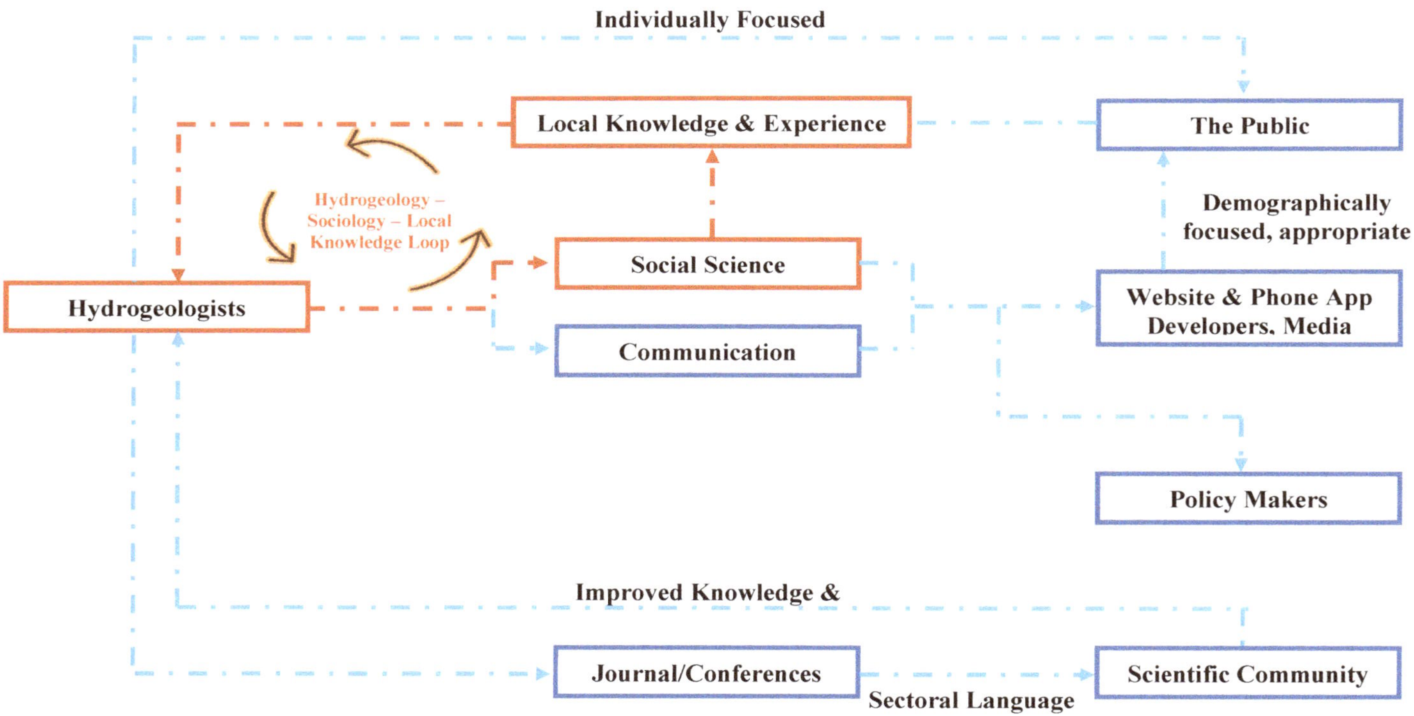

Figure 5. A proposed circular socioeconomy incorporating both "bottom up" (participatory) and "top down" (legislative/policy-driven) elements with hydrogeologists encompassing the communication keystone.

6. Conclusions

As recently noted by Reddy and Syme (2014) [67], sustainable groundwater usage and management likely represents the most difficult challenge within the broader field of water resource management, not only because of its global volumetric significance, but, perhaps more importantly, because of a widespread lack of mechanistic awareness and information, that is, occurrence, transport, and vulnerability to contamination. Similarly, necessary communication and engagement between experts and non-experts within the domain of socio-hydrogeology comprises an inherently higher level of complexity than that encountered within socio-hydrology, and, therefore, cannot employ analogous methods to bridging current gaps between geoscience and society. Previous studies have quite rightly called for an increasingly sociological mentality in attempting to alleviate groundwater-related issues [11,67] and this must continue to be fostered. It is apparent that the "all things to all men" approach may not be appropriate as it places pressure on the hydrogeological community to be societal, sociological, and linguistic experts, in addition to presuming high levels of homogeneity that are not present, socially, geographically, or hydrogeologically. Instead, we recommend that circular socio(hydro)economies comprising all affected and influential actors (hydrogeologists, groundwater users, policymakers), in addition to experts within the social and communications sciences, in order to ensure effective translation and demographically focused message framing. This approach would be particularly effective within medium and high-income countries that are no less reliant on groundwater resources, and no less sensitive to groundwater pressures; however, the demographic and cultural profiles within these regions are vastly different to those in developing countries. As such, effective bi-directional provision of information, experience, guidance, and recommendations should be both regionally and demographically bespoke, accounting not only for hydrogeological setting, but also for regional demographics, socioeconomics, and media preferences. As noted by Hynds et al. (2017) [68], "hydrogeologists possess an inherent understanding of the complex and unpredictable nature of groundwater contamination, and thus in collaboration with microbiologists, epidemiologists, geochemists, medical practitioners, and policy makers have an opportunity to help achieve global public health goals". While this sentiment undoubtedly remains true both now and into the future, it may pertinent to add social, management, mobile technology, and communications experts to this mix.

Author Contributions: P.H., J.O. and S.R. conceptualised the paper. P.H., S.R., J.O., L.A., S.M., S.D. and K.O. analysed the literature and contributed to the writing of the paper. J.O. and P.H. approved the final manuscript.

Funding: This research received no external funding.

Conflicts of Interest: The authors declare no conflict of interest.

References

1. Widstrand, C. Social and economic aspects of water exploitation. *Water Supply Manag.* **1978**, 2.
2. Falkenmark, M. Main problems of water use and transfer of technology. *GeoJournal* **1979**, *3*, 435–443. [CrossRef]
3. Mitchell, M.; Curtis, A.; Sharp, E.; Mendham, E. Directions for social research to underpin improved groundwater management. *J. Hydrol.* **2012**, *448–449*, 223–231. [CrossRef]
4. Sivakumar, B. Socio-hydrology: Not a new science, but a recycled and re-worded Hydrosociology. *Hydrol. Process.* **2012**, *26*, 3788–3790. [CrossRef]
5. Sivapalan, M.; Savinje, H.H.; Blöschl, G. Socio-hydrology: A new science of people and water. *Hydrol. Process.* **2012**, *26*, 1270–1276. [CrossRef]
6. Gober, P.; Wheater, H.S. Socio-hydrology and the science–policy interface: A case study of the Saskatchewan River basin. *Hydrol. Earth Syst. Sci.* **2014**, *18*, 1413–1422. [CrossRef]
7. Sivapalan, M.; Blöschl, G. Time scale interactions and the coevolution of humans and water. *Water Resour. Res.* **2015**, *51*, 6988–7022. [CrossRef]

8. Di Baldassarre, G.; Kooy, M.; Kemerink, J.S.; Brandimarte, L. Towards understanding the dynamic behaviour of floodplains as human-water systems. *Hydrol. Earth Syst. Sci.* **2013**, *17*, 3235–3244. [CrossRef]
9. Pande, S.; Sivapalan, M. Progress in socio-hydrology: A meta-analysis of challenges and opportunities. *Wiley Interdiscip. Rev. Water* **2017**, *4*, e1193. [CrossRef]
10. Burke, J.J.; Sauveplane, C.; Moench, M. Groundwater management and socio-economic responses. *Nat. Resour. Forum* **1999**, *23*. [CrossRef]
11. Re, V. Incorporating the social dimension into hydrogeochemical investigations for rural development: The Bir Al-Nas approach for socio-hydrogeology. *Hydrogeol. J.* **2015**, *23*, 1293–1304. [CrossRef]
12. Elshafei, Y.; Sivapalan, M.; Tonts, M.; Hipsey, M.R. A prototype framework for models of socio-hydrology: Identification of key feedback loops and parameterisation approach. *Hydrol. Earth Syst. Sci.* **2014**, *18*, 2141–2166. [CrossRef]
13. Limaye, S.D. Socio-hydrogeology and low-income countries: Taking science to rural society. *Hydrogeol. J.* **2017**, *25*, 1927–1930. [CrossRef]
14. Loucks, D.P. Debates—Perspectives on socio-hydrology: Simulating hydrologic-human interactions. *Water Resour. Res.* **2015**, *51*, 4789–4794. [CrossRef]
15. Gorelick, S.M.; Zheng, C.M. Global change and the groundwater management challenge. *Water Resour. Res.* **2015**, *51*, 3031–3051. [CrossRef]
16. Whitmee, S.; Haines, A.; Beyrer, C.; Boltz, F.; Capon, A.G.; de Souza Dias, B.F.; Horton, R. Safeguarding human health in the Anthropocene epoch: Report of The Rockefeller Foundation—Lancet Commission on planetary health. *Lancet* **2015**, *386*, 1973–2028. [CrossRef]
17. Famiglietti, J.S. The Global Groundwater Crisis. *Nat. Clim. Chang.* **2014**, *4*, 945–948. [CrossRef]
18. Hynds, P.D.; Misstear, B.D.; Gill, L.W. Unregulated private wells in the Republic of Ireland: Consumer awareness, source susceptibility and protective actions. *J. Environ. Manag.* **2013**, *127*, 278–288. [CrossRef] [PubMed]
19. Moench, M. When the well runs dry but livelihood continues: Adaptive responses to groundwater depletion and strategies for mitigating the associated impacts. In *The Agricultural Groundwater Revolution: Opportunities and Threats to Development*; CABI: Wallingford, UK, 2007; Chapter 9; pp. 173–192.
20. Re, V.; Sacchi, E.; Kammoun, S.; Tringali, C.; Trabelsi, R.; Zouari, K.; Daniele, S. Integrated socio-hydrogeological approach to tackle nitrate contamination in groundwater resources. The case of Grombalia Basin (Tunisia). *Sci. Total Environ.* **2017**, *593*, 664–676. [CrossRef] [PubMed]
21. Guzman Herrador, B.R.; de Blasio, B.F.; MacDonald, E.; Nichols, G.; Sudre, B.; Vold, L. Analytical studies assessing the association between extreme precipitation or temperature and drinking water-related waterborne infections: A review. *Environ. Health Glob. Access Sci. Source* **2015**, *14*, 29. [CrossRef] [PubMed]
22. Ó'hAiseadha, C.; Hynds, P.D.; Fallon, U.B.; O'Dwyer, J. A geostatistical investigation of agricultural and infrastructural risk factors associated with primary verotoxigenic, *E. coli* (VTEC) infection in the Republic of Ireland, 2008–2013. *Epidemiol. Infect.* **2017**, *145*, 95–105.
23. Murphy, H.M.; Prioleau, M.D.; Borchardt, M.A.; Hynds, P.D. Review: Epidemiological evidence of groundwater contribution to global enteric disease, 1948–2015. *Hydrogeol. J.* **2017**, *25*, 981–1001. [CrossRef]
24. Re, V.; Sacchi, E. Tackling the salinity-pollution nexus in coastal aquifers from arid regions using nitrate and boron isotopes. *Environ. Sci. Pollut. Res.* **2017**, *24*, 132–147. [CrossRef] [PubMed]
25. Leduc, C.; Pulido-Bosch, A.; Remini, B. Anthropization of groundwater resources in the Mediterranean region: Processes and challenges. *Hydrogeol. J.* **2017**, *25*, 1529–1547. [CrossRef]
26. Rodrigues-Capítulo, L.; Carretero, S.C.; Kruse, E.E. Comparative study of urban development and groundwater condition in coastal areas of Buenos Aires, Argentina. *Hydrogeol. J.* **2017**, *25*, 1407–1422. [CrossRef]
27. Limaye, S.D. Socio-Hydrogeology and its Application for Promoting Ground Water Management in India. *J. Geol. Geosci.* **2017**, *1*, 1–3.
28. Tringali, C.; Re, V.; Siciliano, G.; Chkir, N.; Tuci, C.; Zouari, K. Insights and participatory actions driven by a socio-hydrogeological approach for groundwater management: The Grombalia Basin case study (Tunisia). *Hydrogeol. J.* **2017**, *25*, 1241–1255. [CrossRef]
29. López-Corona, O.; Padilla, P.; Escolero, O.; Morales-Casique, E. Heuristic Formulation of a Contextual Statistic Theory for Groundwater. *Found. Sci.* **2016**, *23*, 75–83. [CrossRef]

30. Re, V.; Maldaner, C.H.; Gurdak, J.J.; Leblanc, M.; Resende, T.C.; Stigter, T.Y. Topical Collection: Climate-change research by early-career hydrogeologists. *Hydrogeol. J.* **2018**, *26*, 673–676. [CrossRef]
31. Re, V.; Thin, M.M.; Setti, M.; Comizzoli, S.; Sacchi, E. Present status and future criticalities evidenced by an integrated assessment of water resources quality at catchment scale: The case of Inle Lake (Southern Shan state, Myanmar). *Appl. Geochem.* **2018**, *92*, 82–93. [CrossRef]
32. Pahl-Wostl, C. Towards sustainability in the water sector—The importance of human actors and processes of social learning. *Aquat. Sci.-Res. Bound.* **2002**, *64*, 394–411. [CrossRef]
33. Wesselink, A.; Kooy, M.; Warner, J. Socio-hydrology and hydrosocial analysis: Toward dialogues across disciplines. *Wiley Interdiscip. Rev. Water* **2017**, *4*, e1196. [CrossRef]
34. Pew Research Centre. Global Divide on Smartphone Ownership. 2016. Available online: http://www.pewglobal.org/2016/02/22/smartphone-ownership-rates-skyrocket-in-many-emerging-economies-but-digital-divide-remains/technology-report-03-06/ (accessed on 12 September 2017).
35. Troy, T.J.; Konar, M.; Srinivasan, V.; Thompson, S. Moving socio-hydrology forward: A synthesis across studies. *Hydrol. Earth Syst. Sci.* **2015**, *19*, 3667–3679. [CrossRef]
36. Weaver, D.S.; Nejmeh, B.; Vader, D.; Beers, T. The Intelligent Water Project: Bringing Understanding to Water Pumps in Africa. In Proceedings of the 2nd International Conference on Geographical Information Systems Theory, Applications and Management (GISTAM 2016), Rome, Italy, 26–27 April 2016; pp. 211–218.
37. Costanza, R.; de Groot, R.; Braat, L.; Kubiszeqski, I.; Fioramonti, L.; Sutton, P.; Farber, S.; Grasso, M. Twenty years of ecosystems services: How far have we come and how far do we still need to go? *Ecosyst. Serv.* **2017**, *28*, 1–16. [CrossRef]
38. Reed, M.S.; Graves, A.; Dandy, N.; Posthumus, H.; Hubacek, K.; Morris, J.; Prell, C.; Quinn, C.H.; Stringer, L.C. Who's in and why? A typology of stakeholder analysis methods for natural resource management. *J. Environ. Manag.* **2009**, *90*, 1933–1949. [CrossRef] [PubMed]
39. Lienert, J.; Schnetzer, F.; Ingold, K. Stakeholder analysis combined with social network analysis provides fine-grained insights into water infrastructure planning processes. *J. Environ. Manag.* **2013**, *125*, 134–148. [CrossRef] [PubMed]
40. Brunner, N.; Starkl, M.; Sakthivel, P.; Elango, L.; Amirthalingam, S.; Pratap, C.E.; Parimalarenganayaki, S. Policy preferences about managed aquifer recharge for securing sustainable water supply to Chennai city, India. *Water* **2014**, *6*, 3739–3757. [CrossRef]
41. Orr, P.; Colvin, J.; King, D. Involving stakeholders in integrated river basin planning in England and Wales. In *Integrated Assessment of Water Resources and Global Change*; Springer: Dordrecht, The Netherlands, 2006; pp. 331–349.
42. Dungumaro, E.W.; Madulu, N.F. Public participation in integrated water resources management: The case of Tanzania. *Phys. Chem. Earth Parts A/B/C* **2003**, *28*, 1009–1014. [CrossRef]
43. Lennox, J.; Proctor, W.; Russell, S. Structuring stakeholder participation in New Zealand's water resource governance. *Ecol. Econ.* **2011**, *70*, 1381–1394. [CrossRef]
44. Van der Plank, S.; Walsh, B.; Behrens, P. The expected impacts of mining: Stakeholder perceptions of a proposed mineral sands mine in rural Australia. *Resour. Policy* **2016**, *48*, 129–136. [CrossRef]
45. Barthel, R.; Foster, S.; Villholth, K. Interdisciplinary and participatory approaches: The key to effective groundwater management. *Hydrogeol. J.* **2017**, *25*, 1923–1926. [CrossRef]
46. Fishman, R.M.; Seigfried, T.; Raj, P.; Modi, V.; Lall, U. Over-extraction from shallow bedrock versus deep alluvial aquifers: Reliability versus sustainability considerations for India's groundwater irrigation. *Water Resour. Res.* **2011**, *47*. [CrossRef]
47. Brown, R.R.; Keath, N.A. Drawing on social theory for transitioning to sustainable urban water management: Turning the institutional super-tanker. *Aust. J. Water Resour.* **2008**, *12*, 73–83. [CrossRef]
48. Farrelly, M.; Brown, R. Rethinking urban water management: Experimentation as a way forward? *Glob. Environ. Chang.* **2011**, *21*, 721–732. [CrossRef]
49. Rotmans, J.; Kemp, R.; Van Asselt, M. More evolution than revolution: Transition management in public policy. *Foresight* **2001**, *3*, 15–31. [CrossRef]
50. Fiorino, D.J. Environmental policy as learning: A new view of an old landscape. *Public Adm. Rev.* **2001**, *61*, 322–334. [CrossRef]

51. Dresner, M.; Handelman, C.; Braun, S.; Rollwagen-Bollens, G. Environmental identity, pro-environmental behaviors, and civic engagement of volunteer stewards in Portland area parks. *Environ. Educ. Res.* **2015**, *21*, 991–1010. [CrossRef]
52. Carmi, N.; Kimhi, S. Further than the eye can see: Psychological distance and perception of environmental threats. *Hum. Ecol. Risk Assess. Int. J.* **2015**, *21*, 2239–2257. [CrossRef]
53. Armitage, C.J.; Conner, M. Social cognition models and health behaviour: A structured review. *Psychol. Health* **2000**, *15*, 173–189. [CrossRef]
54. Abraham, C.; Sheeran, P.; Johnston, M. From health beliefs to self-regulation: Theoretical advances in the psychology of action control. *Psychol. Health* **1998**, *13*, 569–591. [CrossRef]
55. Kasl, S.V.; Cobb, S. Health behavior, illness behavior and sick role behavior: I. Health and illness behavior. *Arch. Environ. Health Int. J.* **1966**, *12*, 246–266. [CrossRef]
56. Rosenstock, I.M. Historical origins of the health belief model. *Health Educ. Monogr.* **1974**, *2*, 328–335. [CrossRef]
57. Fishbein, M.; Ajzen, I. *Belief, Attitude, Intention and Behavior: An Introduction to Theory and Research*; Addison-Wesley: Reading, MA, USA, 1975.
58. Ajzen, I. *Attitudes, Personality, and Behavior*; McGraw-Hill Education: Berkshire, UK, 2005.
59. Wallston, K.A. Perceived control and health behaviour. In *Cambridge Handbook of Psychology, Health and Medicine*; Cambridge University Press: Cambridge, UK, 1997; Volume 1.
60. Contrada, R.J.; Wright, R.A.; Glass, D.C. Task difficulty, Type A behavior pattern, and cardiovascular response. *Psychophysiology* **1984**, *21*, 638–646. [CrossRef] [PubMed]
61. Asah, S.T.; Blahna, D.J. Motivational functionalism and urban conservation stewardship: Implications for volunteer involvement. *Conserv. Lett.* **2012**, *5*, 470–477. [CrossRef]
62. Crompton, T.; Kasser, T. *Meeting Environmental Challenges: The Role of Human Identity*; WWF-UK: Godalming, UK, 2009; pp. 1–93.
63. Ruddell, B.L.; Wagener, T. Grand Challenges for Hydrology Education in the 21st Century. *J. Hydrol. Eng.* **2013**, *20*, A401400. [CrossRef]
64. Hynds, P.; Naughton, O.; O'Neill, E.; Mooney, S. Efficacy of a national hydrological risk communication strategy: Domestic wastewater treatment systems in the Republic of Ireland. *J. Hydrol.* **2018**, *558*, 205–213. [CrossRef]
65. O'Dwyer, J. Microbiological Contamination of Private Water Wells in the Midwest Region of Ireland: Investigation of Water Quality, Public Awareness and the Application of Logistic Regression in Contaminant Modelling. Ph.D. Thesis, University of Limerick, Limerick, Ireland, 2014.
66. Barthel, R.; Seidl, R. Interdisciplinary Collaboration between Natural and Social Sciences–Status and Trends Exemplified in Groundwater Research. *PLoS ONE* **2017**, *12*, e0170754. [CrossRef] [PubMed]
67. Reddy, R.V.; Syme, G.J. Social sciences and hydrology: An introduction. *J. Hydrol.* **2014**, *518*, 1–4. [CrossRef]
68. Hynds, P.D.; Borchardt, M.; Ibaraki, M. Preface: Hydrogeology and Human Health. *Hydrogeol. J.* **2017**, *25*, 897–902. [CrossRef]

water

MDPI

Review

Artificial Aquatic Ecosystems

Chelsea C. Clifford * and James B. Heffernan

Nicholas School of the Environment, Duke University, P.O. Box 90328, Durham, NC 27708, USA; james.heffernan@duke.edu
* Correspondence: chelseaclifford@gmail.com; Tel.: +1-804-815-9611

Received: 20 June 2018; Accepted: 10 August 2018; Published: 17 August 2018

Abstract: As humans increasingly alter the surface geomorphology of the Earth, a multitude of artificial aquatic systems have appeared, both deliberately and accidentally. Human modifications to the hydroscape range from alteration of existing waterbodies to construction of new ones. The extent of these systems makes them important and dynamic components of modern landscapes, but their condition and provisioning of ecosystem services by these systems are underexplored, and likely underestimated. Instead of accepting that artificial ecosystems have intrinsically low values, environmental scientists should determine what combination of factors, including setting, planning and construction, subsequent management and policy, and time, impact the condition of these systems. Scientists, social scientists, and policymakers should more thoroughly evaluate whether current study and management of artificial aquatic systems is based on the actual ecological condition of these systems, or judged differently, due to artificiality, and consider resultant possible changes in goals for these systems. The emerging recognition and study of artificial aquatic systems presents an exciting and important opportunity for science and society.

Keywords: artificiality; reconciliation ecology; drainage; irrigation; ditches; ponds

1. Introduction

Humans alter geomorphology on an ever-increasing scale [1], one comparable with [2], and in some ways exceeding [3], rates of natural processes. Every change people make to the Earth's surface has the potential to affect the flow and accumulation of water. People have dug ditches, impounded streams and rivers, and otherwise shifted Earth's surface to direct and store water for human use, especially agriculture, for over 5000 years [1]. Today, land use matrices that entail human-engineered waterbodies, such as urban settlements, rice villages, and irrigated cropland, cover significant fractions of terrestrial Earth [4]. Patterns of surface water modification and extent are tightly linked to these land uses [5]. Man-made and -modified aquatic systems have become ubiquitous landscape features [6].

In spite of their commonness, artificial aquatic systems remain poorly understood. Indeed, it is often unclear which waterbodies even belong in the category of "artificial" or "anthropogenic". To date, the limited study of different artificial aquatic systems has been fragmented among various domains of ecology and other environmental sciences, and more often discussed on the margins of natural ecosystems than in conjunction with them, as part of a complete hydroscape [7]. Because science and management does not often focus on artificial aquatic ecosystems, their abundance and extent are poorly quantified, and their ecological statuses and causes thereof poorly understood. As a result, we lack a scientific basis for assessing the ecological value of artificial aquatic systems, or determining how management and policy might improve that value. The ubiquity of artificial aquatic systems, the potential commonalities among them and with natural aquatic ecosystems, and our limited understanding of their central and evolving role in the modern hydroscape all argue for more integrated study of the waterbodies created and transformed by human activity.

In this paper, we propose a framework for aquatic ecosystem artificiality that includes both the intent and magnitude of modification, and argue that policy and management often implicitly use these characteristics to differentiate in their treatment of aquatic systems. We assemble current estimates of the extent of some types of artificial waterbodies in the U.S., and review established knowledge of their ecological condition, including the ecosystem services and disservices that they provide. We argue that the condition of artificial aquatic systems, as for their natural counterparts, likely reflects ecological processes and human decisions both in place and within their watersheds, and that the often poor condition of these systems [8] is not necessarily inherent to their anthropogenic origin. Finally, we posit that scientific undervaluation of artificial aquatic systems may lead scientists, managers, and policymakers to treat artificial waterbodies in ways that perpetuate poor ecological conditions. To manage the rapidly changing and increasingly anthropogenic hydroscape, aquatic scientists need to better inventory its myriad artificial components, evaluate their current structure and function, and link these findings to drivers. Future assessment of artificial aquatic systems that clarifies what their real and perceived values are, and how to purposefully change those values, will require more deliberate, intensive, systematic and mechanistic study, and, perhaps, a shift in our perspective regarding what counts as nature.

2. Degrees and Axes of Artificiality

What does it mean to identify aquatic ecosystems as "artificial"? Designer ecosystems, like rain gardens and ponds, swales, and wetlands conceived and built for water treatment [9], are obviously deliberately constructed for human purposes, often where no waterbody existed before, and thus undisputedly artificial [10]. A stream or lake in a protected watershed far removed from intensive human land use might serve, in a traditional ecological study, as a "completely natural" or "pristine" reference site. However, many, or likely most, ecosystems fall between these extremes. For example, scientists and policymakers often differentiate between a channelized stream and a wholly man-made trench, but either may be colloquially called a "ditch" or "artificial", and the two may look and even function quite similarly, particularly within a highly modified agricultural landscape. All natural waterbodies do not necessarily maintain better structure and function than all artificial ones [11]. Restoration projects similarly blur the bounds of "natural" and "artificial" [10,12]. While restoration typically has the goal of returning an ecosystem to some more natural state [13], the process of restoration necessitates human intervention, which is often sustained through maintenance [14], an implicit acknowledgement that many ecosystems cannot withstand human disturbance without purposeful human assistance. Meanwhile, undirected human actions, like stormwater efflux, can accidentally "restore" natural systems [15–17]. Even mostly or wholly man-made systems, like retention ponds, can "naturalize", or become more biotic and ecosystem-like, in time, without deliberate human effort [18]. Both current character and driving forces behind it are often a combination of human and wild, artificial and natural.

These examples suggest that artificial aquatic ecosystems can be usefully organized along two axes: the degree to which their existence and characteristics depend on human activity, and the degree to which that activity is specifically intended to produce those changes (Figure 1). The first axis spans from moderate alteration of systems with initially geologic origins, to the wholesale creation of waterbodies on formerly dry land, sometimes away from topographic lows. The latter axis ranges from aquatic systems that came to exist as inadvertent or accidental by-products of other human activities on the Earth's surface to deliberate and intentional products of such activity [19]. These two artificiality axes are gradients, not discrete categories, and can be difficult to parse, especially for the many waterbodies with complex, multi-layered histories of modification [20–22]. Depending on the purpose, it may be appropriate to define "artificial" waterbodies broadly or narrowly within this space.

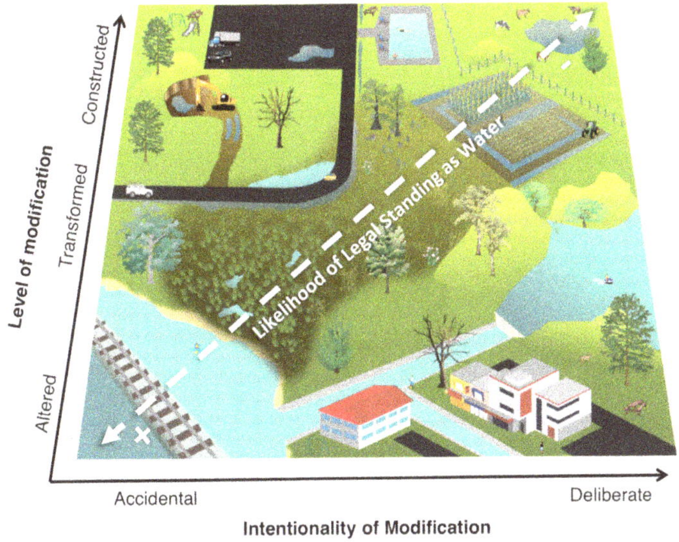

Figure 1. Classification of artificial aquatic systems. Level of deliberateness of modification increases from left to right, and degree of modification increases from bottom to top. These two axes combined yield a third axis, likelihood of legal standing as a water for regulatory purposes in the U.S., running from unlikely among deliberately constructed ecosystems like swimming pools and upland farm ponds in the upper right, to likely among accidentally altered ecosystems, like a lake bisected by a railroad causeway, in the lower left. Other potentially influential characteristics, such as size and permanence, may influence regulation in both natural and artificial systems.

Construction, Transformation, and Alteration

The most clearly artificial aquatic systems are those that humans construct where none existed before. Even the existence of these constructed artificial ecosystems relates to other natural waterbodies because they represent water that could have gone or stayed elsewhere, on the landscape, underground, or to other parts of the hydrologic cycle. Deliberate examples of construction include fountains, many roadside ditches, rain gardens, stormwater treatment areas, many farm ponds, and all designer ecosystems [10]. Conservation-oriented water regulation typically exempts such constructs outright; they usually do not count as water [7,23,24]. Accidental examples include logging ruts, erosional gullies in building sites and agricultural fields, poorly drained impervious surfaces, and even bomb craters [25]. Some of these accidental features represent failures of water conservation regulation or other damaging abstraction from natural water sources. Nonetheless, left unmaintained, in time, such accidental waterbody construction in relative uplands can "naturalize" to a seemingly "wild" ecosystem [25–27]. Waterbodies that humans have constructed accidentally, but that appear relatively free from human intent, are more likely to be regulated than waterbodies that humans have constructed on purpose [23,24].

Transformation occurs when human intervention changes waterbodies from one type to another, fundamentally different in morphology and flow, such as from a lake to a wetland, or a wetland to a stream. Deliberate transformations include ditching of wetlands for agriculture, conversion of wetlands to ponds during development, damming of streams to build reservoirs, piping of creeks, and many restorations. The impoundment of the Everglades behind Tamiami Trail, a road that interrupted sheet flow, is one prominent such example [28]. Similarly, humans accidentally transformed the bed of the Salt River, dried through damming upstream, into wetlands, at stormwater outflows [15]. Transformed waters generally retain their regulatory status [23,24] regardless of intent. Such transformations are often considered degrading and thus may require regulatory permission [23,29,30].

The least modification that might make a waterbody appear artificial is alteration, in which fundamental morphology and flow are retained. Deliberately, people straighten and channelize rivers, harden riverbanks and shorelines, and dredge lakes. Accidentally, sedimentation from agriculture, mining, or construction makes streams and lakes shallower [31]; mill dams clogged river valleys all over the eastern U.S. [32]. Meanwhile, "urban stream syndrome", including incision, flashiness, and other changes largely in response to stormwater drainage, has become a well-known issue in developed areas everywhere [33]. In one particularly grand example of accidental alteration, a railroad causeway divides the Great Salt Lake into mostly independent halves with very different chemistry and community assemblages [34]. Alterations generally do not remove jurisdiction of regulation from waterbodies, and instead likely invoke regulatory oversight [23,29,30]. Of the range of artificial aquatic systems, altered waterbodies are often the easiest to imagine in their "natural" state, the likeliest to be labeled as simply "degraded", the most likely to attract conservation, restoration, and related scientific interest, and, depending on context, possibly the most appropriate for restoration [35–37].

Together, these two axes of artificiality—Degree and intent of modification—may help explain the regulatory protection afforded to various aquatic ecosystems (Figure 1) in U.S. water law [23,29,30]; increasing modification with increasing intent renders waterbodies less likely to be protected [38]. Scientists, regulators, legislators, and other policymakers often do not explicitly acknowledge the value judgement implied by differential treatment of waterbodies according to their type of artificiality. Other traits such as technology, purpose, age, size, and permanence may also figure into value judgments and policy decisions people make about aquatic systems, and so might serve as a basis for further regulatory classification of artificial waterbodies. Some of these attributes, like small size and impermanence, may disproportionately characterize artificial aquatic systems, but also apply to most natural waterbodies [6,39]. Regulatory standards that omit smaller, less permanent waterbodies may do so as much because of their biological, geomorphological, and chemical features [23], or due to their dense distribution inconveniences property and land use considerations [40], as because of their human origins per se.

3. The Ecological Significance of Artificial Aquatic Systems

Understanding the ecological and socio-ecological value of artificial aquatic systems requires that we understand their extent and distribution, their physical and chemical condition and how they relate to biotic communities, and the range of ecosystem services that they provide, but considerable uncertainty surrounds all of these characteristics [8]. Artificial aquatic systems are likely to be ecologically important, due to their extent, which may rival that of natural drainage systems and waterbodies. The ecological functions of artificial systems likely have social significance, often as ecosystem services and disservices, due to their frequent placement near large numbers of people. Moreover, the extent, distribution, and characteristics of artificial waterbodies are likely changing rapidly, in conjunction with those of natural waterbodies. Interdisciplinary understanding of the services and disservices of artificial aquatic systems, the factors that influence them, and their distribution in space and time could foster decisions that increase their ecological value.

3.1. The Extent and Dynamics of Artificial Aquatic Systems

Our understanding of the extent of artificial aquatic systems is piecemeal. Available estimates are largely limited to the U.S. and other developed countries, and largely for intentionally designed aquatic features that are ubiquitous in agricultural, industrial, urban, and recreational land uses, but not their accidental counterparts, including in forests. Even with these incomplete inventories, it is clear that the deliberately constructed or altered fraction of the hydroscape is both large and growing, and must be included in any comprehensive assessment of aquatic resources.

3.1.1. Deliberately Modified Waterbodies

Deliberately constructed and transformed channels constitute a significant portion of the U.S. hydroscape (Figure 2). The U.S. National Hydrography Database includes 5525 km of ditches and canals, or approximately the length of the Missouri River through the Mississippi to the Gulf of Mexico. This aggregation is probably a substantial underestimate, as many smaller ditches do not appear in the database. In 2010, ditches in U.S. agricultural lands occupied about 115,760 km^2 [41], a surface area similar to that of Lakes Superior and Huron combined, or 2.67 times the surface area of all U.S. streams combined [39]. Channelization, another deliberate transformation, has altered the geomorphology of upwards of 26,550 km of rivers and streams in the U.S. [31,42], more than seven times the length of the Mississippi River. Humans have channelized more than 500,000 km of rivers worldwide, and built more than 63,000 km of canals [43–45]. The U.S. also has 6.5 million km of roads [46], many of which have ditches or gutters along both sides that contain water at least occasionally and often for much longer periods. Thus, it is likely that road drainage in the U.S., which effectively serves as urban headwaters, is of comparable length to the 5.3 million km of the country's rivers and streams. The U.S. EPA estimates that 77% of the approximately 1.8 million km of wadable streams in the U.S. are in poor (42%) or fair (25%) condition [47]. Some of these poor and fair streams are likely geomorphically modified enough to fall within our gradients of artificiality. Within this inventory, the remaining degraded "natural" streams, and even perhaps "natural" streams in good condition as well, likely occupy less length than artificial channels. Known lengths of "natural" streams, especially ephemeral and intermittent ones, likewise underestimate their extent [8].

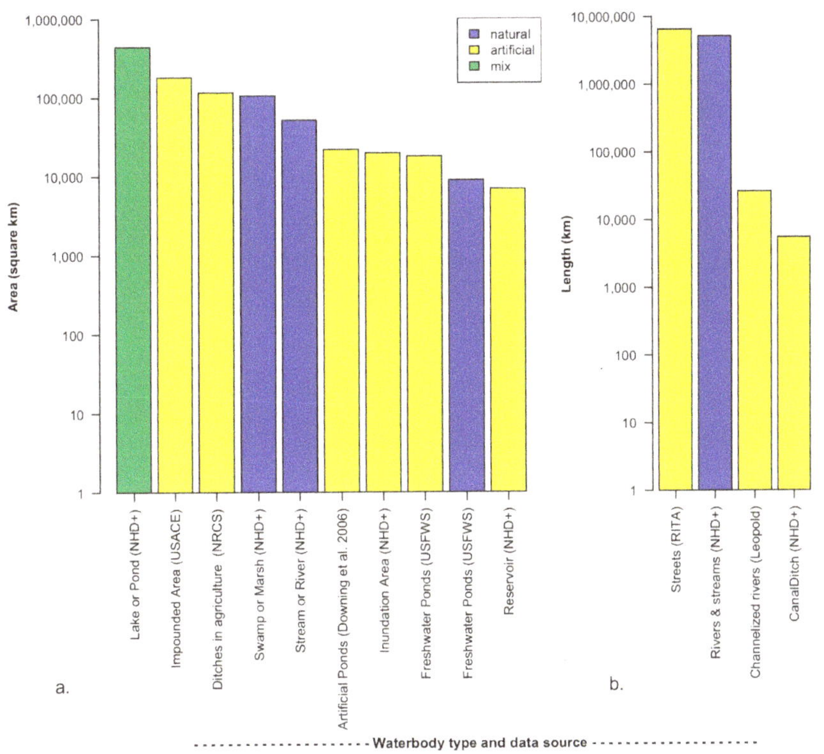

Figure 2. Extent of artificial as compared to natural waterbodies in the U.S. Data drawn from a variety of sources, shown in parentheses. (**a**) areal features; (**b**) linear features. Data are taken from the National Hydrography Dataset (NHD+) [48], U.S. Army Corps of Engineers National Inventory of Dams (USACE) [49], U.S. Department of Agriculture Natural Resources Conservation Service (NRCS) [41], Downing et al. [6], U.S. Fish and Wildlife Service (USFWS) [50], Research and Innovative Technology Administration (RITA) [46], and Leopold [42].

The extent of constructed lakes and ponds are similarly significant in comparison to natural waterbodies (Figure 2). The U.S. has about 22,000 km^2 of deliberately constructed or transformed farm ponds an area similar to that of Lake Michigan; the world as a whole has about 76,830 km^2 of farm ponds [6]. In total, for the U.S., the U.S. Army Corps of Engineers inventories an impounded (deliberately transformed) area in excess of 180,000 km^2 [49], for purposes including irrigation, hydroelectricity, flood control, navigation, water supply, and recreation [49]. The world had approximately 258,570 km^2, or slightly more than the area of the Great Lakes, of impounded water in the mid-2000s [6], before the completion of the Three Gorges Dam in China and other projects. In 2009, the U.S. Fish and Wildlife Service estimated that only 31% of the country's 27,151 km^2 of freshwater ponds were natural. Of the artificial pond area, about the size of Lake Ontario, farm ponds took up about 1.5 times the space as natural ponds; urban ponds occupied about half the area of natural ponds despite the relatively small amount of urban space, and industrial and aquaculture ponds made up significant fractions as well [50]. This estimated aquaculture pond area, of just over a thousand square kilometers, is much smaller than that of many countries'; globally, 1.7 million km^2 of the world's 2.7 million km^2 of irrigated land goes to rice production, and is flooded seasonally, at least [4].

3.1.2. Accidental Waterbodies

The abundance and extent of accidentally created aquatic systems is extremely poorly quantified. Human earth movement has risen in the last 150 years, from a historic background level of less than 5 tons per capita to more than 30 tons per capita annually in the U.S. [1], creating the potential for the formation of local low areas and water accumulation. Moreover, earth movement has become more common in wet spaces [51], where the potential for accidental creation of waterbodies is higher. Water infrastructure, such as stormwater or water supply pipes, can create accidental wetlands wherever leakage occurs [27]. Accidental creation of aquatic ecosystems is perhaps most likely in abandoned areas, where anthropogenic depressions and impoundments that accumulate water may remain, often with minimal human interference. Better information about the density of accidental waterbodies, combined with estimates of the land area over which they might occur, would allow us to estimate their extent, but that information is currently lacking.

3.1.3. Change

The distribution of artificial waterbodies, like and reciprocally with many natural systems, is dynamic in time, owing to both seasonal and event-driven hydrologic change, as well as longer-term changes in land cover. Some of these changes involve the destruction or reduction of natural waterbodies; net effect of growth in artificial waterbodies includes what they replace and from whence they divert water, including those accidental changes that typically go unmeasured. Between 1984 and 2015, North America as a whole, home to 52% of the world's non-ocean permanent surface water, added a net 17,000 km^2 to this area. The area of permanent surface water in the U.S. as a whole grew 0.5%, even as six of its western states lost 33%, or over 6,000 km^2, of their permanent surface water [52]. Farm ponds, in particular, are highly dynamic in use, creation, and abandonment [53]. In commercial and residential developments, ponds and other stormwater features too small to appear in the above analysis, wink in and out of existence too quickly for inventorying, or further scientific study [54].

Emerging technologies and modeling approaches have the potential to improve inventories of small and accidental waterbodies and better characterize the distribution and dynamics of hydroscape change. Advances in remote-sensing technology, such as the increasing availability of high-resolution lidar data, may very soon yield much better maps of small and otherwise over-looked waterbodies over broad extents [55–57].

4. The Condition of Artificial Aquatic Systems and Its Drivers

The perceived poor condition of artificial aquatic systems matches the reality of poor water quality and altered ecological structure in many man-made waterbodies [58,59]. Many artificial waterbodies support species-poor [60] or otherwise undesirable communities or organisms, including disease vectors [58,61,62], and can spread pest species to natural habitats [63,64]. Some have also contributed to, accelerated, or facilitated flow of excess nutrients and other pollutants [65,66], activation of toxicants [58], interrupted desirable species' movement and dispersal [67], increased greenhouse gas emissions [27,68], yielded bad smells [69], and even concealed crime [70]. Other examples of ecosystem disservices proffered by artificial water bodies appear in Table A1. While natural waterbodies can possess the same undesirable characteristics, we are more likely to assume that artificial waterbodies have a negative influence without investigation [71].

Are artificial aquatic systems intrinsically less biologically diverse and less functional than natural ones? It is at least plausible that humans cannot create a waterbody that supports communities as diverse or provides as many ecosystem services. Certainly, when transforming, altering, or removing a functional natural aquatic ecosystem, one should expect a reduction in current ecosystem services provisioning, unless or until scientific study confirms a better outcome possible from the change. One important constraint on artificial aquatic systems is that with their anthropogenic origin comes a severely shortened evolutionary, ecological, and geophysical history [72,73]. To the extent that diversity and other aspects of ecosystem structure depend on slow processes of physical change and community assembly, the recent origin of most artificial systems will likely limit their function.

There are other potential limits on the condition and value of artificial aquatic systems. First, imperfect understanding of how differing designs and constructions affect ecological outcomes, and imperfect ability to reproduce natural structures and conditions, may constrain the most ecologically-motivated projects, as may be the case for many ecosystem restorations [13]. Second, artificial aquatic systems are often embedded in intensively used landscapes, potentially exposing them to anthropogenic stressors and disconnecting them from diverse natural populations [13]. Finally, and perhaps most importantly, many artificial aquatic systems may support limited diversity and ecosystem function because they are not designed or managed to do so; in many cases, their intended function may preclude or limit the provision of other services [27]. As a result, scientists, policy-makers, and the general public have tended to accept that artificial aquatic systems will necessarily and inherently have limited value. However, these assumptions are often not subject to the same critical assessment and process-based explanations that are applied to explanations of the condition of other aquatic systems.

The poor condition of artificial aquatic systems is far from universal, and at least some, perhaps many, artificial aquatic systems also have clear ecological value. Constructed and transformed aquatic systems, whether agricultural, industrial, urban, or recreational, can sustain biodiversity [74], sometimes including rare and desirable species [18]. In Europe, manmade farm ponds serve as primary or important habitat for amphibians [75], birds [76], invertebrates [77–79], plants, and other species [80,81]. Indeed, European conservation proceeds from the assumption that "artificial", "man-made" ponds are not fundamentally ecologically different from "natural ones" [82]. Some species now apparently depend primarily on deliberate artificial aquatic ecosystems for habitat [83–85]; even species new to science continue to emerge from ditches [86–88]. Equally in the U.S., habitats that we presently tend to overlook, such as stormwater treatment wetlands, can sometimes be the best available sites for reproduction of amphibians and other species with specific hydrologic needs [89]. Artificial aquatic systems, whether designed for the purpose or not, have improved water quality in critical watersheds like the Mississippi River basin [90–92]. Additional examples of ecosystem services provided by artificial aquatic systems appear in Table A1. Without more intensive and systematic study, it remains unclear whether good ecological conditions, and the desirable ecosystem services that derive from them, are a negligible, rare, or even commonplace occurrence in artificial aquatic systems. Similarly, their net impacts, and value relative to natural counterparts, remain undetermined.

Making artificial aquatic systems more functional and valuable will require a mechanistic and predictive understanding of their condition and their capacity to provide ecosystems services and disservices. We propose that the science of artificial ecosystems should entertain and evaluate hypotheses about what drives variation among them as well as their differences from their natural counterparts. Like their natural counterparts, the ecological characteristics of artificial aquatic systems are likely to depend on their physical structure, the characteristics of the watershed and landscape in which they are embedded, their age and trajectory over time, and the ongoing interventions of humans for various purposes (Figure 3). While all of these mechanisms are shaped by human design decisions, they also have clear analogs to factors commonly invoked to explain the condition of natural aquatic systems. This re-casting of rationales for why artificial aquatic systems are assumed to be in poor condition as testable alternative mechanisms allows us to consider how different decisions about the design, placement, and longevity of artificial aquatic systems might improve their condition and value. The poor condition and seemingly inherent limitations of artificial aquatic systems could be simply a syndrome of those decisions. In our exploration of these possible causal variables for the ecological condition of artificial aquatic systems here, we focus on this management-oriented way of examining the causal variables, and barely begin to explore the possible interactions between them. We recognize that scientists with different foci could propose other valid testable hypotheses, and indeed invite them to do so, but we consider setting, time, physical design, and subsequent management of artificial waterbodies to be good intellectual places to start trying to understand these ecosystems.

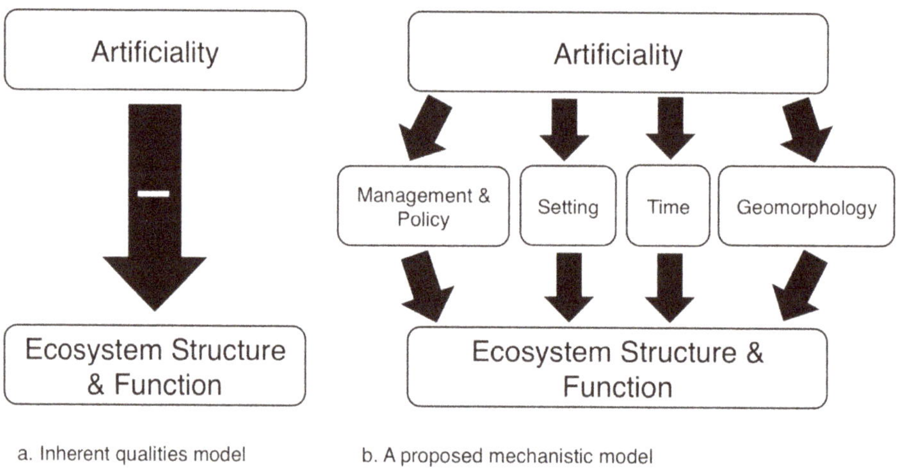

Figure 3. Inherent vs. process-based models of the condition of artificial aquatic systems. (**a**) inherent qualities model; (**b**) process-based model. We suggest that scientists and policy-makers have too readily accepted model (**a**), in which the inherent qualities of artificiality negatively impact ecosystem structure and function, rather than scientifically exploring a mechanistic model such as the one we propose in (**b**), which breaks the influence of artificiality down into multiple processes.

4.1. Setting

Understanding how the watershed setting of artificial aquatic systems affects their ecology is important both because understanding will be essential for better policy and management, and because the effects of setting may obscure the ecological effects of other factors such as design, management, and time. The condition of the watershed and landscape around any waterbody influences its condition [93], and artificial aquatic systems should be no different in this respect. Since humans tend to create artificial aquatic systems in and around heavily modified landscapes with substantial chemical inputs like agricultural fields, roads, and parking lots, artificial aquatic systems such as ditches tend to have lower water quality than their natural counterparts [94–96]. The communities of artificial aquatic systems also tend to reflect the local and regional species pools, yielding, for example, more exotic species in a restoration in a developed area [97].

Available evidence suggests that setting does exert a strong and often overwhelming influence on artificial waterbodies, and that these effects are similar to those observed in natural systems. Water quality of artificial aquatic systems such as ditches responds to catchment land use in much the same way as that of waterbodies of natural origin [8,98], and agricultural land cover impacts reservoirs food webs [99]. In the Salt River in Arizona, level of urbanization explained much of the variance in communities of plants, birds, non-avian reptiles, and amphibians, for reference, restored, and accidentally restored river reaches alike [15]. A study in the Florida panhandle found that natural streams, altered streams, and ditches within the same forested region had similar macroinvertebrate and fish assemblages [100]. Agricultural intensification around fishponds has contributed to the rapid decline in breeding populations of black-headed gulls (*Chroicocephalus ridibundus*) in central France [101]. More such comparisons between artificial and natural waterbodies in similar settings are needed to disentangle the effects of watershed setting from other factors that influence the condition of artificial aquatic systems.

The predictable responses of artificial aquatic systems to their watershed setting has implications for how these systems are managed and how that management could be improved. The importance of watershed land cover for reservoir water quality shapes economically motivated conservation, like New York City's efforts to prevent development in the watersheds of its reservoirs upstate [102]. More generally, stream restoration is more effective in undeveloped than developed catchments [103]. While more studies are needed, the available evidence suggests that the condition of artificial aquatic systems depends strongly on their setting, and that those conditions, and the ecosystems services that depend on them, could be improved by the same watershed-scale policy and management that protects natural waterbodies.

4.2. Time

Most artificial aquatic systems are young, both because most land use change and earth moving has occurred within the past few hundred years [1] and because artificial aquatic systems turn over more quickly than natural ones [93]. Given the timescales over which community assembly occurs in newly formed natural streams and lakes [104], it is likely that limited diversity of some artificial aquatic systems simply reflects their recent origin. Understanding the consequences of recent formation requires that we understand the timescales over which newly created ecosystems develop, whether they arose from anthropogenic or from geologic processes, and potentially attain the characteristics of their older counterparts. At present, we lack both general and system-specific models of these trajectories, as well as criteria with which to judge that an artificial aquatic system has "naturalized".

A relatively limited set of long-term and chronosequence studies indicate that artificial aquatic systems can change in important ways over timescales that are similar to those in natural ecosystems, and that are relevant to decision-making [105–107]. In restored wetlands, for example, ecological structures and functions such as carbon sequestration can improve with time since intervention, and soil characteristics approach natural properties over decades [108]. Re-configured two-stage ditches can achieve soil formation and a geomorphological "quasi-equilibrium" within a decade [109]. Agricultural ditches also undergo a relatively predictable succession of plants and associated invertebrate communities [110]. Accidentally created waterbodies also change over time, often acquiring more 'natural' characteristics. For example, gravel quarries develop more structurally complex and diverse vegetation over several decades [111]. At some point in time, artificial aquatic systems may be difficult to distinguish from natural systems. Many of the small, ephemeral wetlands that sustain populations of amphibians in the Piedmont of the U.S. Southeast are likely legacies of historical human disturbance [26]. Such examples suggest that time eventually erases many signatures of anthropogenic origin, and that this naturalization may change how aquatic systems are perceived and valued.

Better understanding of how time constrains the characteristics of artificial aquatic systems, and the mechanisms by which they evolve, could improve our ability to manage them, individually and as part of the broader aquatic landscape [112]. Properties associated with age may elude newly created waterbodies, and expectations that artificial waterbodies adequately replace natural ones should be tempered accordingly. Goals and expectations of restorations and other interventions might need to reflect the differential responses to the same management technique, as has been observed in young and old artificial aquatic systems [113]. Deliberate management of successional stages has been used to increase the abundance and diversity of desirable species [107]. Wider adoption of such approaches will require better models of succession and its dependence on design, setting, and management.

4.3. Design

Design is a goal-oriented process with multiple stages, including the establishment of goals, a plan to achieve those goals given constraints, implementation (including initial construction and subsequent maintenance), and, ideally, subsequent iterations of goal-setting and redesign [14]. Decisions, whether unconscious or deliberate, at each of these stages have the potential to shape the outcomes of later stages, and ultimately the ecological character of artificial aquatic systems, including their trajectories over time and how they respond to the forcings imposed by their watershed setting. The physical structure and management of accidentally created waterbodies obviously does not depend directly on goal-oriented design, though their structure may reflect design decisions and management regimes of which they are not the object.

4.3.1. Design Goals

Historically, many deliberate artificial aquatic systems have been designed and maintained to provide one or a few services, such as water conveyance and storage [27,114]. The exclusion of many such artificial waterbodies from protection within the U.S. apparently reflects that policymakers and legal frameworks value these systems almost exclusively for their intended, fully human-oriented purposes [23]. Planning for only one or a few ecosystem services, such as water storage and conveyance for flood control, can limit the ability of a deliberate artificial aquatic ecosystem to provide other services, especially when designers overbuild that system for its given purposes [27]. In many cases, the design goal itself can inherently produce a major ecological cost, as in wetland drainage by agricultural ditches [115,116], or can result in unintended disservices arising from synergies and trade-offs in ecosystem services [117,118]. Nonetheless, many designed artificial aquatic systems also provide a range of additional ecosystem services beyond the purpose of their design [15,89].

The designs of aquatic ecosystems, including both newly constructed waterbodies and restoration of degraded systems, increasingly seek to provide a portfolio of ecosystem services and functions through redesign of physical structure as well as changes in management [119,120]. Urban dwellers appreciate open expanses of water in spaces where they go for recreation [121], and even modified or constructed waterbodies can mitigate pollutants and floods, cool the air, and provide spaces for recreational, spiritual, and community-building activities [70]. For example, the Los Angeles River, converted to a concrete flood chute and movie set for car chases in the mid-20th century, has recently become the focus of an ambitious revitalization project to improve water quality and sustain wildlife while also providing a greenway and other recreational opportunities [120]. Similar redesigns of channelized rivers have already demonstrated the benefits of design for a range of ecosystem services [122]. The Landscape Architecture Foundation has endorsed projects throughout the U.S. and around the world with similar methods and goals, specifically including stormwater management, water conservation, water quality, flood protection, and groundwater recharge alongside other environmental, social, and economic goals [22]. Deliberate artificial aquatic ecosystems like these tend to remain primarily human oriented, and not ecologically oriented, in their goals, however. Even restoration designs often explicitly and unapologetically include human-specific concerns, such as ease of maintenance, accessibility, recreational appeal, aesthetics, regulatory standards, finances, and property lines, alongside more ecologically oriented values [123–125].

4.3.2. Planning and Construction

The reduced physical complexity of many artificial aquatic systems, such as concrete-lined channels, obviously limits their value as habitat and potential for improvements in water quality [126]. Restorations that seek to improve these values therefore often focus on the (re-) introduction of heterogeneous structures that are more similar to natural systems [120]. Such designs, and their implementation, can be constrained or flawed in ways that limit their ecological value, including by insufficient scientific understanding of how design features and subsequent management influence eventual outcomes [35,93,127,128]. However, many such systems are also affected by intensive land use and short lifespans [103]; artificial systems whose structure mimics that of natural systems can support similar biotic communities when water quality is high [100].

Conversely, engineering research on designer ecosystems constructed for a very specific subset of aquatic ecosystem services, such as water quality improvement, clearly demonstrates that design plays a role in how effectively these systems achieve their purpose. For example, plant species choice in wastewater treatment wetlands affects speed and removal efficiency of different forms of nitrogen [129]. In wetlands constructed to remove pharmaceuticals from water, design choices of substrate, plants, and regimes of hydrology, temperature, oxygen, and light all affected removal efficiency, which varied from compound to compound in ways apparently related to microbial processes [130,131]. While much variation remained unexplained even in these relatively controlled systems, they do demonstrate that how an ecosystem is constructed affects its ecological behavior.

Physical, legal, and cultural constraints exert strong control on goals and resulting designs. For example, restoration efforts are typically constrained and otherwise impacted by funding, land ownership, and other social and economic variables [13,93,103]; restorations can have a wide range of intended outcomes [97]. Morphology of stream restorations depends in predictable ways upon funding source and legal purposes, and whether the metric for success is stream length (resulting in very sinuous designs) or some other characteristic [124]. Stream restoration in general has tended towards a single-channel, S-shaped, meandering morphology that conforms to longstanding aesthetic concerns [125], reduces maintenance [123], and maximizes mitigation credits, rather than conforming with local natural history [124]. One indication of the limitations of many restoration projects is the finding that accidental aquatic systems can sometimes provide equal or greater services compared to deliberate, designed systems. For example, "accidentally restored" wetlands at stormwater outflows in the dry bed of the Salt River in Arizona had greater wetland plant richness and cover than comparable actively restored sites, though the reverse was true for birds, non-avian reptiles, and amphibians [15].

Changes in goals often dictate substantial changes in the physical structure of artificial aquatic ecosystems. Two-stage ditches, in which miniature floodplains are constructed alongside existing conveyances [132], can significantly reduce concentrations of phosphorus and other nutrients, turbidity, and total suspended sediments [133,134]. Their nutrient removal efficiency compares well with, and can complement, other farmland best practices, like planting cover crops [90]. When properly constructed according to fluvial principles, these ditches can remain functionally stable, without maintenance, for years [109]. Thus, water infrastructure of agricultural landscapes can be designed, and successfully re-built, to achieve a wider range of goals than water conveyance, though additional land area and design and construction effort may be required.

4.3.3. Management and Policy

Ecosystem function and services of artificial waterbodies likely depends on the management they receive after construction as well, just as management matters in natural waterbodies. In reservoirs, the ability of an artificial aquatic system to provide ecosystem services may depend more heavily on ongoing policy and management than on the specifics of the initial design [135,136]. In ditches, management strategies, including dredging, mowing, chemical weeding, burning, and regulation of water depth, can have significant impacts on ditch biodiversity and water quality [137,138]. However, ongoing management and maintenance, like initial design, often does not include these potential outcomes in its considerations, instead opting to continue to focus on relatively few, highly human-oriented goals [126]. Such decisions about ongoing management and policy, however, can, at least, in theory, be revised to reflect changing goals.

In complex landscapes, achieving a portfolio of ecosystem services often requires both structural changes and ongoing active management of artificial aquatic systems. Ditches and canals draining ranchlands in the watershed of Lake Okeechobee were not traditionally managed for water quality or conservation purposes, even though they house large native animals and other species of interest [139]. Establishment of Total Maximum Daily Loads for phosphorus in the lake and its tributaries [140,141] prompted creative responses including regional collaborations among various government agencies, nonprofit conservation organizations, a local scientific research station, and ranchers to raise and actively manage water levels to flood ditched wetlands. This strategy removed phosphorous more cost-effectively than did constructed storm water treatment areas [142], while also increasing wetland vegetated area and vertebrate abundance [143]. Multi-stakeholder management of artificial aquatic systems with ecologically oriented goals could prove a cost-effective way to increase ecosystem services at a similarly regional scale in other locations, perhaps as a complementary tool to traditional restoration.

Accidental waters, which often receive little to no management attention, can provide comparable but non-overlapping ecosystem services to both deliberate and more natural waterbodies. Abandoned features, especially within broader abandoned landscapes and even in the hearts of cities, can provide habitat for urban-avoiders and other organisms that survive best away from humans and human intervention [26]. Abandoned areas can contain accidental artificial waterbodies sustaining both human and nonhuman life, and functioning as little pockets of biodiversity [27]. Accidental urban wetlands can also mitigate nutrient pollution flowing from cities to downstream in natural waterbodies [27]. Two European species of damselfly were believed extinct for decades, until rediscovered, separately, in former industrial and mining areas "not usually explored by biologists". Other neglected artificial habitats in our midst could hold similar surprises. Notably, conservation interventions "focused on returning habitats to a 'natural' state" intended to boost one of those damselfly populations actually backfired [144]. These observations suggest that active intervention, for non-ecological and even ecological goals, can limit the ecological value of artificial aquatic systems.

4.3.4. Monitoring, Learning, and Iteration

One of the criticisms of many restorations is that they require ongoing, often expensive management to avoid reverting to a degraded state, which some scientists consider a failure of resilience. Part of the problem with declaring restoration success or failure is that goals for a specific restoration are often unclear and may change through time [145], but for most aquatic restorations are never evaluated [97,146]. Monitoring protocols often focus on easily quantifiable measures that ensure mitigation credit, rather than landscape-scale and long term ecological contributions [124]. Such monitoring designs may not adequately assess what was lost and what was gained. In all, current practices of stream and wetland restoration may not be well configured for learning and for adapting designs to improve environmental outcomes [14]. More broadly, the exclusion of artificial aquatic systems from policy protections eliminates an important motive for monitoring. In the UK, a recent precipitous decline in farm pond numbers and services in the UK sparked conservation concern and action [147]. The U.S. lacks the monitoring data necessary to characterize trends in its small artificial aquatic systems and to respond accordingly.

Moving forward, adaptive management, designed experiments [148], reconciliation ecology [149], and other ecologically based ways of improving designs may change the outlook for deliberate waterbodies. A future increase in the acceptance of novel ecosystems might allow the creation of new types of waterbodies designed to provide similarly novel suites of ecosystem services [10,150]. Even the tendency for less regulation of more highly and accidentally modified [23,29,30] and smaller waterbodies [7,8], particularly in the U.S., constitutes an opportunity; this quality could make them comparatively easy and low-cost systems in which to study, test, and implement novel ecological design ideas [148,151,152]. Together with the repetitive design and construction of many such waterbodies, like ditches and ponds, the manipulability of artificial aquatic ecosystems makes them prime sites for natural experiments [153] and designed experiments [148]. Irrigation canals can serve as "lotic mesocosms", ideal because of their known histories, predictability, and accessibility [152]. Ditch network structure recently served as a good system within which to model possible alternative stable states in primary producer structure [154]. Science in artificial aquatic ecosystems could contribute substantially to broader ecological theory and practice. While win–win design decisions to support multiple desired ecosystem services and other goals can prove challenging to envision and implement, even in artificial aquatic systems, these waterbodies remain sufficiently understudied that exploration of the many remaining questions around them likely has many win–wins, in terms of furthering both applied and theoretical science, left to yield. We hope that the conceptual structures introduced in this article will assist in future such work.

5. Artificiality and Perception of Ecosystem Value

The concept of artificiality, its associated dichotomy between human and nature, and its connotations for valuation, have deep roots. One of the earliest abstract concepts U.S. children master is the difference between artificial and natural, in terms of origin; they learn to tell whether an object is "made by people or something that people can't make" [155]. Accordingly, Western culture has a long tradition of elaborating upon the natural/artificial dichotomy [156] and including it in value systems [157]. In American history, wild nature provided a divine purpose for European settlers, a spiritual rejuvenation for Romanticists and early conservationists like Muir, a source of strength to manage for technocrats, and a rallying point for complex unity among environmentalists [158]. Today, untouched wilderness "exists nowhere but in the imagination" [157]; every ecosystem is somewhat artificial, yet the concept of pristine wild nature continues to exert a strong pull. A recent psychological study found that subjects preferred environments when told that they were natural [159], perceiving them "less dangerous, cleaner, and more plentiful" than those already exploited by other humans. We argue that research and policy-making about artificial aquatic systems reflects this cultural subordination of artificial things to the natural and wild, inherited from broader contemporary Western culture [159,160].

The past several decades have seen ferocious clashes in environmental philosophy and conservation biology over the role of artificial ecosystems. In the 1980s, prominent ethicists lampooned restoration on the grounds that "faked nature is inferior" in much the same way that an art forgery is inferior because it is "a product of contrivance", lacking "causal continuity with the past" [161], and that man-made natural areas represent "domination, the denial of freedom and autonomy" that defines nature [162]. Some philosophers have since softened this dichotomy, viewing it as a gradient or broadening the criteria that constitute a necessary fidelity to nature [163–165]. While naturalness remains valuable by all standards of environmental ethics, many ethicists increasingly distinguish categories (or dimensions) of naturalness, including "as a physical property of species and ecosystems", such as native biodiversity, and "as a quality of processes that are free of human intervention" [74]. This particular pair of categories, posited as a distillation of values already in wide use, corresponds well with our proposed axes of degree of physical modification and level of intentionality, which U.S. policy tends to reflect [23,29,30].

Discussions of humanity's relation to authentic nature intermingle with and parallel debates within conservation biology and broader environmental science and policy. Proponents of traditional wilderness- and biodiversity-based conservation have reacted with alarm to "new conservation", a loose grouping of movements that include human and socio-economic goals, such as poverty reduction, in their conservation plans [73,166]. While restoration has become a widely accepted practice, novel and designed ecosystems are on the battle lines between "new conservationists" who would like to include them in conservation plans and more traditional conservationists who would not [167]. The difficulty in reconciling these perspectives [168,169] may arise in part from different priorities, i.e., the physical and ecological condition of ecosystems versus their freedom from human intervention, and in part from conflicting views about whether human intervention inherently degrades ecological condition.

Interactions between Perception and Condition in Artificial Aquatic Systems

The presumption that artificial aquatic systems have little ecological value matters because it promotes neglect. People make management decisions about aquatic systems not on the basis of perfect factual knowledge of the state of these systems and their impact on the broader hydroscape, but instead upon how they perceive them [170]. Natural aquatic ecosystems in poor condition often retain perceived potential value, which restoration seeks to regain, no matter how little realized value remains [171]. However, traditionally, scientists and policymakers have regarded artificial ecosystems as relatively low in ecological value [71], regardless of their actual function and services (Figure 4). Conversely, a high-functioning artificial system may be overlooked in conservation planning [18]. For example, the 250 km of canals of the North Poudre Irrigation Company near Fort Collins, Colorado, supported 92% of wetland area in the 23,300-hectare service area through leakage. In spite of the ecosystem services these wetlands provide, this leakage is considered an unacceptably inefficient use of a scarce resource, and may cease as irrigation practices change, without considering the value of lost accidental wetlands [17]. Perceived value influences design and management, which, along with any other more direct impacts of artificiality, in turn influence ecological condition (Figure 5). If, as appears prevalent among the ecologically minded, perceived artificiality downgrades perceived value of aquatic ecosystems, and management and policy decisions reflect this lower valuation in low expectations and low protections for artificial waterbodies, then assumptions of the low quality status of artificial aquatic systems could be self-fulfilling.

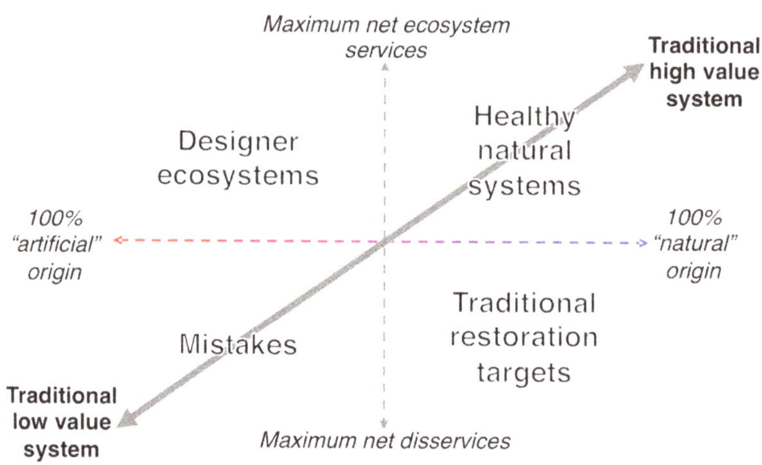

Figure 4. Value axes for aquatic ecosystems. The solid line represents a traditional axis for the value of aquatic ecosystems. The dashed lines parse out artificiality from ecosystem services provisioning along this traditional axis.

The divergent consequences of this positive feedback loop may be illustrated by examining water management policy in the U.S., where the Clean Water Rule prioritized the exclusion of many artificial aquatic systems from jurisdiction under the Clean Water Act [24]. In contrast, the European Union's 1996 Water Framework Directive resolved to gradually expand protection "to all waters, surface waters and groundwater" [172]. In line with this inclusive view of aquatic ecosystems, pond degradation [173] and loss is a stated conservation concern for the EU [147] and NGOs [174]. Freshwater Habitats Trust's Million Ponds Project aims to "to reverse a century of pond loss, ensuring that once again the UK has over one million countryside ponds", and claims more than 1000 ponds created in 2008–2012, housing about 50 rare and declining species [175].

Meanwhile, European researchers continue to explore pond conservation measures [82], including management options that improve habitat quality in existing ponds [105,176]. Similar research and conservation activity is progressing for British and other European ditches [177,178]. Assuming even modest success of such efforts, the condition of artificial aquatic systems in the EU is likely to improve, while the quality of artificial waterbodies in the U.S. is likely to decline. Europe's example suggests that how people regulate, perceive and manage farm ponds, and other artificial aquatic systems does impact conservation outcomes.

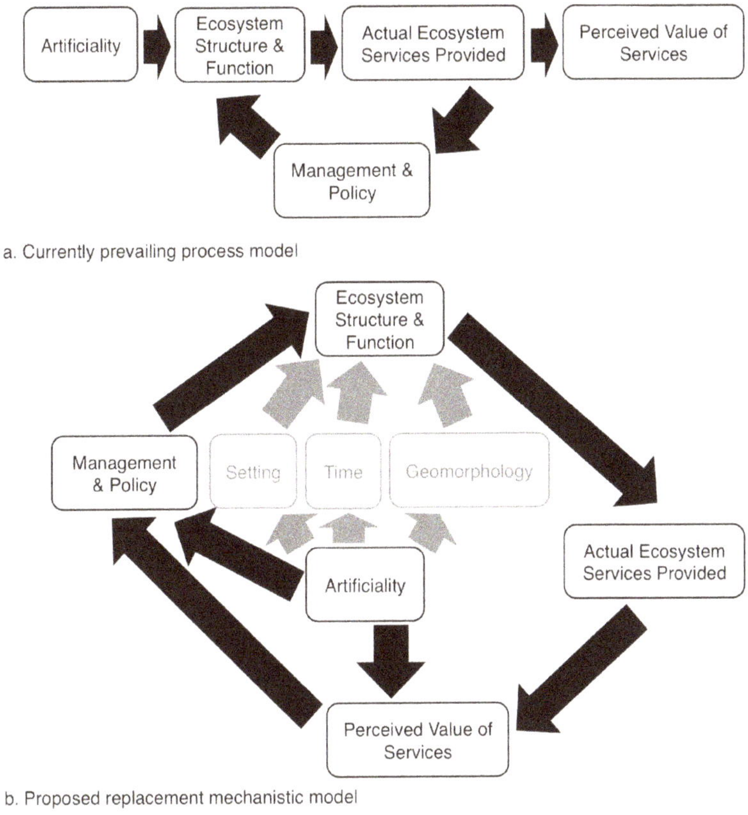

Figure 5. Conceptual diagram of the role of artificiality in the management and services provided by an aquatic ecosystem. (**a**) the currently prevailing process model depicts an approximation of how environmental scientists appear to typically think of the role of artificiality in impacting ecosystems; (**b**) our proposed replacement suggests that, while artificiality may impact ecosystem function directly through mechanisms yet little elucidated, we are more certain that it impacts the perceived value of ecosystem services. Because perception impacts policy, policy affects reality, and reality impacts perception, this proposed replacement process model for the role of artificiality in aquatic ecosystems sets up a positive feedback loop.

Getting the policy, management, and science around these systems right matters not just ecologically, but for social reasons as well. Under-managed artificial waterbodies, particularly environmentally hazardous ones, may often occur in already at-risk communities. Two well-studied examples of 20th-century environmental injustice in the United States, in Anniston, Alabama, and Hyde Park, Georgia, both involved predominantly black communities contaminated and sickened in part by ditches bearing water laced with toxic industrial waste [179,180]. Recently, hog waste lagoons associated with industrial swine facilities in eastern North Carolina have proved resistant to regulation despite repeated flooding during hurricanes and tropical storms and persistent strong detrimental effects on the health and quality of life of neighbors, who are disproportionately black and low in income [181,182]. Thus, what artificial aquatic systems go unregulated may say as much about what we socially undervalue as what we ecologically undervalue. Relatedly, the same accidental wetlands that host birds and remove nitrogen in the Salt River in Phoenix, Arizona, provide somewhat unsafe, legally unauthorized sources of water and cool places to rest for homeless people [15,16,27], which calls into question the design of non-aquatic infrastructure whose functions may have been deputized to or externalized on artificial aquatic ecosystems. When science and policy overlooks artificial aquatic systems, it risks overlooking the people impacted by them as well.

6. Conclusions—Artificial Aquatic Ecosystems in Hybrid Hydroscapes

Artificial aquatic systems comprise a substantial, perhaps predominant, and likely enduring component of the modern hydroscape. Because the sheer extent of artificial aquatic ecosystems may, by some measures, increasingly rival that of natural systems, they have the potential to play an important role in both conservation and in the provision of ecosystem services within these hybrid aquatic landscapes. The premise underlying reconciliation ecology [149] is the insufficient extent of relatively undisturbed habitats to preserve anything but a fraction of extant species. In some regions, it may be difficult to enact any sufficiently wide-reaching biodiversity conservation policy without inclusion of artificial systems [183]. Because artificial aquatic systems are interwoven with, rather than separate from, natural elements of the hydroscape, improvements in the condition of artificial systems may benefit natural waterbodies as well [75], or may degrade natural waterbodies through abstraction; the net effect of their creation must account for all of the above. Thus, plans to improve land and water management should target artificial aquatic systems as well as those of natural origin [183].

To realize greater socio-ecological benefits from artificial aquatic systems, we need to understand not just their current value, but their possible provisioning of ecosystem services. This understanding will require, first and foremost, better assessments of the extent and condition of artificial aquatic systems. Improving that condition will require that we suspend our conventional assumption that artificial aquatic systems are intrinsically inferior; instead, we need more hypothesis-driven study that evaluates the factors, such as watershed setting, physical structure and design, time, and management, that influence their ecological condition. We will need to move beyond this initial exploration to more thoroughly consider interactions among these drivers and alternative ways of framing the mechanisms underlying artificiality (e.g., physical vs. biological), first conceptually and then through well-controlled studies.

Because the very way we perceive artificial aquatic systems may affect their ultimate condition and value, effective management of the modern hybrid hydroscape may require reconsidering cultural norms about the concept of artificiality, even undoing our deeply held notions about a human/nature dichotomy. Environmental scientists, and our cross-disciplinary collaborators, must first take on such efforts in support of our own work, but can also play a role in helping policy-makers and others meet these challenges.

Author Contributions: Both C.C. and J.B.H. contributed substantially to all aspects of this work. Conceptualization, C.C. and J.B.H.; Investigation, C.C. and J.B.H.; Data Curation, C.C. and J.B.H.; Writing—Original Draft Preparation, C.C. and J.B.H.; Writing—Review and Editing, C.C. and J.B.H.; Visualization, C.C. and J.B.H.; Supervision, J.B.H.; Funding Acquisition, C.C. and J.B.H.

Funding: This research was funded by a National Science Foundation Graduate Research Fellowship and Duke University's Nicholas School for the Environment.

Acknowledgments: Thanks to the Duke River Center (labs of James Heffernan, Martin Doyle, Brian McGlynn, and Emily Bernhardt), Duke MediaLab 2014 participants, Dean Urban, Monica Palta, Nancy Grimm, Erick Carlson, Lauren McPhillips, Alex Webster, Danielle Purifoy, Cari Ficken, Marissa Lee, Priscilla Wald, Matthew Taylor, and Daniel Richter for contributing to the development of these ideas. Vector files used in Figure 1 courtesy of the Integration and Application Network, University of Maryland Center for Environmental Science (ian.umces.edu/symbols/).

Conflicts of Interest: The authors declare no conflict of interest. The funders had no role in the design of the study; in the collection, analyses, or interpretation of data; in the writing of the manuscript, and in the decision to publish the results.

Appendix A

Table A1. Documented ecosystem services and disservices of artificial aquatic systems. This list, while incomplete, provides examples of supporting, provisioning, and cultural services and disservices, including biodiversity, for a variety of ecosystems around the world. Actual monetary valuation of ecosystem services and disservices of artificial aquatic ecosystems, particularly of net effects of waterbodies and comparisons with natural waterbodies, remain lacking, and constitute an area inviting further study.

Ecosystem Service/Disservice	Waterbody Type	Location
Supporting		
Biodiversity retention		
Rare damselfly habitat	Agricultural ditches	Czech Republic [144]
Mite diversity	Agricultural ditch and depression	Slovakia [184]
Diverse macroinvertebrate habitat	Agricultural ditches and peat lakes	The Netherlands [185]
Stickleback genetic diversity	Agricultural ditches	Japan [186]
Reduced fish diversity	Drainage ditches and dredged streams	Estonia [60]
Endangered turtle habitat	Mined peat bogs and drainage ditches	Canada [187]
Amphibian habitat and breeding area	Anthropogenic small isolated wetlands	USA [188]
	Rice paddies	Japan [67], Brazil [189]
	Carp aquaculture ponds	Poland [190]
Barriers to amphibian dispersal	Roadside ditches	Japan [67]
River fragmentation	Reservoirs & dammed rivers	Global [45]
Bird habitat	Open water salt marsh management and mosquito ditches	USA [191]
	Integrated marsh management	USA [192]
	Rice fields	Philippines [193], Brazil [189], China, Japan [194], France [195]
	Bomb craters	Hungary [25]
	Carp aquaculture ponds	Poland [190]
Macroinvertebrate/ zooplankton habitat	Bomb craters	Hungary [25]
	Rice fields	Brazil [189]
	Urban and agricultural ponds	UK [196]
Wetland plant dispersal	Agricultural ditches	Netherlands [197,198]
Wetland/aquatic plant habitat	Paddies and ditches	China [199,200], Brazil [189]
	Drainage ditches	China [201]
	Fen restoration and ditch	UK [202]
	Open water salt marsh management	USA [191]
	Bomb craters	Hungary [25]
Reduced plant diversity	Fish ponds and managed fens	Czech Republic [203]
Wetland habitat loss	Forestry drainage ditches	Southeast Asia [68]
	Shrimp aquaculture ponds	Mexico, Central America, Indonesia [204]
	Agricultural drainage	USA [205], Global [206]
	Rice fields	Brazil [189]
	Fish ponds and managed fens	Czech Republic [203]
Instream habitat loss	River channelization	Global [31,95]
Nutrient cycling		
Habitat for common collector-gatherers	Channelized agricultural headwater streams	USA [207]
Habitat for common aquatic vegetation	Agricultural ditches and peat lakes	Netherlands [185]
Soil erosion	Roadside ditches and culverts	USA [208]
	Peatland forestry ditches	Finland [209]
Reduced soil bulk density and mineral content	Salt marsh ditches	USA [116]
Groundwater recharge	Agricultural drainage ditches	Netherlands [210], China [115]
Lowered groundwater table	Forestry drainage ditches	Southeast Asia [68]
	Open water salt marsh management (sometimes)	USA [191]
	Hot springs swimming pools and baths	Turkey [211]

Table A1. *Cont.*

Ecosystem Service/Disservice	Waterbody Type	Location
Water overuse	Rice fields	USA [212]
	Impoundments and abstractions	Global [95]
	Mining and industrial diversion	Global [45]
	Swimming pools and golf courses	Turkey [211], Global [213]
Provisioning		
Fisheries		
Dispersal corridors for fish and shrimp	Paddy irrigation ditches	Taiwan [214]
Fish and mussel habitat and nursery	Irrigation ditch	Japan [215]
Nekton habitat (prey fish and shrimp)	Integrated marsh management	USA [192]
	Open water salt marsh management	USA [191]
Fish habitat	Agricultural ditches	Japan [216]
	Constructed wetlands, recycle pits, and ditches	
Hunting	Rice fields	USA [212]
	Abandoned ditches	USA [personal observation]
Animal aquaculture		
Catfish and prawns	Embankment ponds	USA [217]
Shrimp	Shrimp aquaculture ponds	Mexico, Central America, Indonesia [204]
Carp	Aquaculture ponds	Poland [190]
Duck	Integrated Rice-Duck Farms	China, Japan [194]
Crops		
Rice	Paddies / fields	China [194,199], USA [212], Philippines [193], Brazil [189], Japan [194], France [195]
Vegetables	Rice fields with High Diversity Vegetation Patches	Philippines [193]
Biofuel	Cutaway peatland, reed canary grass field and ditches	Finland [218]
Timber	Forestry drainage	Southeast Asia [68]
Regulating		
Pest control		
Dispersal corridors for diverse, mostly predaceous spiders and ground beetles	Agricultural drainage ditches	Belgium [219]
Habitat for frogs, spiders, dragonfly larvae	Paddy ditches	China [199]
Mosquito reduction	Salt marsh mosquito ditches and managed ponds	USA [191]
Reduced invasive plants	Integrated marsh management	USA [192]
Insectivorous birds	Rice fields	Philippines [193]
Weed/invertebrate control/spreading by ducks	Integrated Rice-Duck Farms	China, Japan [194]
Habitat for pest fish	Irrigation ditches	Japan [220]
Movement of invasive predator fish	Irrigation canals	USA [221]
Disease vector		
Fecal bacteria export	Roadside ditches	USA [222]
	Urban ditches and pond	USA [65]
Intestinal parasites	Open sewage	USA [223]
Liver flukes	Irrigation ditches	Southeast Asia [62]
Schistosomiasis	Paddies, ditches, ponds	China [224]
Malaria (mosquitoes)	Puddles, urban farms, construction sites, drains, ditches	Ghana [61]
Pollination	Rice fields with High Diversity Vegetation Patches	Philippines [193]
Pollutant removal	Paddy fields, ditches, and reservoirs	China [225]
Denitrification	Paddy ditches	China [199]
	Traditional and ecological agricultural drainage ditches	China [226]
	Agricultural drainage ditches	USA [227]
	Restored wetlands and two-stage ditches	USA [90]
Soil sorption of P	Traditional and ecological agricultural drainage ditches	China [226], USA [228]
P efflux	Agricultural drainage ditches	UK [202], Germany [229]
	Tile drains and ditches	USA, Canada, Sweden, New Zealand [230]
Plant uptake of nutrients	Aquaculture drainage ditches	USA [217]
	Traditional and ecological agricultural drainage ditches	China [226]
	Restored wetlands and two-stage ditches	USA [90]
	Vegetated agricultural drainage ditches	USA [231,232]
Nutrient export	Agricultural drainage ditches	Germany [229], China [233]
	Rice paddies	China [234]
	Roadside ditches	USA [235]
	Urban ditches and pond	USA [65]
Algal blooms and hypoxia	Urban ditches and pond	USA [65]
Organic matter and C retention	Agricultural drainage ditches	USA [228]
Greenhouse gas emissions	Rice paddies and drained peat	Southeast Asia [68]
	Reed canary grass field and ditches in drained peat	Finland [218]
	Drained peatlands	UK [236]
	Shrimp aquaculture ponds	Mexico, Central America, Indonesia [204]
DOC efflux	Drained peatlands	UK [236]

Table A1. *Cont.*

Ecosystem Service/Disservice	Waterbody Type	Location
Sediment/solids retention	Ecological drainage system (wetlands, ditches, ponds)	China [237]
	Agricultural drainage ditches	USA [228]
Sediment/solids export	Roadside ditches	USA [222]
	Peatland forestry ditches	Finland [209]
Salt export	Agricultural drainage ditches	China [233]
Organic pollutant attenuation	Vegetated agricultural ditch	Mexico [238]
Pesticide degradation	Stagnant ditches	Netherlands [239]
Antibiotic export	Agricultural ditches	Germany [66]
Hormone export	Tile drains and ditches	USA [240]
Bad smell	Industrial ditch	Taiwan [69]
Flood control	Drainage ditches	Netherlands [210]
Increased hydrologic flashiness	Roadside ditches	USA [235], Greece [241]
Flooding	Flooding irrigation	Mexico [242]
Cultural		
Scientific model system	Irrigation canal	USA [152]
	Agricultural drainage ditches	Netherlands [154]
Bird-watching, photography	Road borrow pit, reservoir	USA [243]
Sport	Canals	Netherlands [244]
Source of conflict	Impounded rivers	Global [95]

References

1. Hooke, R.L. On the history of humans as geomorphic agents. *Geology* **2000**, *28*, 843–846. [CrossRef]
2. Haff, P.K. Hillslopes, rivers, plows, and trucks: Mass transport on Earth's surface by natural and technological processes. *Earth Surf. Process. Landf.* **2010**, *35*, 1157–1166. [CrossRef]
3. Wilkinson, B.H. Humans as geologic agents: A deep-time perspective. *Geology* **2005**, *33*, 161–164. [CrossRef]
4. Ellis, E.C.; Ramankutty, N. Putting people in the map: Anthropogenic biomes of the world. *Front. Ecol. Environ.* **2008**, *6*, 439–447. [CrossRef]
5. Steele, M.K.; Heffernan, J.B.; Bettez, N.; Cavender-Bares, J.; Groffman, P.M.; Grove, J.M.; Hall, S.; Hobbie, S.E.; Larson, K.; Morse, J.L.; et al. Convergent surface water distributions in U.S. Cities. *Ecosystems* **2014**, *17*, 685–697. [CrossRef]
6. Downing, J.A.; Prairie, Y.T.; Cole, J.J.; Duarte, C.M.; Tranvik, L.J.; Striegl, R.G.; McDowell, W.H.; Kortelainen, P.; Caraco, N.F.; Melack, J.M.; et al. The global abundance and size distribution of lakes, ponds, and impoundments. *Limnol. Oceanogr.* **2006**, *51*, 2388–2397. [CrossRef]
7. Sayer, C.D. Conservation of aquatic landscapes: Ponds, lakes, and rivers as integrated systems. *Wiley Interdiscip. Rev. Water* **2014**, *1*, 573–585. [CrossRef]
8. Biggs, J.; von Fumetti, S.; Kelly-Quinn, M. The importance of small waterbodies for biodiversity and ecosystem services: Implications for policy makers. *Hydrobiologia* **2017**, *793*, 3–39. [CrossRef]
9. Anderson, B.S.; Phillips, B.M.; Voorhees, J.P.; Cahn, M. Vegetated treatment systems for removing contaminants associated with surface water toxicity in agriculture and urban runoff. *J. Vis. Exp.* **2017**, *8*. [CrossRef] [PubMed]
10. Higgs, E. Novel and designed ecosystems. *Restor. Ecol.* **2017**, *25*, 8–13. [CrossRef]
11. Crail, T.D.; Gottgens, J.F.; Krause, A.E. Fish community response to evolving channel complexity in an agricultural headwater system. *J. Soil Water Conserv.* **2011**, *66*, 295–302. [CrossRef]
12. Palmer, M.A.; Bernhardt, E.S. Hydroecology and river restoration: Ripe for research and synthesis. *Water Resour. Res.* **2006**, *42*. [CrossRef]
13. Palmer, M.A.; Filoso, S. Restoration of ecosystem services for environmental markets. *Science* **2009**, *325*, 575–576. [CrossRef] [PubMed]
14. Ross, M.R.V.; Bernhardt, E.S.; Doyle, M.; Heffernan, J.B. Designer ecosystems: Incorporating design approaches into applied ecology. *Annu. Rev. Environ. Resour.* **2015**, *40*, 419–443. [CrossRef]
15. Bateman, H.L.; Stromberg, J.C.; Banville, M.J.; Makings, E.; Scott, B.D.; Suchy, A.; Wolkis, D. Novel water sources restore plant and animal communities along an urban river. *Ecohydrology* **2015**, *8*, 792–811. [CrossRef]
16. Palta, M.; du Bray, M.V.; Stotts, R.; Wolf, A.; Wutich, A. Ecosystem services and disservices for a vulnerable population: Findings from urban waterways and wetlands in an American desert city. *Hum. Ecol.* **2016**, *44*, 463–478. [CrossRef]

17. Sueltenfuss, J.P.; Cooper, D.J.; Knight, R.L.; Waskom, R.M. The creation and maintenance of wetland ecosystems from irrigation canal and reservoir seepage in a semi-arid landscape. *Wetlands* **2013**, *33*, 799–810. [CrossRef]
18. Chester, E.T.; Robson, B.J. Anthropogenic refuges for freshwater biodiversity: Their ecological characteristics and management. *Biol. Conserv.* **2013**, *166*, 64–75. [CrossRef]
19. Kueffer, C.; Kaiser-Bunbury, C.N. Reconciling conflicting perspectives for biodiversity conservation in the Anthropocene. *Front. Ecol. Environ.* **2013**, *12*, 131–137. [CrossRef]
20. Hupy, J.P.; Koehler, T. Modern warfare as a significant form of zoogeomorphic disturbance upon the landscape. *Geomorphology* **2012**, *157*, 169–182. [CrossRef]
21. Meade, R.; Wheeler, B.D. Illtyd Pools, Brecon: Description and conservation. *New Phytol.* **1990**, *115*, 187–199. [CrossRef]
22. Landscape Architecture Foundation (LAF). *Landscape Performance Series*; LAF: Washington, DC, USA, 2016; Available online: http://landscapeperformance.org/ (accessed on 14 December 2016).
23. Zollitsch, B.; Christie, J. *Report on State Definitions, Jurisdiction and Mitigation Requirements in State Programs for Ephemeral, Intermittent and Perennial Streams in the United States*; Association of State Wetland Managers: Windham, ME, USA, 2014.
24. U.S. Ecological Protection Agency; U.S. Army Corps of Engineers. *Definition of 'Waters of the United States' under the Clean Water Act*; U.S. Army Corps of Engineers and U.S. Environmental Protection Agency: Washington, DC, USA, 2015.
25. Vad, C.F.; Pentek, A.L.; Cozma, N.J.; Foldi, A.; Toth, A.; Toth, B.; Bode, N.A.; Mora, A.; Ptacnik, R.; Acs, E.; et al. Wartime scars or reservoirs of biodiversity? The value of bomb crater ponds in aquatic conservation. *Biol. Conserv.* **2017**, *209*, 253–262. [CrossRef] [PubMed]
26. Hawley, J. Isolated and Ephemeral Wetlands of Southern Appalachia: Biotic Communities and Environmental Drivers across Multiple Temporal and Spatial Scales. Ph.D. Thesis, Clemson University, Clemson, SC, USA, 2014.
27. Palta, M.M.; Grimm, N.B.; Groffman, P.M. "Accidental" urban wetlands: Ecosystem functions in unexpected places. *Front. Ecol. Environ.* **2017**, *15*, 248–256. [CrossRef]
28. Boler, R.B.; Sikkema, D.A. *Tamiami Trail Modifications: Next Steps Project Summary of Findings and Draft Environmental Impact Statement*; National Park Service: Homestead, FL, USA, 2010.
29. Zollitsch, B.; Christie, J. *Status and Trends Report on State Wetland Programs in the United States*; Association of State Wetland Managers: Windham, ME, USA, 2016.
30. Hough, P.; Robertson, M. Mitigation under Section 404 of the Clean Water Act: Where it comes from, what it means. *Wetl. Ecol. Manag.* **2009**, *17*, 15–33. [CrossRef]
31. Gregory, K.J. The human role in changing river channels. *Geomorphology* **2006**, *79*, 172–191. [CrossRef]
32. Walter, R.C.; Merritts, D.J. Natural streams and the legacy of water-powered mills. *Science* **2008**, *319*, 299–304. [CrossRef] [PubMed]
33. Walsh, C.J.; Roy, A.H.; Feminella, J.W.; Cottingham, P.D.; Groffman, P.M.; Morgan, R.P. The urban stream syndrome: Current knowledge and the search for a cure. *J. N. Am. Benthol. Soc.* **2009**, *24*, 706–723. [CrossRef]
34. Naftz, D.; Angeroth, C.; Kenney, T.; Waddell, B.; Darnall, N.; Silva, S.; Perschon, C.; Whitehead, J. Anthropogenic influences on the input and biogeochemical cycling of nutrients and mercury in Great Salt Lake, Utah, USA. *Appl. Geochem.* **2008**, *23*, 1731–1744. [CrossRef]
35. Wohl, E.; Lane, S.N.; Wilcox, A.C. The science and practice of river restoration. *Water Resour. Res.* **2015**, *51*, 5974–5997. [CrossRef]
36. Hallett, L.M.; Standish, R.J.; Hulvey, K.B.; Gardener, M.R.; Suding, K.N.; Starzomski, B.M.; Murphy, S.D.; Harris, J.A. Towards a conceptual framework for novel ecosystems. In *Novel Ecosystems*; John Wiley & Sons: Hoboken, NJ, USA, 2013; pp. 16–28.
37. Hobbs, R.J. Degraded or just different? Perceptions and value judgements in restoration decisions. *Restor. Ecol.* **2016**, *24*, 153–158. [CrossRef]
38. Crifasi, R.R. Reflections in a stock pond: Are anthropogenically derived freshwater ecosystems natural, artificial, or something else? *Environ. Manag.* **2005**, *36*, 625–639. [CrossRef] [PubMed]
39. Downing, J.A.; Cole, J.J.; Duarte, C.M.; Middelburg, J.J.; Melack, J.M.; Prairie, Y.T.; Kortelainen, P.; Striegl, R.G.; McDowell, W.H.; Tranvik, L.J. Global abundance and size distribution of streams and rivers. *Inland Waters* **2012**, *2*, 229–236. [CrossRef]

40. Rabkin, J.A. Against the EPA, absurdity is no defense symposium: The federal leviathan: Is there any area of modern life to which federal government does not extend: The thirty-second annual Federalist Society national student symposium on law and public policy-2013 section II: Environmental and property laws. *Harv. J. L. Pub. Policy* **2014**, *37*, 41–50.
41. The United States Department of Agriculture (USDA). *Summary Report: 2012 National Resources Inventory*; Natural Resources Conservation Service and Center for Survey Statistics and Methodology, Iowa State University: Washington, DC, USA; Ames, IA, USA, 2015.
42. Leopold, L.B. A reverence for rivers. *Geology* **1977**, *5*, 429–430. [CrossRef]
43. Naiman, R.J.; Magnuson, J.J.; Stanford, J.A.; McKnight, D.M. *The Freshwater Imperative: A Research Agenda*; Island Press: Washington, DC, USA, 1995.
44. Gleick, P.H.; Singh, A.; Shi, H. *Threats to the World's Freshwater Resources*; Pacific Institute for Studies in Development, Environment, and Security: Oakland, CA, USA, 2001.
45. Meybeck, M. Global analysis of river systems: From Earth system controls to Anthropocene syndromes. *Philos. Trans. R. Soc. Lond. Biol. Sci.* **2003**, *358*, 1935–1955. [CrossRef] [PubMed]
46. Research and Innovative Technology Administration (RITA). *Public Road and Street Mileage in the United States by Type of Surface*; U.S. Department of Transportation, Research and Innovative Technology Administration, Bureau of Transportation Statistics: Washington, DC, USA, 2014. Available online: http://www.rita.dot.gov/bts/sites/rita.dot.gov.bts/files/publications/national_transportation_statistics/html/table_01_04.html (accessed on 18 June 2018).
47. Paulsen, S.G.; Stoddard, J.L.; Holdsworth, S.; Mayio, A.; Tarquinio, E. *Wadeable Streams Assessment: A Colllaborative Survey of the Nation's Streams*; U.S. Environmental Protection Agency: Washington, DC, USA, 2004.
48. *National Hydrography Dataset*; U.S. Geological Survey: Washington, DC, USA, 2015.
49. *National Inventory of Dams*; U.S. Army Corps of Engineers: Washington, DC, USA, 2013.
50. Dahl, T.E. *Status and Trends of Wetlands in the Conterminous United States 2004 to 2009*; U.S. Department of the Interior; Fish and Wildlife Service: Washington, DC, USA, 2011.
51. Wohl, E.; Lininger, K.B.; Baron, J. Land before water: The relative temporal sequence of human alteration of freshwater ecosystems in the conterminous United States. *Anthropocene* **2017**, *18*, 27–46. [CrossRef]
52. Pekel, J.F.; Cottam, A.; Gorelick, N.; Belward, A.S. High-resolution mapping of global surface water and its long-term changes. *Nature* **2016**, *540*, 418–422. [CrossRef] [PubMed]
53. Berg, M.D.; Popescu, S.C.; Wilcox, B.P.; Angerer, J.P.; Rhodes, E.C.; McAlister, J.; Fox, W.E. Small farm ponds: Overlooked features with important impacts on watershed sediment transport. *J. Am. Water Resour. Assoc.* **2016**, *52*, 67–76. [CrossRef]
54. Oertli, B. Editorial: Freshwater biodiversity conservation: The role of artificial ponds in the 21st century. *Aquat. Conserv. Mar. Freshw. Ecosyst.* **2018**, *28*, 264–269. [CrossRef]
55. Rapinel, S.; Hubert-Moy, L.; Clément, B.; Nabucet, J.; Cudennec, C. Ditch network extraction and hydrogeomorphological characterization using lidar-derived DTM in wetlands. *Hydrol. Res.* **2015**, *46*, 276–290. [CrossRef]
56. Bailly, J.S.; Lagacherie, P.; Millier, C.; Puech, C.; Kosuth, P. Agrarian landscapes linear features detection from lidar: Application to artificial drainage networks. *Int. J. Remote Sens.* **2008**, *29*, 3489–3508. [CrossRef]
57. Cazorzi, F.; Dalla Fontana, G.; De Luca, A.; Sofia, G.; Tarolli, P. Drainage network detection and assessment of network storage capacity in agrarian landscape. *Hydrol. Process.* **2013**, *27*, 541–553. [CrossRef]
58. Herzon, I.; Helenius, J. Agricultural drainage ditches, their biological importance and functioning. *Biol. Conserv.* **2008**, *141*, 1171–1183. [CrossRef]
59. *National Lakes Assessment: A Collaborative Survey of the Nation's Lakes*; U.S. Environmental Protection Agency: Washington, DC, USA, 2009.
60. Rosenvald, R.; Jarvekulg, R.; Lohmus, A. Fish assemblages in forest drainage ditches: Degraded small streams or novel habitats? *Limnologica* **2014**, *46*, 37–44. [CrossRef]
61. Mattah, P.A.D.; Futagbi, G.; Amekudzi, L.K.; Mattah, M.M.; de Souza, D.K.; Kartey-Attipoe, W.D.; Bimi, L.; Wilson, M.D. Diversity in breeding sites and distribution of *Anopheles* mosquitoes in selected urban areas of southern Ghana. *Parasit. Vectors* **2017**, *10*, 15. [CrossRef] [PubMed]
62. Wang, Y.C. Examining landscape determinants of *Opisthorchis viverrini* transmission. *EcoHealth* **2012**, *9*, 328–341. [CrossRef] [PubMed]

63. Maheu-Giroux, M.; de Blois, S. Landscape ecology of *Phragmites australis* invasion in networks of linear wetlands. *Landsc. Ecol.* **2007**, *22*, 285–301. [CrossRef]
64. O'Connor, J.H.; Rothermel, B.B. Distribution and population characteristics of African Jewelfish and Brown Hoplo in modified wetlands in south Florida. *Am. Midl. Nat.* **2013**, *170*, 52–65. [CrossRef]
65. Mallin, M.A.; McIver, M.R. Pollutant impacts to Cape Hatteras National Seashore from urban runoff and septic leachate. *Mar. Pollut. Bull.* **2012**, *64*, 1356–1366. [CrossRef] [PubMed]
66. Bailey, C.; Spielmeyer, A.; Frings, R.M.; Hamscher, G.; Schuttrumpf, H. From agricultural fields to surface water systems: The overland transport of veterinary antibiotics. *J. Soils Sedim.* **2015**, *15*, 1630–1634. [CrossRef]
67. Naito, R.; Sakai, M.; Morimoto, Y. Negative effects of deep roadside ditches on *Pelophylax porosa brevipoda* dispersal and migration in comparison with *Hyla japonica* in a rice paddy area in Japan. *Zool. Sci.* **2012**, *29*, 599–603. [CrossRef] [PubMed]
68. Hergoualc'h, K.A.; Verchot, L.V. Changes in soil CH_4 fluxes from the conversion of tropical peat swamp forests: A meta-analysis. *J. Integr. Environ. Sci.* **2012**, *9*, 31–39. [CrossRef]
69. Mao, I.F.; Chen, M.R.; Wang, L.; Chen, M.L.; Lai, S.C.; Tsai, C.J. Method development for determining the malodor source and pollution in industrial park. *Sci. Total Environ.* **2012**, *437*, 270–275. [CrossRef] [PubMed]
70. Perring, M.P.; Manning, P.; Hobbs, R.J.; Lugo, A.E.; Ramalho, C.E.; Standish, R.J. Novel urban ecosystems and ecosystem services. In *Novel Ecosystems*; John Wiley & Sons: Hoboken, NJ, USA, 2013; pp. 310–325.
71. Salomon Cavin, J. Beyond prejudice: Conservation in the city. A case study from Switzerland. *Biol. Conserv.* **2013**, *166*, 84–89. [CrossRef]
72. Robert, A.; Fontaine, C.; Veron, S.; Monnet, A.C.; Legrand, M.; Clavel, J.; Chantepie, S.; Couvet, D.; Ducarme, F.; Fontaine, B.; et al. Fixism and conservation science. *Conserv. Biol.* **2017**, *31*, 781–788. [CrossRef] [PubMed]
73. Miller, B.; Soule, M.E.; Terborgh, J. 'New conservation' or surrender to development? *Anim. Conserv.* **2014**, *17*, 509–515. [CrossRef]
74. Ridder, B. An exploration of the value of naturalness and wild nature. *J. Agric. Environ. Ethics* **2007**, *20*, 195–213. [CrossRef]
75. Reid, N.; Dingerkus, S.K.; Stone, R.E.; Buckley, J.; Beebee, T.J.C.; Marnell, F.; Wilkinson, J.W. Assessing historical and current threats to common frog (*Rana temporaria*) populations in Ireland. *J. Herpetol.* **2014**, *48*, 13–19. [CrossRef]
76. Sebastian-Gonzalez, E.; Sanchez-Zapata, J.A.; Botella, F. Agricultural ponds as alternative habitat for waterbirds: Spatial and temporal patterns of abundance and management strategies. *Eur. J. Wildl. Res.* **2010**, *56*, 11–20. [CrossRef]
77. Ruggiero, A.; Cereghino, R.; Figuerola, J.; Marty, P.; Angelibert, S. Farm ponds make a contribution to the biodiversity of aquatic insects in a French agricultural landscape. *C. R. Biol.* **2008**, *331*, 298–308. [CrossRef] [PubMed]
78. Cereghino, R.; Ruggiero, A.; Marty, P.; Angelibert, S. Biodiversity and distribution patterns of freshwater invertebrates in farm ponds of a south-western French agricultural landscape. *Hydrobiologia* **2008**, *597*, 43–51. [CrossRef]
79. Abellan, P.; Sanchez-Fernandez, D.; Millan, A.; Botella, F.; Sanchez-Zapata, J.A.; Gimenez, A. Irrigation pools as macroinvertebrate habitat in a semi-arid agricultural landscape (SE Spain). *J. Arid. Environ.* **2006**, *67*, 255–269. [CrossRef]
80. Davies, B.; Biggs, J.; Williams, P.; Whitfield, M.; Nicolet, P.; Sear, D.; Bray, S.; Maund, S. Comparative biodiversity of aquatic habitats in the European agricultural landscape. *Agric. Ecosyst. Environ.* **2008**, *125*, 1–8. [CrossRef]
81. Biggs, J.; Williams, P.; Whitfield, M.; Nicolet, P.; Weatherby, A. 15 years of pond assessment in Britain: Results and lessons learned from the work of pond conservation. *Aquat. Conserv. Mar. Freshw. Ecosyst.* **2005**, *15*, 693–714. [CrossRef]
82. Cereghino, R.; Boix, D.; Cauchie, H.M.; Martens, K.; Oertli, B. The ecological role of ponds in a changing world. *Hydrobiologia* **2014**, *723*, 1–6. [CrossRef]
83. Kadoya, T.; Suda, S.; Washitani, I. Dragonfly crisis in Japan: A likely consequence of recent agricultural habitat degradation. *Biol. Conserv.* **2009**, *142*, 1899–1905. [CrossRef]
84. Canessa, S.; Oneto, F.; Ottonello, D.; Arillo, A.; Salvidio, S. Land abandonment may reduce disturbance and affect the breeding sites of an endangered amphibian in northern Italy. *Oryx* **2013**, *47*, 280–287. [CrossRef]

85. Casas, J.J.; Sanchez-Oliver, J.S.; Sanz, A.; Furne, M.; Trenzado, C.; Juan, M.; Paracuellos, M.; Suarez, M.D.; Fuentes, F.; Gallego, I.; et al. The paradox of the conservation of an endangered fish species in a Mediterranean region under agricultural intensification. *Biol. Conserv.* **2011**, *144*, 253–262. [CrossRef]
86. Taylor, C.A.; Robison, H.W. A new burrowing crayfish of the genus *Fallicambarus* Hobbs, 1969 (decapoda: Cambaridae) from the red river drainage of the southcentral united states. *Zootaxa* **2016**, *4144*, 575–583. [CrossRef] [PubMed]
87. Zang, D.K. *Cerasus Laoshanensis* (Rosaceae), a new species from Shandong, China. *Ann. Bot. Fenn.* **2017**, *54*, 135–137. [CrossRef]
88. Gan, Q.L.; Li, X.W. *Impatiens baokangensis* (Balsaminaceae), a new species from Hubei, China. *Ann. Bot. Fenn.* **2016**, *53*, 145–148. [CrossRef]
89. Brand, A.B.; Snodgrass, J.W. Value of artificial habitats for amphibian reproduction in altered landscapes. *Conserv. Biol.* **2010**, *24*, 295–301. [CrossRef] [PubMed]
90. Roley, S.S.; Tank, J.L.; Tyndall, J.C.; Witter, J.D. How cost-effective are cover crops, wetlands, and two-stage ditches for nitrogen removal in the Mississippi river basin? *Water Resour. Econ.* **2016**, *15*, 43–56. [CrossRef]
91. Perez-Gutierrez, J.D.; Paz, J.O.; Tagert, M.L.M. Seasonal water quality changes in on-farm water storage systems in a south-central US agricultural watershed. *Agric. Water Manag.* **2017**, *187*, 131–139. [CrossRef]
92. Cooper, C.M.; Moore, M.T.; Bennett, E.R.; Sammie Smith, J.; Farris, J.L. Alternative environmental benefits of agricultural drainage ditches. *Verh. Int. Verein Limnol.* **2002**, *28*, 1678–1682. [CrossRef]
93. Palmer, M.A. Reforming watershed restoration: Science in need of application and applications in need of science. *Estuar. Coasts* **2009**, *32*, 1–17. [CrossRef]
94. Stewart, J.S.; Wang, L.Z.; Lyons, J.; Horwatich, J.A.; Bannerman, R. Influences of watershed, riparian-corridor, and reach-scale characteristics on aquatic biota in agricultural watersheds. *J. Am. Water Resour. Assoc.* **2001**, *37*, 1475–1487. [CrossRef]
95. Dudgeon, D.; Arthington, A.H.; Gessner, M.O.; Kawabata, Z.; Knowler, D.J.; Leveque, C.; Naiman, R.J.; Prieur-Richard, A.H.; Soto, D.; Stiassny, M.L.; et al. Freshwater biodiversity: Importance, threats, status and conservation challenges. *Biol. Rev. Camb. Philos. Soc.* **2006**, *81*, 163–182. [CrossRef] [PubMed]
96. Burcher, C.L.; Valett, H.M.; Benfield, E.F. The land-cover cascade: Relationships coupling land and water. *Ecology* **2007**, *88*, 228–242. [CrossRef]
97. Suding, K.N. Toward an era of restoration in ecology: Successes, failures, and opportunities ahead. *Annu. Rev. Ecol. Evol. Syst.* **2011**, *42*, 465–487. [CrossRef]
98. Veraart, A.J.; Dimitrov, M.R.; Schrier-Uijl, A.P.; Smidt, H.; de Klein, J.J.M. Abundance, activity and community structure of denitrifiers in drainage ditches in relation to sediment characteristics, vegetation and land-use. *Ecosystems* **2017**, *20*, 928–943. [CrossRef]
99. Bremigan, M.T.; Soranno, P.A.; Gonzalez, M.J.; Bunnell, D.B.; Arend, K.K.; Renwick, W.H.; Stein, R.A.; Vanni, M.J. Hydrogeomorphic features mediate the effects of land use/cover on reservoir productivity and food webs. *Limnol. Oceanogr.* **2008**, *53*, 1420–1433. [CrossRef]
100. Simon, T.N.; Travis, J. The contribution of man-made ditches to the regional stream biodiversity of the new river watershed in the Florida panhandle. *Hydrobiologia* **2011**, *661*, 163–177. [CrossRef]
101. Francesiaz, C.; Guilbault, E.; Lebreton, J.D.; Trouvilliez, J.; Besnard, A. Colony persistence in waterbirds is constrained by pond quality and land use. *Freshw. Biol.* **2017**, *62*, 119–132. [CrossRef]
102. Mehaffey, M.H.; Nash, M.S.; Wade, T.G.; Ebert, D.W.; Jones, K.B.; Rager, A. Linking land cover and water quality in New York City's water supply watersheds. *Environ. Monit. Assess.* **2005**, *107*, 29–44. [CrossRef] [PubMed]
103. Bernhardt, E.S.; Palmer, M.A. Restoring streams in an urbanizing world. *Freshw. Biol.* **2007**, *52*, 738–751. [CrossRef]
104. Brown, L.E.; Milner, A.M. Rapid loss of glacial ice reveals stream community assembly processes. *Glob. Chang. Biol.* **2012**, *18*, 2195–2204. [CrossRef]
105. Sayer, C.; Andrews, K.; Shilland, E.; Edmonds, N.; Edmonds-Brown, R.; Patmore, I.; Emson, D.; Axmacher, J. The role of pond management for biodiversity conservation in an agricultural landscape. *Aquat. Conserv. Mar. Freshw. Ecosyst.* **2012**, *22*, 626–638. [CrossRef]
106. Arthaud, F.; Vallod, D.; Robin, J.; Wezel, A.; Bornette, G. Short-term succession of aquatic plant species richness along ecosystem productivity and dispersal gradients in shallow lakes. *J. Veg. Sci.* **2013**, *24*, 148–156. [CrossRef]

107. Drake, C.M.; Stewart, N.F.; Palmer, M.A.; Kindemba, V.L. *The Ecological Status of Ditch Systems: An Investigation into the Current Status of the Aquatic Invertebrate and Plant communities of Grazing Marsh Ditch Systems in England and Wales*; Buglife-The Invertebrate Conservation Trust: Peterborough, UK, 2010.
108. Ballantine, K.; Schneider, R. Fifty-five years of soil development in restored freshwater depressional wetlands. *Ecol. Appl.* **2009**, *19*, 1467–1480. [CrossRef] [PubMed]
109. D'Ambrosio, J.L.; Ward, A.D.; Witter, J.D. Evaluating geomorphic change in constructed two-stage ditches. *J. Am. Water Resour. Assoc.* **2015**, *51*, 910–922. [CrossRef]
110. Palmer, M.A.; Drake, M.; Stewart, N. *A Manual for the Survey and Evaluation of the Aquatic Plant and Invertebrate Assemblages of Grazing Marsh Ditch Systems*; Buglife-The Invertebrate Conservation Trust: Peterborough, UK, 2013.
111. Rehounkova, K.; Prach, K. Spontaneous vegetation succession in gravel-sand pits: A potential for restoration. *Restor. Ecol.* **2008**, *16*, 305–312. [CrossRef]
112. Oosting, H.J. An ecological analysis of the plant communities of Piedmont, North Carolina. *Am. Midl. Nat.* **1942**, *28*, 1–126. [CrossRef]
113. Zhang, Z.D.; Brown, R.; Bauer, J.; Bedard-Haughn, A. Nutrient dynamics within drainage ditches under recent, medium, and long-term drainage in the black soil zone of southeastern Saskatchewan. *Geoderma* **2017**, *289*, 66–71. [CrossRef]
114. Kaushal, S.; McDowell, W.; Wollheim, W.; Johnson, T.; Mayer, P.; Belt, K.; Pennino, M. Urban evolution: The role of water. *Water* **2015**, *7*, 4063–4087. [CrossRef]
115. Li, M.; Li, L.X.; Liao, H.S.; Huang, Z.W.; Huang, W.D.A. The influence of drainage on wetland degradation in Zoige plateau. *Disaster Adv.* **2012**, *5*, 659–666.
116. Vincent, R.E.; Burdick, D.M.; Dionne, M. Ditching and ditch-plugging in New England salt marshes: Effects on hydrology, elevation, and soil characteristics. *Estuar. Coasts* **2013**, *36*, 610–625. [CrossRef]
117. Bennett, E.M.; Peterson, G.D.; Gordon, L.J. Understanding relationships among multiple ecosystem services. *Ecol. Lett.* **2009**, *12*, 1394–1404. [CrossRef] [PubMed]
118. Power, A.G. Ecosystem services and agriculture: Tradeoffs and synergies. *Philos. Trans. R. Soc. Lond. Biol. Sci.* **2010**, *365*, 2959–2971. [CrossRef] [PubMed]
119. Hale, R.; Baxter, C.; Burnham, M.; Marsh, K. *Co-Producing Urban River Imaginaries over Time: An Idaho River and Its community*; Society for Freshwater Science: Raleigh, NC, USA, 2017.
120. Landers, J. Environmental engineering-Los Angeles aims to combine river restoration, urban revitalization. *Civ. Eng.* **2007**, *77*, 11–13.
121. Polat, A.T.; Akay, A. Relationships between the visual preferences of urban recreation area users and various landscape design elements. *Urban For. Urban Green.* **2015**, *14*, 573–582. [CrossRef]
122. Smith, R.F.; Hawley, R.J.; Neale, M.W.; Vietz, G.J.; Diaz-Pascacio, E.; Herrmann, J.; Lovell, A.C.; Prescott, C.; Rios-Touma, B.; Smith, B.; et al. Urban stream renovation: Incorporating societal objectives to achieve ecological improvements. *Freshw. Sci.* **2016**, *35*, 364–379. [CrossRef]
123. Lave, R. *Fields and Streams: Stream Restoration, Neoliberalism, and the Future of Environmental Science*; University of Georgia Press: Athens, GA, USA, 2012.
124. Doyle, M.W.; Singh, J.; Lave, R.; Robertson, M.M. The morphology of streams restored for market and nonmarket purposes: Insights from a mixed natural-social science approach. *Water Resour. Res.* **2015**, *51*, 5603–5622. [CrossRef]
125. Podolak, K.; Kondolf, G.M. The line of beauty in river designs: Hogarth's aesthetic theory on capability brown's eighteenth-century river design and twentieth-century river restoration design. *Landsc. Res.* **2016**, *41*, 149–167. [CrossRef]
126. Dollinger, J.; Dages, C.; Bailly, J.S.; Lagacherie, P.; Voltz, M. Managing ditches for agroecological engineering of landscape. A review. *Agron. Sustain. Dev.* **2015**, *35*, 999–1020. [CrossRef]
127. Palmer, M.A.; Menninger, H.L.; Bernhardt, E. River restoration, habitat heterogeneity and biodiversity: A failure of theory or practice? *Freshw. Biol.* **2010**, *55*, 205–222. [CrossRef]
128. Bernhardt, E.S.; Palmer, M.A. River restoration: The fuzzy logic of repairing reaches to reverse catchment scale degradation. *Ecol. Appl.* **2011**, *21*, 1926–1931. [CrossRef] [PubMed]
129. Kumwimba, M.N.; Zhu, B. Effectiveness of vegetated drainage ditches for domestic sewage effluent mitigation. *Bull. Environ. Contam. Toxicol.* **2017**, *98*, 682–689. [CrossRef] [PubMed]

130. Verlicchi, P.; Zambello, E. How efficient are constructed wetlands in removing pharmaceuticals from untreated and treated urban wastewaters? A review. *Sci. Total Environ.* **2014**, *470*, 1281–1306. [CrossRef] [PubMed]
131. Li, Y.F.; Zhu, G.B.; Ng, W.J.; Tan, S.K. A review on removing pharmaceutical contaminants from wastewater by constructed wetlands: Design, performance and mechanism. *Sci. Total Environ.* **2014**, *468*, 908–932. [CrossRef] [PubMed]
132. Powell, G.E.; Ward, A.D.; Mecklenburg, D.E.; Jayakaran, A.D. Two-stage channel systems: Part 1, a practical approach for sizing agricultural ditches. *J. Soil Water Conserv.* **2007**, *62*, 277–286.
133. Davis, R.T.; Tank, J.L.; Mahl, U.H.; Winikoff, S.G.; Roley, S.S. The influence of two-stage ditches with constructed floodplains on water column nutrients and sediments in agricultural streams. *J. Am. Water Resour. Assoc.* **2015**, *51*, 941–955. [CrossRef]
134. Mahl, U.H.; Tank, J.L.; Roley, S.S.; Davis, R.T. Two-stage ditch floodplains enhance n-removal capacity and reduce turbidity and dissolved P in agricultural streams. *J. Am. Water Resour. Assoc.* **2015**, *51*, 923–940. [CrossRef]
135. Cordell, H.K.; Bergstrom, J.C. Comparison of recreation use values among alternative reservoir water level management scenarios. *Water Resour. Res.* **1993**, *29*, 247–258. [CrossRef]
136. Branche, E. The multipurpose water uses of hydropower reservoir: The share concept. *C. R. Phys.* **2017**, *18*, 469–478. [CrossRef]
137. Twisk, W.; Noordervliet, M.A.W.; ter Keurs, W.J. The nature value of the ditch vegetation in peat areas in relation to farm management. *Aquat. Ecol.* **2003**, *37*, 191–209. [CrossRef]
138. Needelman, B.A.; Kleinman, P.J.A.; Strock, J.S.; Allen, A.L. Drainage ditches: Improved management of agricultural drainage ditches for water quality protection: An overview. *J. Soil Water Conserv.* **2007**, *62*, 171–178.
139. Smith, C.; Boughton, E.H.; Pierre, S. *Pomacea maculata* (island apple snail) invasion in seasonal wetlands on Florida ranchland: Association with plant-community structure and aquatic-predator abundance. *Southeast. Nat.* **2015**, *14*, 561–576. [CrossRef]
140. Fluck, R.C.; Fonyo, C.; Flaig, E. Land-use-based phosphorus balances for lake Okeechobee, Florida, drainage basins. *Appl. Eng. Agric* **1992**, *8*, 813–820. [CrossRef]
141. *Final Total Maximum Daily Load for Biochemical Oxygen Demand, Dissolved Oxygen, and Nutrients in the Lake Okeechobee tributaries Osceola, Polk, Okeechobee, Highlands, Glade, and Martin, Florida*; U.S. Environmental Protection Association: Atlanta, GA, USA, 2008.
142. Compilation of Benefits and Costs of STA and Reservoir Projects in the South Florida Water Management District. Available online: http://www.fresp.org/pdfs/Compilation%20of%20STA%20and%20REZ%20Benefits%20Costs%20HandS%2011_2011.pdf (accessed on 27 September 2014).
143. Bohlen, P.J.; Boughton, E.; Fauth, J.E.; Jenkins, D.; Kiker, G.; Quintana-Ascencio, P.F.; Shukla, S.; Swain, H.M. *Assessing Trade-Offs Among Ecosystem Services in a Payment-for-Water Services Program on Florida Ranchlands*; Archbold Biological Station: Lake Placid, FL, USA, 2014.
144. Harabis, F.; Dolny, A. Necessity for the conservation of drainage systems as last refugia for threatened damselfly species, *Coenagrion ornatum*. *Insect. Conserv. Divers.* **2015**, *8*, 143–151. [CrossRef]
145. Baker, S.; Eckerberg, K. Ecological restoration success: A policy analysis understanding. *Restor. Ecol.* **2016**, *24*, 284–290. [CrossRef]
146. Palmer, M.; Allan, J.D.; Meyer, J.; Bernhardt, E.S. River restoration in the twenty-first century: Data and experiential future efforts. *Restor. Ecol.* **2007**, *15*, 472–481. [CrossRef]
147. Jeffries, M.J. Ponds and the importance of their history: An audit of pond numbers, turnover and the relationship between the origins of ponds and their contemporary plant communities in south-east Northumberland, UK. *Hydrobiologia* **2012**, *689*, 11–21. [CrossRef]
148. Felson, A.J.; Pickett, S.T.A. Designed experiments: New approaches to studying urban ecosystems. *Front. Ecol. Environ.* **2005**, *3*, 549–556. [CrossRef]
149. Rosenzweig, M.L. *Win-Win Ecology: How the Earth's Species Can Survive in the Midst of Human Enterprise*; Oxford University Press: Oxford, UK, 2003.
150. Hobbs, R.J.; Higgs, E.; Harris, J.A. Novel ecosystems: Implications for conservation and restoration. *Trends Ecol. Evol.* **2009**, *24*, 599–605. [CrossRef] [PubMed]

151. De Meester, L.; Declerck, S.; Stoks, R.; Louette, G.; Van de Meutter, F.; De Bie, T.; Michels, E.; Brendonck, L. Ponds and pools as model systems in conservation biology, ecology and evolutionary biology. *Aquat. Conserv. Mar. Freshw. Ecosyst.* **2005**, *15*, 715–725. [CrossRef]
152. Koetsier, P.; McCauley, L.M.M. An irrigation canal as a lotic mesocosm: Examining the relationship between macroinvertebrate benthos and drift. *West. N. Am. Nat.* **2015**, *75*, 259–270. [CrossRef]
153. Diamond, J.M. Ecology: Laboratory, field and natural experiments. *Nature* **1983**, *304*, 586–587. [CrossRef]
154. Van Gerven, L.P.A.; Kuiper, J.J.; Janse, J.H.; Janssen, A.B.G.; Jeuken, M.; Mooij, W.M.; de Klein, J.J.M. How regime shifts in connected aquatic ecosystems are affected by the typical downstream increase of water flow. *Ecosystems* **2017**, *20*, 733–744. [CrossRef]
155. Petrovich, O. Preschool children's understanding of the dichotomy between the natural and the artificial. *Psychol. Rep.* **1999**, *84*, 3–27. [CrossRef]
156. Chakrabarty, D. The climate of history: Four theses. *Crit. Inq.* **2009**, *35*, 197–222. [CrossRef]
157. Bensaude-Vincent, B.; Newman, W.R. Introduction: The artificial and the natural: State of the problem. In *The Artificial and the Natural*; Bensaude-Vincent, B., Newman, W.R., Eds.; The MIT Press: Cambridge, UK, 2007; pp. 1–19.
158. Purdy, J. *After Nature: A Politics for the Anthropocene*; Harvard University Press: Cambridge, MA, USA, 2015.
159. McMahan, E.A.; Cloud, J.M.; Josh, P.; Scott, M. Nature with a human touch: Human-induced alteration negatively impacts perceived naturalness and preferences for natural environments. *Ecopsychology* **2016**, *8*, 54–63. [CrossRef]
160. Rozin, P.; Spranca, M.; Krieger, Z.; Neuhaus, R.; Surillo, D.; Swerdlin, A.; Wood, K. Preference for natural: Instrumental and ideational/moral motivations, and the contrast between foods and medicines. *Appetite* **2004**, *43*, 147–154. [CrossRef] [PubMed]
161. Elliot, R. Faking nature. In *Environmental Ethics*; Blackwell Publishing: Malden, MA, USA, 1982; pp. 381–389.
162. Katz, E. The big lie: Human restoration of nature. In *Environmental Ethics*; Blackwell Publishing: Malden, MA, USA, 1983; pp. 390–397.
163. Callicott, J.B. A critique of an alternative to the wilderness idea. In *Environmental Ethics*; Blackwell Publishing: Malden, MA, USA, 1995; pp. 437–443.
164. Light, A.; Thompson, A.; Higgs, E.S. Valuing novel ecosystems. In *Novel ecosystems*; John Wiley & Sons: Hoboken, NJ, USA, 2013; pp. 257–268.
165. Higgs, E. *Nature by Design*; Massachusetts Institute of Technology: Boston, MA, USA, 2003.
166. Soule, M. The "New Conservation". *Conserv. Biol.* **2013**, *27*, 895–897. [CrossRef] [PubMed]
167. Hobbs, R.J.; Arico, S.; Aronson, J.; Baron, J.S.; Bridgewater, P.; Cramer, V.A.; Epstein, P.R.; Ewel, J.J.; Klink, C.A.; Lugo, A.E.; et al. Novel ecosystems: Theoretical and management aspects of the new ecological world order. *Glob. Ecol. Biogeogr.* **2006**, *15*, 1–7. [CrossRef]
168. Tallis, H.; Lubchenco, J. A call for inclusive conservation. *Nature* **2014**, *515*, 27–28. [CrossRef] [PubMed]
169. Matulis, B.S.; Moyer, J.R. Beyond inclusive conservation: The value of pluralism, the need for agonism, and the case for social instrumentalism. *Conserv. Lett.* **2017**, *10*, 279–287. [CrossRef]
170. Gelcich, S.; O'Keeffe, J. Emerging frontiers in perceptions research for aquatic conservation. *Aquat. Conserv. Mar. Freshw. Ecosyst.* **2016**, *26*, 986–994. [CrossRef]
171. Francis, R.A.; Hoggart, S.P.G. Waste not, want not: The need to utilize existing artificial structures for habitat improvement along urban rivers. *Restor. Ecol.* **2008**, *16*, 373–381. [CrossRef]
172. Introduction to the New EU Water Framework Directive. Available online: http://ec.europa.eu/environment/water/water-framework/info/intro_en.htm (accessed on 19 August 2014).
173. Declerck, S.; De Bie, T.; Ercken, D.; Hampel, H.; Schrijvers, S.; Van Wichelen, J.; Gillard, V.; Mandiki, R.; Losson, B.; Bauwens, D.; et al. Ecological characteristic's of small farmland ponds: Associations with land use practices at multiple spatial scales. *Biol. Conserv.* **2006**, *131*, 523–532. [CrossRef]
174. European Pond Conservation Network. Available online: https://freshwaterhabitats.org.uk/research/epcn/ (accessed on 12 August 2018).
175. Million Ponds Project. Available online: http://www.freshwaterhabitats.org.uk/projects/million-ponds/ (accessed on 19 August 2014).
176. Canals, R.M.; Ferrer, V.; Iriarte, A.; Carcamo, S.; San Emeterio, L.; Villanueva, E. Emerging conflicts for the environmental use of water in high-valuable rangelands. Can livestock water ponds be managed as artificial wetlands for amphibians? *Ecol. Eng.* **2011**, *37*, 1443–1452. [CrossRef]

177. Clarke, S.J. Conserving freshwater biodiversity: The value, status and management of high quality ditch systems. *J. Nat. Conserv.* **2015**, *24*, 93–100. [CrossRef]
178. Shaw, R.F.; Johnson, P.J.; Macdonald, D.W.; Feber, R.E. Enhancing the biodiversity of ditches in intensively managed UK farmland. *PLoS ONE* **2015**, *10*, 19. [CrossRef] [PubMed]
179. Spears, E.G. *Baptized in PCBS: Race, Pollution, and Justice in an All-American Town*; UNC Press Books: Chapel Hill, NC, USA, 2014.
180. Checker, M. *Polluted Promises: Environmental Racism and the Search for Justice in a Southern Town*; NYU Press: New York, NY, USA, 2005.
181. Setzer, C.; Domino, M.E. Medicaid outpatient utilization for waterborne pathogenic illness following hurricane Floyd. *Public Health Rep.* **2004**, *119*, 472–478. [CrossRef] [PubMed]
182. Kim, J.; Goldsmith, P.; Thomas, M.H. Economic impact and public costs of confined animal feeding operations at the parcel level of Craven County, North Carolina. *Agric. Hum. Values* **2010**, *27*, 29–42. [CrossRef]
183. Rosenzweig, M.L. Reconciliation ecology and the future of species diversity. *Oryx* **2003**, *37*, 194–205. [CrossRef]
184. Luptacik, P.; Miklisova, D.; Kovac, L. Diversity and community structure of soil Oribatida (Acari) in an arable field with alluvial soils. *Eur. J. Soil Biol.* **2012**, *50*, 97–105. [CrossRef]
185. Verdonschot, R.C.M.; Keizer-vlek, H.E.; Verdonschot, P.F.M. Biodiversity value of agricultural drainage ditches: A comparative analysis of the aquatic invertebrate fauna of ditches and small lakes. *Aquat. Conserv Mar. Freshw. Ecosyst.* **2011**, *21*, 715–727. [CrossRef]
186. Ishiyama, N.; Koizumi, I.; Yuta, T.; Nakamura, F. Differential effects of spatial network structure and scale on population size and genetic diversity of the ninespine stickleback in a remnant wetland system. *Freshw. Biol.* **2015**, *60*, 733–744. [CrossRef]
187. Yagi, K.T.; Litzgus, J.D. The effects of flooding on the spatial ecology of spotted turtles (*Clemmys guttata*) in a partially mined peatland. *Copeia* **2012**, 179–190. [CrossRef]
188. Pitt, A.L.; Baldwin, R.F.; Lipscomb, D.J.; Brown, B.L.; Hawley, J.E.; Allard-Keese, C.M.; Leonard, P.B. The missing wetlands: Using local ecological knowledge to find cryptic ecosystems. *Biodivers. Conserv.* **2012**, *21*, 51–63. [CrossRef]
189. Maltchik, L.; Stenert, C.; Batzer, D.P. Can rice field management practices contribute to the conservation of species from natural wetlands? Lessons from Brazil. *Basic Appl. Ecol.* **2017**, *18*, 50–56. [CrossRef]
190. Kloskowski, J.; Nieoczym, M. Management practices to enhance wildlife diversity of man-made fish ponds: The importance of the hydroperiod. *Knowl. Manag. Aquat. Ecosyst.* **2015**, 15. [CrossRef]
191. James-Pirri, M.J.; Erwin, R.M.; Prosser, D.J.; Taylor, J.D. Responses of salt marsh ecosystems to mosquito control management practices along the Atlantic coast (USA). *Restor. Ecol.* **2012**, *20*, 395–404. [CrossRef]
192. Rochlin, I.; James-Pirri, M.J.; Adamowicz, S.C.; Dempsey, M.E.; Iwanejko, T.; Ninivaggi, D.V. The effects of integrated marsh management (IMM) on salt marsh vegetation, nekton, and birds. *Estuar. Coasts* **2012**, *35*, 727–742. [CrossRef]
193. Horgan, F.G.; Ramal, A.F.; Villegas, J.M.; Almazan, M.L.P.; Bernal, C.C.; Jamoralin, A.; Pasang, J.M.; Orboc, G.; Agreda, V.; Arroyo, C. Ecological engineering with high diversity vegetation patches enhances bird activity and ecosystem services in Philippine rice fields. *Reg. Environ. Chang.* **2017**, *17*, 1355–1367. [CrossRef]
194. Pernollet, C.A.; Simpson, D.; Gauthier-Clerc, M.; Guillemain, M. Rice and duck, a good combination? Identifying the incentives and triggers for joint rice farming and wild duck conservation. *Agric. Ecosyst. Environ.* **2015**, *214*, 118–132. [CrossRef]
195. Pernollet, C.A.; Cavallo, F.; Simpson, D.; Gauthier-Clerc, M.; Guillemain, M. Seed density and waterfowl use of rice fields in Camargue, France. *J. Wildl. Manag.* **2017**, *81*, 96–111. [CrossRef]
196. Hill, M.J.; Ryves, D.B.; White, J.C.; Wood, P.J. Macroinvertebrate diversity in urban and rural ponds: Implications for freshwater biodiversity conservation. *Biol. Conserv.* **2016**, *201*, 50–59. [CrossRef]
197. Soomers, H.; Karssenberg, D.; Soons, M.B.; Verweij, P.A.; Verhoeven, J.T.A.; Wassen, M.J. Wind and water dispersal of wetland plants across fragmented landscapes. *Ecosystems* **2013**, *16*, 434–451. [CrossRef]
198. Van Dijk, W.F.A.; van Ruijven, J.; Berendse, F.; de Snoo, G.R. The effectiveness of ditch banks as dispersal corridor for plants in agricultural landscapes depends on species' dispersal traits. *Biol. Conserv.* **2014**, *171*, 91–98. [CrossRef]

199. Zhu, L.; Jiang, C.; Xie, X.; Xie, M.; Hu, X.; Cheng, J. Ecological and landscape effects of ditches on farmland ecosystem. In Proceedings of the Bioinformatics and Biomedical Engineering (ICBBE), Beijing, China, 11–13 June 2009.
200. Zhang, Y.-T.; Zhang, G.-F.; Li, Y.; Li, L.; Yu, L.-P. Diversity of aquatic vascular plants in the Yangtze delta. *Plant Sci. J.* **2012**, *30*, 238–249. [CrossRef]
201. Zhao, Y.; Li, X.R.; Zhang, Z.S.; Hu, Y.G.; Wu, P. Species composition and species richness in the Hetao irrigation region drainage ditches, Northern China. *Arid Land Res. Manag.* **2013**, *27*, 167–177. [CrossRef]
202. Hill, C.R.; Robinson, J.S. Phosphorus flux from wetland ditch sediments. *Sci. Total Environ.* **2012**, *437*, 315–322. [CrossRef] [PubMed]
203. Navrátilová, J.; Hájek, M.; Navrátil, J.; Hájková, P.; Frazier, R.J. Convergence and impoverishment of fen communities in a eutrophicated agricultural landscape of the Czech Republic. *Appl. Veg. Sci.* **2017**, *20*, 225–235. [CrossRef]
204. Henriksson, P.J.G.; Jarvio, N.; Jonell, M.; Guinée, J.B.; Troell, M. The devil is in the details—The carbon footprint of a shrimp. *Front. Ecol. Environ.* **2018**, *16*, 10–11. [CrossRef]
205. Wiebusch, R.K.; Lant, C.L. Policy drivers of US wetland conversion rates, 1955–2009. *Soc. Nat. Resour.* **2017**, *30*, 16–30. [CrossRef]
206. Zedler, J.B.; Kercher, S. Wetland resources: Status, trends, ecosystem services, and restorability. *Annu. Rev. Environ. Resour.* **2005**, *30*, 39–74. [CrossRef]
207. Stone, M.L.; Whiles, M.R.; Webber, J.A.; Williard, K.W.J.; Reeve, J.D. Macroinvertebrate communities in agriculturally impacted Southern Illinois streams: Patterns with riparian vegetation, water quality, and in-stream habitat quality. *J. Environ. Qual.* **2005**, *34*, 907–917. [CrossRef] [PubMed]
208. Katz, H.A.; Daniels, J.M.; Ryan, S. Slope- area thresholds of road- induced gully erosion and consequent hillslope- channel interactions. *Int. J. Circuit Theory Appl.* **2013**, *41*, 285–295. [CrossRef]
209. Stenberg, L.; Tuukkanen, T.; Finer, L.; Marttila, H.; Piirainen, S.; Klove, B.; Koivusalo, H. Ditch erosion processes and sediment transport in a drained peatland forest. *Ecol. Eng.* **2015**, *75*, 421–433. [CrossRef]
210. Bohne, B.; Storchenegger, I.J.; Widmoser, P. An easy to use calculation method for weir operations in controlled drainage systems. *Agric. Water Manag.* **2012**, *109*, 46–53. [CrossRef]
211. Egresi, I. Tourism and sustainability in Turkey: Negative impact of mass tourism development. In *Alternative Tourism in Turkey: Role, Potential Development and Sustainability*; Egresi, I., Ed.; Springer International Publishing Ag: Cham, Switzerland, 2016; Volume 121, pp. 35–53.
212. Elphick, C.S.; Taft, O.; Lourenco, P.M. Management of rice fields for birds during the non-growing season. *Waterbirds* **2010**, *33*, 181–192. [CrossRef]
213. Gossling, S.; Peeters, P.; Hall, C.M.; Ceron, J.P.; Dubois, G.; Lehmann, L.; Scott, D. Tourism and water use: Supply, demand, and security. An international review. *Tourism Manag.* **2012**, *33*, 1–15. [CrossRef]
214. Chen, R.S.; Wang, K.L.; Wu, C.Y. A preliminary study on the improvement of the ecological corridor in paddy fields. *Irrig. Drain.* **2015**, *64*, 115–123. [CrossRef]
215. Sato, M.; Kumagai, M.; Azuma, N. Life history of *Tanakia lanceolata* in irrigation channel network of the iwaki river basin in northern japan. *Trans. Jpn. Soc. Irrig. Drain. Rural Eng.* **2012**, *80*, 1–11.
216. Ishiyama, N.; Sueyoshi, M.; Watanabe, N.; Nakamura, F. Biodiversity and rarity distributions of native freshwater fish in an agricultural landscape: The importance of beta diversity between and within water-body types. *Aquat. Conserv. Mar. Freshw. Ecosyst.* **2016**, *26*, 416–428. [CrossRef]
217. Flora, C.; Kroger, R. Use of vegetated drainage ditches and low-grade weirs for aquaculture effluent mitigation: I. Nutrients. *Aquac. Eng.* **2014**, *60*, 56–62. [CrossRef]
218. Hyvonen, N.P.; Huttunen, J.T.; Shurpali, N.J.; Lind, S.E.; Marushchak, M.E.; Heitto, L.; Martikainen, P.J. The role of drainage ditches in greenhouse gas emissions and surface leaching losses from a cutaway peatland cultivated with a perennial bioenergy crop. *Boreal Environ. Res.* **2013**, *18*, 109–126.
219. Decleer, K.; Maes, D.; Van Calster, H.; Jansen, I.; Pollet, M.; Dekoninck, W.; Baert, L.; Grootaert, P.; Van Diggelen, R.; Bonte, D. Importance of core and linear marsh elements for wetland arthropod diversity in an agricultural landscape. *Insect. Conserv. Divers.* **2015**, *8*, 289–301. [CrossRef]
220. Onikura, N.; Nakajima, J. Age, growth and habitat use of the topmouth gudgeon, *Pseudorasbora parva* in irrigation ditches on northwestern Kyushu Island, Japan. *J. Appl. Ichthyol.* **2013**, *29*, 186–192. [CrossRef]

221. Cowley, D.E.; Wissmar, R.C.; Sallenave, R. Fish assemblages and seasonal movements of fish in irrigation canals and river reaches of the Middle Rio Grande, New Mexico (USA). *Ecol. Freshw. Fish* **2007**, *16*, 548–558. [CrossRef]
222. Falbo, K.; Schneider, R.L.; Buckley, D.H.; Walter, M.T.; Bergholz, P.W.; Buchanan, B.P. Roadside ditches as conduits of fecal indicator organisms and sediment: Implications for water quality management. *J. Environ. Manag.* **2013**, *128*, 1050–1059. [CrossRef] [PubMed]
223. McKenna, M.L.; McAtee, S.; Bryan, P.E.; Jeun, R.; Ward, T.; Kraus, J.; Bottazzi, M.E.; Hotez, P.J.; Flowers, C.C.; Mejia, R. Human intestinal parasite burden and poor sanitation in rural Alabama. *Am. J. Trop. Med. Hyg.* **2017**, *97*, 1623–1628. [CrossRef] [PubMed]
224. Yu, Q.; Wan, X.X.; Liu, Q.; Cao, C.L.; Bao, Z.P.; Zhu, H.Q.; Zhong, B.; Guo, J.G. Cost-effectiveness evaluation and investigation of control measure changes in areas of schistosomiasis transmission control in hilly regions of mountain areas I: Epidemiological investigation and analysis of prevalence factors of schistosomiasis. *Chin. J. Schistosomiasis Control* **2012**, *24*, 250–254.
225. Fan, X.Y.; Cui, B.S.; Zhao, H.; Zhang, Z.M. The changes of wetland network pattern associated with water quality in the Pearl River Delta, China. *Clean-Soil Air Water* **2012**, *40*, 1064–1075. [CrossRef]
226. Fu, D.F.; Gong, W.J.; Xu, Y.; Singh, R.P.; Surampalli, R.Y.; Zhang, T.C. Nutrient mitigation capacity of agricultural drainage ditches in tai lake basin. *Ecol. Eng.* **2014**, *71*, 101–107. [CrossRef]
227. Powell, K.L.; Bouchard, V. Is denitrification enhanced by the development of natural fluvial morphology in agricultural headwater ditches? *J. N. Am. Benthol. Soc.* **2010**, *29*, 761–772. [CrossRef]
228. Usborne, E.L.; Kroger, R.; Pierce, S.C.; Brandt, J.; Goetz, D. Preliminary evidence of sediment and phosphorus dynamics behind newly installed low-grade weirs in agricultural drainage ditches. *Water Air Soil Pollut.* **2013**, *224*, 11. [CrossRef]
229. Kahle, P.; Schonemann, S.; Lennartz, B. Effect of vegetated buffer strips on nitrate inputs into surface waters in drained lowland catchments. *Hydrol. Wasserbewirtsch.* **2013**, *57*, 60–68.
230. Kleinman, P.J.A.; Smith, D.R.; Bolster, C.H.; Easton, Z.M. Phosphorus fate, management, and modeling in artificially drained systems. *J. Environ. Qual.* **2015**, *44*, 460–466. [CrossRef] [PubMed]
231. Tyler, H.L.; Moore, M.T.; Locke, M.A. Potential for phosphate mitigation from agricultural runoff by three aquatic macrophytes. *Water Air Soil Pollut.* **2012**, *223*, 4557–4564. [CrossRef]
232. Tyler, H.L.; Moore, M.T.; Locke, M.A. Influence of three aquatic macrophytes on mitigation of nitrogen species from agricultural runoff. *Water Air Soil Pollut.* **2012**, *223*, 3227–3236. [CrossRef]
233. Li, Q.K.; Hu, Y.W.; Luo, L.G. Source characteristics analysis of discharge and pollutants in typical drainage ditch of Qingtongxia irrigation district. *Huanjing Kexue* **2012**, *33*, 1579–1586.
234. Zhang, Z.Y.; Kong, L.L.; Zhu, L.; Xia, J.H.; Patricia, X. Fate characteristics of nitrogen in runoff from a small agricultural watershed on the south of Huaihe river in China. *Environ. Earth Sci.* **2012**, *66*, 835–848. [CrossRef]
235. Buchanan, B.P.; Falbo, K.; Schneider, R.L.; Easton, Z.M.; Walter, M.T. Hydrological impact of roadside ditches in an agricultural watershed in central New York: Implications for non-point source pollutant transport. *Hydrol. Process.* **2013**, *27*, 2422–2437. [CrossRef]
236. Peacock, M.; Jones, T.G.; Airey, B.; Johncock, A.; Evans, C.D.; Lebron, I.; Fenner, N.; Freeman, C. The effect of peatland drainage and rewetting (ditch blocking) on extracellular enzyme activities and water chemistry. *Soil Use Manage.* **2015**, *31*, 67–76. [CrossRef]
237. Shan, B.Q.; Li, N.; Tang, W.Z. Construct of Yangtze-Huai River rural areas ecological drainage system and its retention effect on pollutants. *Chin. J. Environ. Sci.* **2012**, *33*, 3797–3803.
238. Moeder, M.; Carranza-Diaz, O.; Lopez-Angulo, G.; Vega-Avina, R.; Chavez-Duran, F.A.; Jomaa, S.; Winkler, U.; Schrader, S.; Reemtsma, T.; Delgado-Vargas, F. Potential of vegetated ditches to manage organic pollutants derived from agricultural runoff and domestic sewage: A case study in Sinaloa (Mexico). *Sci. Total Environ.* **2017**, *598*, 1106–1115. [CrossRef] [PubMed]
239. Adriaanse, P.I.; Boesten, J.; Crum, S.J.H. Estimating degradation rates in outdoor stagnant water by inverse modelling with TOXSWA: A case study with prosulfocarb. *Pest Manag. Sci.* **2013**, *69*, 755–767. [CrossRef] [PubMed]
240. Gall, H.E.; Sassman, S.A.; Jenkinson, B.; Lee, L.S.; Jafvert, C.T. Comparison of export dynamics of nutrients and animal-borne estrogens from a tile-drained midwestern agroecosystem. *Water Res.* **2015**, *72*, 162–173. [CrossRef] [PubMed]

241. Soulis, K.X.; Dercas, N.; Papadaki, C.H. Effects of forest roads on the hydrological response of a small-scale mountain watershed in Greece. *Hydrol. Process.* **2015**, *29*, 1772–1782. [CrossRef]
242. Fernández Tejedo, I. Fragilidad de un espacio productivo: Cambio climático e inundaciones en el bajío, siglo xviii. *Tzintzun* **2012**, *55*, 107–156.
243. Henderson, M. Venice area Audubon rookery. *Outdoor Photographer*, 10 May 2017.
244. Milbank, D. Leapin ljeppers get on the stick or wind up all wet—For Frisians in The Netherlands, vaulting murky canals is a hot spectator sport. *Wall Street Journal*, 6 July 1994; A1.

MDPI
St. Alban-Anlage 66
4052 Basel
Switzerland
Tel. +41 61 683 77 34
Fax +41 61 302 89 18
www.mdpi.com

Water Editorial Office
E-mail: water@mdpi.com
www.mdpi.com/journal/water

www.ingramcontent.com/pod-product-compliance
Lightning Source LLC
Chambersburg PA
CBHW040817250926
51850CB00007BA/1195
9783039363407